MW00447651

ANNALS OF COMMUNISM

Each volume in the series Annals of Communism will publish selected and previously inaccessible documents from former Soviet state and party archives in a narrative that develops a particular topic in the history of Soviet and international communism. Separate English and Russian editions will be prepared. Russian and Western scholars work together to prepare the documents for each volume. Documents are chosen not for their support of any single interpretation but for their particular historical importance or their general value in deepening understanding and facilitating discussion. The volumes are designed to be useful to students, scholars, and interested general readers.

EXECUTIVE EDITOR OF THE ANNALS OF COMMUNISM SERIES
Jonathan Brent, Yale University Press

PROJECT MANAGER
Vadim A. Staklo

AMERICAN ADVISORY COMMITTEE

Ivo Banac, Yale University
Zbigniew Brzezinski, Center for Strategic and International Studies
William Chase, University of Pittsburgh
Friedrich I. Firsov, former head of the Comintern research group at RGASPI
Sheila Fitzpatrick, University of Chicago
Gregory Freeze, Brandeis University
John L. Gaddis, Yale University
J. Arch Getty, University of California, Los Angeles
Jonathan Haslam, Cambridge University

Robert L. Jackson, Yale University
Norman Naimark, Stanford University
Gen. William Odom (deceased), Hudson Institute and Yale University
Daniel Orlovsky, Southern Methodist University
Timothy Snyder, Yale University
Mark Steinberg, University of Illinois, Urbana-Champaign
Strobe Talbott, Brookings Institution
Mark Von Hagen, Arizona State University
Piotr Wandycz, Yale University

RUSSIAN ADVISORY COMMITTEE

K. M. Anderson, Moscow State University
N. N. Bolkhovitinov, Russian Academy of Sciences
A. O. Chubaryan, Russian Academy of Sciences
V. P. Danilov, Russian Academy of Sciences
A. A. Fursenko, secretary, Department of History, Russian Academy of Sciences (head of the Russian Editorial Committee)

V. P. Kozlov
N. S. Lebedeva, Russian Academy of Sciences
S. V. Mironenko, director, State Archive of the Russian Federation (GARF)
O. V. Naumov, director, Russian State Archive of Social and Political History (RGASPI)
E. O. Pivovar, Moscow State University
V. V. Shelokhaev, president, Association ROSSPEN
Ye. A. Tyurina, director, Russian State Archive of the Economy (RGAE)

The Voice of the People

Letters from the Soviet Village
1918–1932

C. J. Storella and A. K. Sokolov

Documents Compiled by S. V. Zhuravlev, V. V. Kabanov, T. P. Mironova,
T. V. Sorokina, A. K. Sokolov, and E. V. Khandurina

Text Preparation and Commentary by C. J. Storella, A. K. Sokolov,
S. V. Zhuravlev, and V. V. Kabanov

Documents translated by C. J. Storella

Yale

UNIVERSITY PRESS

New Haven & London

Copyright © 2013 by Yale University.
All rights reserved.
This book may not be reproduced, in whole or in part, including illustrations, in any form
(beyond that copying permitted by Sections 107 and 108 of the U.S. Copyright Law and
except by reviewers for the public press), without written permission from the publishers.

Yale University Press books may be purchased in quantity for educational, business,
or promotional use. For information, please e-mail sales.press@yale.edu (U.S. office)
or sales@yaleup.co.uk (U.K. office).

Designed by James J. Johnson.
Set in Sabon type by Westchester Book Group.
Printed in the United States of America.

Library of Congress Cataloging-in-Publication Data

Storella, C. J. (Carmine John)
The voice of the people : letters from the Soviet village, 1918–1932 / C. J. Storella and
A. K. Sokolov ; documents compiled by S. V. Zhuravlev . . . [et al.]; text preparation and
commentary by C. J. Storella . . . [et al.]; documents translated by C. J. Storella.
p. cm. — (Annals of communism)
Includes bibliographical references and index.
ISBN 978-0-300-11233-7 (cloth : alk. paper)
1. Land reform—Soviet Union—History—Sources.   2. Peasants—Soviet Union—
Correspondence.   3. Krest'ianskaia gazeta.   4. Peasants—Soviet Union—Social
conditions—Sources.   5. Villages—Soviet Union—History—Sources.
6. Collectivization of agriculture—Social aspects—Soviet Union—History—
Sources.   7. Communism—Social aspects—Soviet Union—History—Sources.
8. Soviet Union—Rural conditions—Sources.   9. Soviet Union—Economic
policy—1917–1928—Sources.   10. Soviet Union—Economic policy—
1928–1932—Sources.   I. Sokolov, A. K.   II. Title.
HD1333.S65S76 2012
333.3'14709042—dc23
2012022661

A catalogue record for this book is available from the British Library.

This paper meets the requirements of ANSI/NISO Z39.48–1992 (Permanence of Paper).

10 9 8 7 6 5 4 3 2 1

Yale University Press gratefully acknowledges the financial support given for this publication by the John M. Olin Foundation, the Lynde and Harry Bradley Foundation, the Historical Research Foundation, Roger Milliken, the Rosentiel Foundation, Lloyd H. Smith, Keith Young, the William H. Donner Foundation, Joseph W. Donner, Jeremiah Milbank, and the David Woods Kemper Memorial Foundation.

Contents

Acknowledgments

This book proved a difficult undertaking and would not have come to fruition without the help of a number of friends and colleagues.

Jonathan Harris, William Chase, and Wendy Z. Goldman read all or parts of various versions of the manuscript, providing, as always, insightful and helpful comments and, when necessary, the occasional kick in the pants.

Thanks are also due to Lynne Viola, Orysia Karapinka, Gregory Freeze, and ChaeRan Freeze for their availability, advice, and encouragement.

Both editors wish to acknowledge the significant contributions of Jeffrey Burds, who, with Andrei Sokolov, originally conceived this project and did much early and important work on it.

S. V. Zhuravlev, V. V. Kabanov, T. P. Mironova, T. V. Sorokina, and E. V. Khandurina ably assisted Dr. Sokolov in selecting and compiling the letters and documents that make up this volume. Professors Kabanov and Zhuravlev also contributed to the text that accompanied the publication of these documents in Russia. Selected parts of that text have been incorporated in the present volume.

Jonathan Brent and Vadim Staklo of Yale University Press deserve high praise for their patience and commitment to this project. Heartfelt thanks as well to Mary Pasti for her attention to detail and her engagement with the many historical issues raised in this document collection. The book is much the better for her editorial labors.

In the course of translating these documents, Yury Starostine, L. R. Vaintraub, Natalia Basovskaia, and Aleksei Kilichenkov provided very helpful advice and answers to all my questions. I am particularly grateful to Naum Kats, my friend and colleague at Carnegie-Mellon University, for generously and enthusiastically giving of his time to wrestle with especially opaque and troublesome passages, and for always supplying the espresso.

Thanks are also due to Robert Hayden, Eileen O'Malley, and the staff at the University of Pittsburgh's Center for Russian and East European Studies for their many considerations over the course of this project.

Finally, I would like to thank my wife, Irene Kugler, for her loving support and unlimited forbearance.

C. J. STORELLA

Note on Transliteration and Translation

The Library of Congress transliteration system is used in citations, but a modified version is employed in the text. Hard (") and soft signs (') are omitted, but in certain combinations soft signs may appear as *i* (e.g., Zinoviev, not Zinov'ev). The vowel *ë* is transliterated as *yo* unless preceded by *sh, ch,* or *shch*. Other changes include:

> In the initial position:
> > E = Ye (Yevseev, not Evseev)
> > Ia = Ya (Yanenko, not Ianenko)
> > Iu = Yu (Yury, not Iurii)
> In the final position:
> > ii = y (Trotsky, not Trotskii)
> > iia = ia (Izvestia, not Izvestiia)

Plurals of Russian words are often rendered with the English *s*.

Since these translations are not presented as facsimiles of the original letters, in the interest of providing a more readable text we have minimized the use of bracketed words (e.g., "comrade" instead of "c[omrade]" or "com[rade]" for *t.* or *tov.*) and occasionally spelled out an acronym (e.g., "Komsomol" instead of "VKLSM"). For the same reason, we have not endeavored to approximate every Russian spelling, grammar, or punctuation error in English and have, on the contrary, made minor corrections in tense, singulars and plurals, subject-verb agreement, and punctuation, though making exceptions to preserve the flavor of the original.

Note on the Documents

Russian archival documents are cited and numbered by collection (*fond* or f.), inventory (*opis'* or op.), file (*delo* or d.), page (*list* or l., plural ll.), and sometimes part (*chast* or ch.). The majority of documents in the present volume were selected from the vast collection of unpublished readers' letters to *Krestianskaia gazeta* preserved in the Russian State Archive of the Economy (RGAE). Unpublished letters to other newspapers (e.g., *Pravda, Bednota*) have also been included. In addition, documents from the following archives have been used: the Russian State Archive of Social-Political History (RGASPI), the State Archive of the Russian Federation (GARF), and the Central State Archive of Social Movements of Moscow City (TsGAOD g. Moskvy).

Some letters written to a variety of institutions and publications were compiled in summaries (*svodkis*). These are primarily typed copies of excerpts from letters written to different publications in late 1929 and early 1930 and are preserved in *fond* 7486s of RGAE. Unfortunately, the specific addressee and precise date of composition are not always indicated. According to Andrei Sokolov, the original letters are most likely located in the archives of the secret police, OGPU.

Some letters and documents were too long to reproduce in full. Editorial excisions are indicated by ellipses in brackets: [...]. Within letters and documents, text inserted by the editors or the translator is likewise in brackets. Where the compilers of the typed copy from which our version is taken could not decipher a writer's handwriting, they indicated this with ellipses, in which case the ellipses are labeled "[words not clear in original]."

The organization of the book, and hence the order of the documents, is broadly chronological. It follows rural developments in the 1920s from the introduction of the NEP through the early phase of agricultural collectivization. Each chapter is organized thematically. To follow the evolution of peasant thinking on a given subject over time it was sometimes necessary to violate strict chronological order by including letters written earlier or later in the decade. Chapter 1 provides background on the revolution and the civil war. Since *Krestianskaia gazeta* did not exist at this time, the few letters included in this chapter are taken primarily from those written directly to Party and government leaders. By and large, chapters 2 and 3 address economic issues and the developing relationship between town and countryside. Chapters 4 and 5 are devoted to various aspects of village life pertaining to daily existence and the exercise of political power at the local level. Chapter 6 presents letters indicating how the peasants understood the aims and benefits of socialist construction. Chapter 7, the lengthiest, covers the lead up to and the implementation of wholesale collectivization and, more than the other chapters, contains letters from svodkis. A list of the documents is included at the back of the book.

The Voice of the People

Introduction

An important link between the government and the peasant is the *Krestyanskaya Gazeta,* or "Peasants' Gazette," a little paper which appears twice a week and is entirely addressed to a peasant audience. Its circulation fluctuates with the season, but has been as high as a million. Perhaps the most interesting feature of this paper is the number of letters which it receives from its readers. They pour in at the rate of 1500 or 2000 a day, written on leaves torn out of school notebooks, on bits of wrapping paper, often scrawled and misspelled, but giving convincing proof that the peasant has awakened to a point where he has ideas and wishes to express them.

—WILLIAM HENRY CHAMBERLIN, 1930

In the early hours of 26 October 1917, shortly after seizing power from the Provisional Government in the Russian capital, Petrograd, Bolshevik Party leaders hurriedly hammered out a proclamation transferring control over private farmlands to the peasants who worked them. Determined to get their Decree on Land approved at the All-Russian Congress of Soviets then in session, the exhausted yet triumphal insurrectionists scribbled the sweeping terms of the new law down on paper in fervent, all-but-illegible scrawl. According to an eyewitness, the drafters' poor penmanship actually prevented the Bolshevik Party chief, Vladimir Lenin, from reading out the decree to the boisterous gathering of people's deputies. After several false starts, the man poised to head the government of the largest country on earth abandoned hope of deciphering the document and ceded the honor of announcing its momentous contents to an unnamed, keener-eyed delegate who had stepped from the crowd to help him. Not long afterward, amid the prevailing commotion and with hardly a word of discussion, the decree passed with only one vote against it.[1]

Besides illustrating the impromptu and decisive manner by which great upheavals sometimes resolve major issues, this minor episode from the October Revolution's first full day has something appealing about it. The land question loomed large for the new Soviet government. Only the vote calling on the belligerents in the Great War to immediately lay down their arms carried sufficient weight and urgency to precede it in the congress's order of business that morning. As the Bolsheviks well knew, the

Provisional Government's failure to undertake meaningful land reform during its eight-month existence had inspired peasant support for its overthrow and the transfer of state authority to the popularly elected councils known as soviets. Without that support, the new government would also be short-lived, which explains the Bolsheviks' expeditious and, from the peasants' point of view, positive resolution of the matter. The Decree on Land effectively answered the peasants' age-old demand to control the land they seasonally plowed, planted, and harvested. Therefore, it seems entirely appropriate that the terms of the decree, which affected so many faceless millions, was announced by one of the anonymous people's emissaries who packed the Smolny Institute that eventful day.

This book is a collection of letters from a few among the millions directly affected by the decree who, like Lenin's surrogate reader, sensed that the Bolsheviks were having difficulty seeing rural matters as clearly as they might. To help sharpen that vision, they took up pen or pencil to provide the new authorities with firsthand information about their work, their surroundings, and their lives. Like Lenin at the congress tribune, the Soviet leadership had welcomed, even encouraged, this assistance for the majority of letters in this collection were solicited—but never published—by the editorial board of Krestianskaia gazeta, or The Peasant Gazette, a newspaper that began life as a weekly in late 1923, six years after the October Revolution.[2] The peasants who wrote to the paper were not content to read from a prepared text, but took full advantage of the opportunity afforded them to communicate their frustrations and disappointments in their own fashion, to point fingers and name names, and to articulate alternative visions of the life they believed the revolution had made possible not only in their villages or even just in the countryside but throughout the former Russian empire and beyond.

"Krestianskaia Gazeta: *The Friend and Defender of the Peasantry*"

Following the civil war, Lenin's well-known prerevolutionary emphasis on the organizational centrality of the press found expression in the Communist Party's creation of newspapers, *Krestianskaia gazeta* among them, aimed at specific social groups like peasants, women, and soldiers. The Communist Party congress held in May 1924 called on the press to strengthen its ties to workers and peasants and defined the press's primary role as explicating the "fundamental questions of life and existence" for them. This strategy was part of a broad campaign that Matthew Lenoe has designated a "mass enlightenment project," the ultimate aim of which was to create the "new Soviet man." In place of the emphasis on agitation and mobilization that had marked the press of the civil war era, during much of the 1920s Soviet newspapers assumed the role of popular educators whose goal was to disseminate values and promote practices that would make good Soviet citizens of their readers. To this end, a crucial function of the press in these years entailed familiarizing these readers with Bolshevism's official language and habituating them to perceive the world and to organize their thoughts about it in terms established by the Party.[3]

Such a psychological transformation among the peasantry could not succeed without greatly increasing the influence of the towns on the countryside. The Central Committee of the Party intended the indifferently entitled *Krestianskaia gazeta* to be a means of extending this influence to the villages and far-flung hamlets where communist cadres were scarce in order to facilitate propagandizing among the peasantry along lines that accorded with Party and state interests. The rural press in particular was instructed to display an "attentive attitude" toward peasant letters and peasant complaints about local officials and to use stories on agronomy to illuminate political and economic issues.[4] As in the period of underground activism, when professional revolutionaries took it upon themselves to infuse workers with a proletarian class consciousness, now, with the help of newspapers, the urban vanguard would disseminate its political and cultural values among the rustic masses in the hopes of begetting a new Soviet peasantry.

Newspaper readers' letters provided Party leaders a way to judge the success of these efforts. Letters supplemented the more formal channels of information on which central authorities relied, adding direct anecdotal accounts of peasant and working-class life and concerns to the picture of domestic conditions provided in police dispatches, statistical compilations, and the reports that local officials regularly sent to Moscow. This information was all the more instructive for being expressed in people's own words. As a result, more than other sources of information, letters had the virtue of providing the central leadership with a sense of public opinion on a variety of questions and policies, as well as a means by which to gauge the "mood" of the population. Given the paucity of rural Party cadres and institutions, and the critical role that agricultural production played in the life of the nation, peasant letters were a particularly important source of information for the leadership. Moreover, in the 1920s, before effective state rule had been established in the countryside, both the peasants and the central leadership shared a common interest in bringing instances of misrule to light. It is not surprising, then, that Moscow encouraged and paid close attention to the revelations of official misconduct and other irregularities that readers' letters often contained.[5]

Through the selection process, editorial boards controlled which let-ters appeared in their newspapers and, consequently, were read by the nation. But Moscow did more than simply encourage individuals to write to newspapers so it could publish those letters it deemed correct and use-ful. To organize letter writing itself, the Communist Party also provided official support for the formation of, first, a workers', then a peasants' "correspondent" movement. Through congresses, conferences, written instructions, and guidelines, central officials endeavored to shape the form and content of the discourse that newspaper correspondents em-ployed. To further encourage letters, *Krestianskaia gazeta* supplied its correspondents with preprinted forms on which they could write their letters and where instructions admonished them to write about what they knew firsthand and to "write only the truth." Anyone who wrote to a newspaper might be considered a worker correspondent or village cor-respondent (*rabkor* or *sel'kor,* respectively), and by the mid-1920s, the movement claimed hundreds of thousands of adherents. Thus, news-paper letter writing provided the center with more than a source of pop-ular opinion; it also served to organize "advanced" (i.e., correct-thinking) workers and peasants into extensions of the Party throughout the na-tion's factories and villages.[6]

We might expect, then, a high degree of conformity in how and what rural readers wrote to newspapers. As Jeffrey Brooks has noted, and as a large number of letters in this collection confirm, dispensing with offi-cial terminology and expressions proved impossible for many lower-class newspaper readers, and this, no doubt, served the regime's purposes.[7] But, given the sharply conflicting interests that underlay peasant-state relations, there is also a danger of overstating Bolshevik successes on this particular sector of the cultural front. Although several writers of letters in this collection identify themselves as village correspondents, and a few even provide their official correspondent numbers, most do not, and even among those that do, hostility toward the regime and its policies is not always absent. As for the rest, the language, tone, and lessons im-parted in these letters are often far from what Moscow would have con-sidered comfortable or correct. Bolshevik efforts at cultural organization and production were serious and consequential, but, as in many other areas of life in the 1920s, the official attempt to organize letter writing was also highly susceptible to spontaneity, not to mention the peasants' evidently sincere desire to "write the truth" about what they knew.

In the 1930s, after forced collectivization had strengthened the Party's hold on the countryside, peasants had fewer means at their disposal to counter the Party's highly developed cultural offensive. Although the overall situation remained chaotic, the combination of political repression, increased censorship, extensive surveillance, terror, and the state's wholesale cooptation of the remaining quasi-independent institutions imposed a new level of regimentation on society, narrowing the possibilities for autonomous activity and demanding a high degree of conformity. Yet, even in this confined and often dangerous atmosphere, individuals and groups found ways to blunt, counter, and otherwise resist the logic of the official discourse and the actions of the state.[8]

Such resistance was no less evident earlier, in the pre-collectivization countryside, when the regime tolerated a modicum of open debate and cultivated good relations with the peasantry. Coming between the civil war—when the Bolsheviks pursued policies designed to aggravate class differences within the village—and the all-out military-type crusade against the countryside that was wholesale agricultural collectivization, the years of the New Economic Policy, or NEP (1921–1929) were relatively calm. Rural state and Party institutions, still in the process of formation, remained weak and played a less prominent role in the day-to-day lives of the peasantry than they did later, since traditional peasant institutions like the commune and the village assembly continued to function. In what amounted to a transitional situation, where the new had yet to completely supplant the old, peasants could be less guarded and feel more encouraged to express themselves freely in the direct fashion that was their wont.

Paradoxically, Moscow also contributed to the atmosphere of relatively open expression by initiating well-publicized campaigns targeting, for example, corruption, financial waste or mismanagement, privileged lifestyles, vice, or groups identified as anti-proletarian or antisocialist. Through these campaigns the regime sought, among other things, to direct popular anger and to establish the terms of popular discourse. In this, to a certain extent, it succeeded. Typically, letters addressing campaign themes increased, as did denunciations of alleged *kulaks* (wealthy peasants), derelict local officials, and other campaign targets. But inviting criticism from below is always risky; leaders and led are usually reading from different scripts and acting from different motivations. Once the dam is opened, criticism, and the dissatisfactions and emotions behind it, can be difficult to channel. A number of letters included here show that letter writers were not always content to remain within the bounds of specified topics and often used the sanctioned discussion of a particular issue as a point of departure to make more general criticisms of policy, ideology, and even powerful state and Party figures.

Whatever the Party leaders' goals, the peasants seemed to have immediately recognized in the new gazette a forum or public space where they could express their views on a variety of pressing issues—this even though the paper was a Communist Party publication run by Party members, the same Party bent on getting ever more revenue and grain from the countryside. Nonetheless, as we see in the remark above by the Brooklyn-born journalist and early Sovietologist William Henry Chamberlin, who spent many years in Soviet Russia during the 1920s, letters from villages and settlements near and far poured into *Krestianskaia gazeta*'s Moscow office. The quantity of correspondence may be considered evidence that the peasants took the newspaper's simple title to heart, that they perceived it to be, as one of its slogans claimed, "the friend and defender of the peasantry." Though not a paper by the peasants, it was a paper for the peasants. In addition to keeping them abreast of news and political developments, it regularly addressed issues directly relevant to the peasants' daily life and work. As the volume of mail suggests, however, the peasants' enthusiasm for the paper was kindled most by the role it played as a conduit, as a way for them to communicate their thoughts and concerns to an audience beyond the village, an audience that not only included fellow peasants scattered across the country but the nation's political leadership. Having the opportunity to address the leadership was especially important for peasants in the 1920s; though finding themselves now ruled by a state that defined itself as "worker-peasant," they did not enjoy the status or full array of constitutional rights granted their alleged partners in this arrangement, the working class.[9]

As this discrimination indicates, the regime did not deem the peasants ready for full integration into the social order it was building. On the contrary, Party leaders considered the outlook and economic practices of much of the peasantry both a symbolic and a material threat to that social order. In large part, this view can be attributed to the Bolsheviks' understanding of their rule as a "dictatorship of the proletariat" through which the working class, as the bearer of political consciousness and progress, exercised its dominance and influence over the rest of society. In the Party's class analysis, most peasants were or aspired to be small-property owners and, as a result, were gripped by a "petty-bourgeois" mentality that made them a vacillating and unreliable ally of the working class in its historical mission to attain socialism. Above all else, this mentality placed the peasants at odds with the Bolsheviks' ultimate objective of socializing agricultural production by means of collective farming. On this basis, Party members were inclined to view the peasants with suspicion and to vilify many of their activities and ambitions as self-serving and exploitive rather than socially useful—a view Party members would hold as long as peasants practiced individual, not socialized, cultivation and remained actually or potentially the capitalist "other."

It became apparent following the revolution that the low technological level of Russian agriculture and the regime's dependence on the output of individual farming made the immediate realization of widespread collective farming—whether through peaceful or forcible means—impracticable, and required instead the adoption of a different course of action. In response, Party ideologists, statisticians, and organizers sought to construct a distinct class of Soviet peasants that would provide a base of support for Bolshevik policies in the thousands of settlements that made up rural Russia.[10] Pursuit of this goal involved identifying and separating from the mass those elements within the peasantry whose economic interests were deemed to coincide with the Party's long-range plans for agricultural socialization. Consequently, individuals and households found themselves located along a continuum with, at the poles, what may, without too much distortion, be termed "good" and "bad" peasants. Wealth, as measured by the quantity of land or livestock possessed, served as the chief determinant of where an individual fell on the continuum. Party doctrine classified landless and poor peasants favorably as "proletarian" and "semi-proletarian," respectively. "Rich" peasants were dubbed *kulaks,* literally "fists," a term that previously had referred to village moneylenders or peasants of an entrepreneurial bent but that after the revolution acquired the added pejorative political connotation of "anti-proletarian." The vast majority of so-called middle peasants occupied a position somewhere between the two extremes of "rural proletariat" and "rural capitalist." From the Party's standpoint, the primary goal of the worker-peasant alliance amounted to winning over the middle peasants—often the most hardworking and productive farmers—by freeing them of their petty-bourgeois attachment to individual farming and convincing them to support radical agricultural reorganization.

As this suggests, while peasants may have been located in Soviet society, they were not fully part of it. Further, they could not achieve full integration into that society until they renounced their own interests and adopted the outlook and values dictated by the state. The peasants recognized the situation and to one degree or another resented it—often bitterly. Their resentment frequently took the form of resisting, in word and deed, Party organizers' attempts to split the peasantry into distinct strata of rich (*kulak*), middle (*seredniak*), and poor (*bedniak*) and of questioning the very idea of a "worker-peasant" state, at least as Party leaders conceived it. Lynne Viola, based on her study of peasant letters published in the newspaper *Bednota* (Poor Peasants), shows that peasants took a much more nuanced approach to village social divisions than Party members did, especially as regards the definition of who was a kulak. Peasants were more likely to apply the "kulak" label to those who violated the rural "moral economy" or disregarded practices considered "just" than to those who simply had, through their own labors, honestly accumulated a degree of wealth. Peasants' subordinate status also placed them at odds with the dominant Soviet political discourse, which sought to define the interests and goals of the state as, ultimately, identical with the best interests of those who labored, workers and peasants alike. As a result, the peasantry, as much as any social group, was well placed to adopt a critical attitude toward the claims and actions of the Soviet state and the Communist Party, an attitude they did not hesitate to express in their letters.[11]

In the letters, many correspondents, consciously or not, offered a conception of peasant identity that challenged the official portrait of the class. In contrast to the narrow, greedy, exploitative, rapacious petty capitalist of Party dogma, they extolled the virtues of hardworking agriculturalists who, through their backbreaking labors, fed the entire nation. In this view, it was they who had provided the critical mass for revolutionary victory and counterrevolutionary defeat, often sacrificing and suffering greatly as a result, and who now manned the Red Army in further defense of the revolutionary citadel. Successful peasants—to the extent that they had prospered through hard work and intelligence—strengthened the nation by contributing to its overall wealth and therefore deserved the thanks and support of the state. That they instead should be subject to repressive or discriminatory measures seemed, truly, to indicate a world turned upside down. Through this type of dialogue, letter writers undertook a rectification project of sorts, engaging in a defense of peasant interests that simultaneously questioned the Party's competence to define the class on its terms and provided a rationale that might eliminate the obstacles to the peasantry's full acceptance into the ranks of Soviet citizenship. Some of those who adopted a more antagonistic position took the opportunity to reject membership in the Soviet polity altogether. Letters written in either vein may be considered a form of popular opposition to the hegemonic claims of the state.[12]

By the time *Krestianskaia gazeta* began publication at the end of 1923, the Bolsheviks had eliminated oppositional political parties, and most government organizations—including the soviets—had come under Communist Party domination. In this highly circumscribed political environment, letter writing provided one of the few remaining means by which citizens could gain the ear of higher authority. Letter writing redirected the flow of newspaper and other official communications, turning readers from passive recipients of news, information, and government pronouncements into active participants in the public discourse and, by extension, in the construction of the new society. It particularly enabled literate and semiliterate peasants (and, through surrogate writers, illiterate peasants) to transcend the limits imposed on them by geographical isolation and the provincialism of village culture, allowing even the lowliest *muzhik* (male peasant) to give notice of his views on topics great and small. Through a letter to an editor, the nonproletarian, "pettybourgeois" peasant could hold some hope that his (or her) views would get the same consideration as those of a factory worker or city dweller. In this respect, a large degree of enfranchisement and class assertiveness also attached to peasant letter-writing.

Krestianskaia gazeta's letter column provided peasants a rostrum from which to challenge official affronts to their dignity, discuss national and local issues, and expose wrongdoings perpetrated by officials or other peasants. It was a part of what Jeffrey Brooks has called the "interactive" sphere of Soviet journalism that marked the 1920s, when newspapers tried to gain mass support for the regime by encouraging rational discussion of issues.[13] Like the *cahiers de doléances* during the French Revolution, the letters served as expressions of class interest vis-à-vis other groups in society. But unlike the cahiers, they were not formal petitions. While some were appeals submitted in the traditional fashion—collectively, in the name of a group or village—most were submitted by individuals and were often written in the familiar, colloquial style of everyday speech. Peasants did not limit themselves in their writings to class concerns narrowly defined. On the contrary, they commented on a vast array of political, economic, and social questions, as well as matters of local or personal interest. Like the peasant cahiers, which historians have described as particularly "outspoken," the letters are remarkable in their candor.[14] Peasants did not, it seems, consider any subject beyond the pale of discussion or criticism.

This is not to suggest that the letters provide a pure, undisguised version of what James C. Scott calls the "hidden transcript"—the "offstage" discourse that subordinate groups engage in for and among themselves at sites of their own making.[15] The very act of writing to authority precludes such an eventuality. The peasants certainly employed, to a lesser or greater degree, various devices in their writing that muted or shaded their more venturous thoughts; rarely are exchanges between the powerful and the less powerful completely transparent. But readers of these letters are bound to be struck by how so many of them seem devoid of self-censorship. Most letters are signed, almost all express dissatisfaction, a fair number emit anger, even rage; the language of many, though certainly not all, is plain and direct. This helps explain why so many letters remained unpublished—they did not fit the officially sanctioned, if unwritten, definition of the conversation that should occur between rulers and ruled. Ultimately, the regime's priorities determined which letters would be published and which not. This did not, however, deter peasants from speaking their mind. Even when the Party leadership attempted to set the terms of acceptable political discourse more explicitly by publishing provocative letters only to denounce them as examples of counterrevolutionary writing and thinking, this, in itself, did not seem to discourage peasants from writing how and what they wanted.[16] Throughout the period under review, peasants expressed themselves boldly on all manner of topics.

This said, it must be noted that these letters confront the reader with a variety of styles and discursive strategies, making general or categorical characterizations difficult. The peasants' inclination to express their thoughts forthrightly was often tempered by the recognition that their appeals would be most effective if framed in a manner that would gain them a sympathetic hearing with the powers that be. This meant, in many cases, adopting the new idiom fostered by the regime as part of a discursive strategy. Sometimes its adoption was no more than a formal or ingratiating gesture, as when a peasant headed a letter—in newspaper-column fashion—with one of the ubiquitous Party slogans of the day (or, frequently, a slogan or epigram drawn from another source or devised by the writer), often copied out in capital letters, before launching into whatever complaint or commentary had motivated him or her to write. A number of letters provide examples of peasants employing official expressions or emulating the style and vocabulary of newspapers and government pronouncements. This may be seen as evidence of the Party's success in shaping discourse, but it may also be a tactical ploy on the writers' part: mimicking "correct" or "appropriate" speech to give weight to their words or simply to display loyalty—heartfelt or not. Today's readers must recognize, therefore, that this form of discourse can be highly ambiguous.[17]

At other times, by contrast, writers drew on the regime's standard phraseology not to show subordination but, ironically, to deliver a defiant message. A good example is provided by references to the Bolsheviks' designation of the new state as "worker-peasant," an appellation the peasants had ample reason to question. Yet, when it suited their purposes, peasants took the formulation seriously and at face value to prove that the state and its agents were failing to live up to the letter or the spirit of their own precepts, especially as these concerned the muzhik. Peasant letter-writers employed such official clichés as a cover behind which they could comment on the unfairness of their second-class status, expose the abuses perpetrated by officials at all levels of the hierarchy, even ridicule the absurdities of Communist Party policy, and do it all in the name of the Bolsheviks' declared principles. Embedded in this sort of criticism is the more subversive, usually (but not always) unstated charge that the Soviet state was neither "peasant," "worker," nor even "socialist," and that communists used these terms hypocritically to conceal their own privileged position while they continued to exploit the laboring classes.

This is but one among many possible examples illustrating how peasants turned the new Soviet lingua franca against its originators. So frequently did peasants undertake to refute egalitarian- or democratic-sounding phrases and neologisms in order to give the lie to Bolshevik distortions regarding rural affairs that they must have derived a visceral pleasure from the practice. By that means, the muzhiks could serve notice to the country's new masters that they were not fools and knew, in the words of the Russian proverb, "where the dog was buried." The sentiment can be expressed in less poetic terms by noting that as a result of their particular material interests and standing in the new order, peasants did not find the symbols or ideological rationales generated by the Bolsheviks to justify aspects of their rule either mystifying or impenetrable.[18]

Critics of authoritarian regimes, Russian and Soviet writers in particular, have always found in irony an effective rhetorical weapon, so it is hardly surprising that peasants should frequently draw on this fundamental trope to frame their reproofs. But peasants did not limit their written discourse to ironical statements nor habilitate their missives in Soviet officialese. Peasant language is rich and expressive, and letter writers freely mined the plentiful discursive cache to give life and force to their comments. In their letters the reader encounters adages, the language of the Bible and religious belief, vulgarities and profanities, and literary and historical references, in addition to a creative, if not always standard, use of the vernacular.[19] Perhaps because of their inferior status, peasants also felt free to compare life under the Bolsheviks unfavorably with that under the tsars, thereby challenging an axiomatic proposition of the regime that Communist Party rule was primarily dedicated to improving the condition of the formerly exploited masses. In short, the war of words between the new regime and the peasantry was not a one-way affair. Peasants had at their disposal a ready dialogic arsenal that they could use to counter the symbolic and dialectical salvos let loose from the cities and from on high.

The Peasants and the State

The concerns of this book, then, are the concerns of the Russian peasantry as it confronted a new and revolutionary political regime intent on transforming the life and work to which it had been accustomed for generations. Inevitably, at its core, this book is about the relationship of state and society. Since many letters were commentaries on state action or inaction, the aims and goals of Party and state policy are emphasized throughout. If the letters reveal anything about the peasants' relationship to the Soviet state, it is that the relationship was complex and not easily reducible to simple narratives of opposition or cooperation, resistance or capitulation. History and experience supplied peasants with cause to fear and shun the state, yet peasants were not inveterate anarchists opposed to each and every state intervention in village life. On the contrary, they expected, and in many cases demanded, that Soviet power provide rural dwellers with economic, agronomic, cultural, social-welfare, and other forms of assistance. The peasants felt that this sort of aid was their due as supporters of revolutionary change, as laborers, as providers of food for the nation. In fact, the lack of such assistance tended to raise peasant hackles.

Nevertheless, even when they applauded Bolshevik initiatives—like the revival of the free market in grain in 1921—peasants understood that the ultimate goal of Party policy was their further integration into the state economic system and the appropriation of more of what they produced. Throughout the years covered by this book, 1918–1932, the Party leadership worked to establish an effective state presence in the historically undergoverned countryside to ensure uninterrupted access to peasant "surpluses" and to facilitate social and political change. Neither task was easy. In the wake of the tsarist state's demise in February 1917, rural authority had collapsed. Very quickly its agents—landlords, police, and gentry land captains—found themselves under assault and then supplanted by popular, locally based assemblies and committees of the peasants' own creation. The peasants resisted the Provisional Government's efforts to influence the formation and activities of newly established institutions like district-level (*volost*) committees. By the summer of 1917, this attitude led peasants in many regions to establish what amounted to autonomous governments based on the principles of localism and grass-roots democracy.[20] After overthrowing the Provisional Government in October, the Bolsheviks initiated efforts to erect a central state apparatus of their own, a process that did not significantly penetrate the countryside until the spring of 1918, when severe food shortages in the cities launched the Party on the path of bringing the peasantry—its institutions and its food supplies—under state control.

The ruling regime's campaign to establish Soviet administrative control over the countryside proceeded in fits and starts until the completion of the bloody first stage of collectivization, a decade-and-a-half-long struggle between state and peasant that one historian has called the greatest of all European peasant wars. During this period, Bolshevik state-building continuously clashed with peasant traditions, institutions, and self-interest. As the sides confronted one another, both employed a variety of tactics. Compromise and concessions occurred on an almost daily basis. When these failed, each side had recourse to greater or lesser shows of violence. Recognizing their disadvantage in this form of struggle, however, peasants were just as likely to engage in less direct forms of resistance to what they viewed as unreasonable state demands or encroachments.[21]

In short, as in much else and by necessity, peasants took a pragmatic stance vis-à-vis the state, calling for and expecting its help when it suited their purposes, evading or fighting it when they found its actions threatening. Bolshevik grand strategy, for its part, was obvious from the very beginning: to fracture peasant unity along the lines of internal economic disparities in wealth and landholding in order to separate the poorer strata from the better-off and to win over the former to the side of the state. That is, it aimed to foster class discord among the peasants themselves.

At the root of the relationship between regime and peasant lay the Bolsheviks' revolutionary project. To realize their socialist dreams, the Bolsheviks believed that they first had to turn Russia into an advanced industrial colossus—a condition the country was nowhere near at the outset of the 1920s. In the last quarter-century of its existence, the empire ruled by the Romanov tsars had undergone a rapid and tempestuous industrial revolution. Private entrepreneurs and the state had combined to build a strong heavy-industrial sector, and Russia had come to rank among the world leaders in many important industrial categories. Economic modernization also resulted in the swelling of the empire's leading cities by migrants seeking work; factories in the burgeoning industrial centers rivaled in size the largest to be found anywhere in Europe or North America. This economic progress, however, was more than offset by the condition of the rest of the country, which remained caught in a web of underdevelopment, populated as it was by peasants practicing a tradition-bound, mostly nonmechanized, labor-intensive form of agriculture. The First World War, the revolution, and the subsequent civil war destroyed any chance that Russian agriculture would follow Russian industry and be modernized through capitalist means. These combined cataclysms also destroyed much of the industrial base laid during the previous three decades. The Bolsheviks saw their party as the engine of historical change, and socialism as the historically predetermined goal of that change. Once securely ensconced in power, they set themselves the task of resuscitating and completing by other means the economic modernization that the tsarist state and Russian capitalism had begun—a Herculean task that demanded a regular and uninterrupted flow of food supplies from the countryside to the cities.

To ensure this flow, the Party hoped to socialize agricultural production by remaking the Russian countryside through the amalgamation of small, individual peasant holdings into larger, more efficient, and more productive collective units. But, as the Bolsheviks were only too aware, their long-term program of land nationalization and agricultural collectivization placed them directly at odds with the peasants, whose long-held desire was to own the land they farmed. In Marxist terms, capitalism had failed to prepare Russia for socialism by eradicating the feudal legacy of peasant agriculture. As a result, the Bolsheviks now confronted millions of agriculturalists who, having just wrested control of the land from their landlords, were not inclined to conjoin their hard-won holdings in collective farms. On the contrary, as the Polish communist Rosa Luxemburg declared in her early critique of the Bolshevik revolution, the rural revolution had created an enormous landowning class, "a new and powerful layer of popular enemies of socialism in the countryside, enemies whose resistance will be much more dangerous and stubborn than that of the noble large landowners."[22] To succeed in their mission, therefore, the Bolsheviks, who by and large concurred with Luxemburg's characterization of the peasantry, had to find some means of circumventing history in order to extend the socialist revolution into the villages and settlements of rural Russia.

From a Marxist standpoint, Luxemburg's observation was self-evident. The old regime's collapse and the redistribution of land among the peasantry had, indeed, created millions of small-property holders who could not be expected to acquiesce in a radical reorganization of their lives and work in the name of economic efficiency or historical necessity. The Bolsheviks themselves recognized the reality of this situation first in October 1917, when they refrained from pursuing full socialization of the land (to Luxemburg's chagrin), and then again after the civil war, when they turned to the market to induce the peasants to provide food to the cities. Despite these concessions to private farming, however, Lenin never abandoned hope that the peasant economy could be socialized—to think otherwise would have been akin to admitting that the revolution had run into a brick wall. To this end, in the early 1920s he proposed that generous state support be provided for the formation of agricultural cooperatives. In Lenin's evaluation, cooperatives, an organizational form he deemed "most acceptable" to the peasants, would demonstrate the benefits of socialist production to them and facilitate their transition to collective farming. Lenin recognized, however, that encouraging peasants to join these ventures would require more than state investment in a cooperative network. To rouse broad peasant enthusiasm for cooperative farming, Lenin believed that material aid had to be accompanied by a program of rural enlightenment that would work to overcome muzhik habits and muzhik attachment to tradition by disseminating scientific knowledge and instilling the desire to learn. Only by addressing the mental as well as the material foundations of rural life, he argued, could the possibility of widespread socialist agriculture be kept alive. Through a wide expansion of literacy, the introduction of efficient agricultural techniques, and the cultivation among the peasants of the "habit of book-reading," Lenin hoped to remake the individual peasant consciousness and create, in effect, a rural version of the new Soviet man. Ideally, this strategic coupling of economic assistance and consciousness-raising would free peasants' from the yoke of their prejudices and backwardness—their *nekulturnost*—convincing them, ultimately, to accept the superiority of collective over individual farming.

Although Lenin's advocacy of cooperation retained a good deal of utopianism, it was a response to the peasants' rejection of communist attempts to organize collective farms during the civil war and, as such, a concession to reality. It reflected two related sides of Lenin's intellectual makeup that came to the fore as revolution-induced expectations of immediate change subsided: that of the positivist, devoted to progress through scientific advance, and that of the Europeanized Russian *intelligent,* convinced of the transformative power of (Western) culture. Proposed at a time when the domestic economy was in shambles and Soviet Russia isolated internationally, cooperation offered the possibility of furthering change in the desired direction peacefully, on the basis of available internal resources. In place of coercive measures that risked inciting the peasants to active opposition, inducements would be offered that would, nonetheless, still make possible the capture and revolutionary transformation of the peasant economy. Having been rebuffed in early efforts to carry through a socialist revolution in the countryside, Party workers would now build a cooperative network by abandoning frontal attacks on rural institutions and traditions and providing the peasants instead with the means "to grow into socialism." It was this program that Lenin bequeathed his successors. Ambitious in its goals, it was, for all that, a program that emphasized patience. As Lenin modestly calculated, it would require an entire "historical epoch" to fully succeed.[23]

What is noteworthy about this program in the present context is that, as seen in a number of letters, whatever peasants thought of cooperative or collective agriculture, many agreed that the state should actively improve the material and intellectual conditions of the village. One of the benefits of publishing a collection of peasant letters is that it allows the reader to question just how implacable an enemy of socialism the small-holding peasant really was. The letters that touch on this subject indicate that the answer to this question is far from black or white. "If socialism became the dominant language of 1917," Orlando Figes and Boris Kolonitskii write in their study of language use during the Russian Revolution, "then it was largely because it provided the peasants with an idiom in which to formulate their own revolutionary ideals." As many letters collected here show, in like fashion, peasants continued expressing their aspirations in a socialist dialect during the 1920s even after the war-communism experience had shown them the extremist measures Bolsheviks were willing to employ to get what they wanted from the countryside. These letters suggest that the peasants' use of socialist language is only partly explainable by the regime's ability to set the terms of discourse or the peasants' desire to ingratiate themselves with the country's new rulers. After the revolution, the language of socialism continued to provide peasants with a means of articulating their demands because fundamental socialist principles intersected at many points with their own notions of what constituted a just society. And, as Figes and Kolonitskii suggest, where these intersections were unclear or even non-existent, or where a socialist concept was simply not understood, peasants did not hesitate to make interpretations advantageous to their own interests—all in all, neatly reversing the tactics that urban propagandists had used to gain peasant support during the revolution when they presented socialist doctrine in religious or other terms designed to appeal to peasant sensibilities.[24]

An important role in the peasants' relation to socialism must be assigned to their experiences since 1914. By removing Russia from the war, the socialist revolution had ended the bloodletting in which the principal sufferers were peasant soldiers, and returned millions of peasant fathers and sons from the front lines to their homes and families. Just as decisively, it had eliminated the social and legal obstacles to peasant land ownership and, at least in the peasants' mind, the obstacles to self-government. Thus, from the outset, and in a very real way, socialism meant peace, land, and liberty, all of which the peasants held in near-sacred regard. The commonplace that peasants ceased to be revolutionary after gaining control of land expresses an important but only partial truth. Certainly, the peasants remained committed to the family farm, the commune, the village assembly, and the regularities of the church calendar as the foundations of their social and economic life. But support of tradition should not be taken as evidence that peasants were essentially backward-looking. After the revolution they, too, were interested in a sort of socialist construction. Their letters are full of calls for the state to carry out changes in the countryside, frequently justified as extensions or fulfillments of the socialist revolution whether or not they comported with the Bolshevik program. In this way, peasants drew on an older paternalistic ideal to define, to their benefit, both socialism and the terms of their relation with the new state.[25]

The anger and bewilderment that many peasants expressed at their second-class status ("stepsons" as opposed to "true sons" of the revolution) are perfectly understandable. To the peasants, the postrevolutionary land redistribution had been a signal act of social justice, and they saw no necessary contradiction between landownership and the basic tenets of socialism as they understood them or, more accurately perhaps, as they wanted them understood. In any case, if socialism was a morally and ethically just form of economic organization that privileged laborers over the exploiters of labor, then why, the peasants reasoned, should their revolutionary gains be seen as incompatible with it? Likewise, why should avowed socialists like the Bolsheviks not endeavor to improve the working peasants' lot? In this sense, for many peasants, landholding was not a contradiction but a precondition of socialism, and they reasonably expected that the future development of the revolution would result in continued improvements in their economic situation. Thus, while most peasants, except those in the direst of circumstances, questioned the advantages of collectivized farming, they were not unreceptive to socialism's ethical claims or the material benefits it promised to deliver.

Such generalizations must be qualified. The peasant class was not homogeneous, and neither were the interests and convictions of its members. Economic differences marked the class coming to the fore at various times and over various issues. For all its crudity and limitations, the Party's division of the peasantry into rich, middle, and poor strata— categories that carried political as well as economic ramifications that the peasants accepted and rejected as it suited their purposes—does help explain different reactions among the peasants to Bolshevik policies and to the changes they portended. Understandably, given the precarious terms of rural existence, the peasant was to a large degree *Homo economicus;* but economics alone did not determine peasant attitudes to the new regime. Regional variations, religious commitment, ethnic identity, and, importantly, personal experience helped shape reactions to the revolution and to the social and political relations that emerged from it. Civil war veterans understood the meaning of the revolution and the role of the Party differently than did their fellow villagers who had not served in the ranks. The young, more so than their elders, could see the new order as providing an alternative to traditional village life and the means for individual advancement. Men and women viewed through different lenses the revolution's promise to rearrange domestic life by altering gender roles, childrearing practices, and the definition of marriage itself. No class-wide consensus on all the transformations set in motion by the revolution was possible. Indeed, Chamberlin's observation that the peasants had "awakened" barely hints at the variety of opinions they held or the fervor with which they expressed them.

The Voice of the People

Seen in this light, letters from individuals provide, at best, subjective responses to Bolshevik policies and actions; they must be used carefully and critically in coming to any general or systematic understanding of state-peasant relations. For every opinion or judgment expressed by one peasant another will voice a different or even contradictory view. The historian must avoid the temptation to consider one or another of these views as more or less valid or representative. True as this might be, the editors believe that the accumulated weight of the subjective and anecdotal evidence contained in these letters adds a great deal to our understanding of rural dwellers' relation to the revolution and the state and will be of interest equally to social, political, and cultural historians of the Soviet Union. In a very immediate way, these letters place real human beings at the center of the political and economic debates that so defined the era. Historically, peasants have been the least literary of social groups; written evidence of their thoughts and intentions has been the exception rather than the rule. The spread of literacy among the peasantry in the late nineteenth and early twentieth centuries, however, and the establishment of a state that claimed to act on behalf of toilers, encouraged an unprecedented outpouring of compositions that shed tremendous light on peasant mentalities. The wealth of detail on economic, political, local, and daily-life questions provided in these letters vividly supplements the scholarly literature on the subject and helps expose both peasant perceptions of state policy and actions and peasant responses.

The story of the Bolshevik Party's efforts to "solve" the problem of Russia's agricultural backwardness and the tragic results has been told many times by contemporary observers and historians.[26] The uniqueness of the present book lies in its angle of vision and most especially in its voice, which is that of the peasants' themselves. In other words, this is history from below as recorded in the words of the common people. It is a history that until the collapse of the Soviet Union had been largely inaccessible, since the letters presented here, contrary to most of their authors' wishes and intentions, were not published when they were written and remained locked away in the archives of the Communist Party and Soviet state. Their publication in English is a continuation of the historical project made possible by the collapse of Soviet communism. Since 1991, historians have been able to replace the monologue of congress resolutions, governmental decrees, laws, and political speeches that for so long provided the basic source material for Soviet historical studies with a dialogue in which Soviet citizens are given their chance to comment on the events and the times through which they were living.[27]

And there is no end of comment. The more than 150 documents included here represent only a tiny fraction of the hundreds of thousands of letters that flooded Soviet newspaper offices each year during the 1920s. In 1927, *Krestianskaia gazeta* alone received over 700,000 letters.[28] The sheer volume of correspondence guaranteed that most letters would never see publication in broadsheets. The letters touch on every facet of life in the Soviet countryside from the mundane to the momentous. Much of the correspondence centers on matters of personal or local importance, although writers usually take it for granted that their problems and those of their particular village or hamlet have important implications for the condition of the nation at large and provide a barometer of the success or failure of the Bolsheviks' socialist experiment. Since the new regime claimed to be the champion of those who labored, this is as it should be. The laborers whose letters are included in this collection, however, were peasants and not the Communist Party's preferred industrial workers. The correspondence occurs, that is, between individuals from the overwhelming majority class of the country and a ruling elite dedicated to a radical reconstruction of that class as a precondition to its own success. To the peasants, the complexities and contradictions inherent in this situation were fundamental, and they frequently took it upon themselves to inform the Bolsheviks that their policies revealed a basic ignorance of agriculture and rural life and hindered rather than encouraged the development of the nation's resources.

The opinions, political and otherwise, expressed in the letters range from sympathy for the new regime to calls for its eternal damnation. Many letters express support for the state's efforts to modernize village life by disseminating up-to-date agricultural techniques or alleviating the hardships of daily life. Requests that more be done to help eliminate rural backwardness and "darkness" are common. A number of letters offer details of agricultural policy and reveal a nuanced understanding of land use, taxes, and other issues. On the other hand, hostile letters are in ample supply. In these, the Bolsheviks and their rural representatives are ridiculed for their stupidity, damned for their cruelty and their atheism, or exposed as corrupt parasites no better, and often worse, than the old-regime officials they had replaced. Comment was not limited to matters of immediate or local interest, however. According to one study, in any given year *Krestianskaia gazeta* devoted as much as 16 percent of its space to international news, and foreign affairs and high policy are important topics of concern in the letters.[29] Whatever the views of the writers, most of the letters, even the highly critical ones, are signed, indicating the importance their authors attached to the views expressed and their belief that they had every right to express them.

A large number of letters identify corrupt officials and expose abuses of power. Such letters were not necessarily written to be published but to call governmental attention to a problem and to plead for intervention from above. A number of the letters in this collection are, in fact, just such appeals or petitions for Moscow's help in righting a local injustice or rescuing an individual in dire straits. Petitioning higher authority for the redress of grievances is a very old tradition in Russia. Historically, this device had been employed, especially by peasants, to leapfrog over local authorities or landowners considered dishonest or oppressive in order to speak directly to the sovereign, who was believed to be sympathetic to individual subject's plight and powerful enough to deal effectively with local power wielders. Historians generally consider that this popular belief in the sovereign's essential goodness, known as "naive monarchism" or "the myth of the good tsar," was shot down during the massacre of workers who marched on the tsar's palace in St. Petersburg on 9 January 1905—the infamous Bloody Sunday. Whatever the truth of this as far as Nicholas II and the Romanovs are concerned, the practice of petitioning Moscow continued throughout the Soviet era, no doubt encouraged by the Communist Party's designation of the new state as "worker-peasant" and by Party members' efforts to portray themselves as the people's representatives. Although the monarch was gone, individual leaders, as well as newspapers, regularly received many appeals for help.

Though written to public individuals and institutions, as opposed to family or acquaintances, the letters collected here are personal; to characterize them as "not personal" would be, to a certain extent, to misrepresent them. In many cases, their authors poured into them their deeply held feelings about what was occurring in their villages, on their farms, and even in their homes. Many also express well-thought-out—if not always well-informed—views about the course of the revolution and the nature of the new society that was or, in their opinion, should be coming into existence. For uneducated or semiliterate people, little is more personal than opinions on "important" questions that they have struggled to formulate and articulate before a wider audience. The value that such individuals tended to place on the written word added significance to an act—writing to authority—that already carried great weight. The writers' sense of responsibility may have been heightened by the realization that letter writing provided one of the few opportunities for a citizen to participate in the political process. The regime certainly hoped that inviting letters would help give it more legitimacy in the eyes of those who previously had had little say in matters that directly affected them.

Though subjective and often idiosyncratic, the letters share many traits. Among the more immediately obvious are the apologies for poor handwriting and bad grammar. Writer after writer begs forgiveness for awkward or clumsy phraseology and misspellings. Typically these self-effacing confessions lead to an explanation of how the difficulties encountered in life had prevented the writer from acquiring a proper education. That these declarations were sincere there can be no doubt. In a society where attaining literacy was a major accomplishment and a thorough education a rare thing, the inadequacies felt by the uneducated before the educated were tangible. Russia had long been a society rent by a deep cultural gulf between its educated elite and its mass of uneducated inhabitants. For many of our correspondents, the very act of putting their thoughts on paper could have been accomplished only after great struggle. By the same token, in a political system that glorified the lower classes, a humble apology for ignorance helped establish the author's proletarian bona fides and could be expected to promote a sympathetic hearing.

Detailed biographies, especially elucidations of work history, revolutionary activity, or civil war service and the hardships these entailed for the writer and the writer's family are also frequently encountered. Here, too, the authors were establishing their class credentials. Often, supplying a biography was also a method for legitimizing the author's right to speak to an issue assumed to require a degree of experience or expertise to deal with, particularly farming. In other cases, the writer seeks to explain or atone for a past transgression—committed, in some cases, by a parent or even a grandparent—such as having engaged in some type of commercial or profit-making activity. That the act may have occurred years before the revolution was of little importance, since a person's past now went a long way toward determining legal status, access to an education, and acceptance into communist and Soviet organizations.

The proliferation of publications directed at a peasant audience in the 1920s encouraged this outpouring of correspondence and was a direct

result of the Communist Party's efforts to propagandize among the peasant class and to gain its political sympathies. In March 1921 the Party
officially ended the forcible requisitioning of peasants' grain and introduced the New Economic Policy. Under this new arrangement peasants,
on fulfilling their tax obligations to the state, were granted the right to
sell their remaining grain on the private market. The NEP introduced a
new era in which state and private economic sectors existed side by side
and fiscal and market mechanisms replaced the coercive measures used
to get peasant grain during the civil war. In turn, the new economic measures required a new way of propagandizing among the peasants.

On the eve of the civil war, the Party had begun publishing the
newspaper *Bednota,* which, as its name indicates, targeted the village
poor, whose support Party activists sought to cultivate. Under the NEP,
however, the Party looked to broaden its appeal beyond the poorest
peasants—"the village proletariat"—and rural communist activists to
the numerically larger and economically better-off middle-peasant stratum. In April 1923 the Twelfth Communist Party Congress called for the
creation of a cheap mass-circulation weekly newspaper written in language accessible to the peasantry. The result was *Krestianskaia gazeta*
which began limited publication in November 1923. Yakov Arkadeevich
Yakovlev, a leader of the Central Committee's press department and a
Party expert on rural affairs, organized the newspaper and served as its
first editor. (In 1924, Yakovlev also assumed editorial duties at *Bednota.*)
From the beginning, the Party leadership placed great hopes on the paper's
ability to improve the Party's relations with the peasants and instructed
lower-level Party and Komsomol (communist youth) organizations to see
to its widest possible distribution. The February 1924 Central Committee
plenary session called for circulation to reach 200,000 by the spring and
placed special emphasis on the paper's distribution among teachers and
Red Army soldiers (most of whom were peasants). A Central Committee
circular sent to Party organizations after the Thirteenth Communist
Party Congress at the end of May over the names of Yakovlev and the
Central Committee secretary, A. A. Andreev, styled *Krestianskaia gazeta*
"the most important mode of communication between the peasantry
and the working class and the means of carrying the Party's influence
into the countryside." Echoing the instructions of the Congress, the circular also advanced the slogan "One newspaper for every ten peasant
households by spring 1925." This meant a combined circulation of
2,000,000 newspapers (from central to local) aimed at the peasant audience. Indeed, *Krestianskaia gazeta*'s circulation steadily grew beginning
in 1924. Between January 1925 and March 1926, circulation increased
from 600,000 to 1,000,000 copies. In October 1928, *Krestianskaia*

gazeta began appearing twice a week, and by the end of 1929—on the very eve of wholesale collectivization—circulation had reached 1,400,000.[30]

The cost to the peasant was low—initially fifteen kopeks for a one-month subscription (four issues). To encourage subscriptions, the paper offered such bonuses as color portraits of Lenin, Trotsky, Kalinin, and other revolutionary leaders and a printed collection of laws pertaining to peasants. A six-month subscriber received, in addition, a daily calendar. *Krestianskaia gazeta* answered peasant demands for an accessible newspaper attuned to their interests and practical needs, and this accounts for its popularity. Its editors encouraged letters from readers, and even a cursory reading of these suggests that, absent any independent means of political expression, peasants saw the paper as a sort of advocate or intercessor on behalf of their interests with the Soviet government in addition to a valuable source of information. A mass newspaper, as opposed to a political newspaper, *Krestianskaia gazeta* mixed news, entertainment, and practical advice in its columns and fulfilled the Party's requirement for a paper the peasants would want to read. Its success encouraged the publication of more peasant-oriented reading matter. During the 1920s villagers were hungry for information, so within a year of the appearance of the first issue of *Krestianskaia gazeta,* its publishing house began printing other periodicals aimed at specific segments of the rural audience. Ultimately, these included the satirical *Lapot* (Bast Shoe), *Derevenskii kommunist* (The Village Communist), *Krestianka* (The Peasant Woman), *Druzhnye rebiata* (Friendly Fellows) for children, and *Sam sebe agronom* (Be Your Own Agronomist).

The newspaper's archives for the 1920s contain millions of letters from across the country. Because giving them all space in print was simply impossible, editorial decisions had to be made about which letters to print and which to consign to the archive. Letters written in response to one or another issue composing the current official agenda stood a better chance of placement in the newspaper than those addressing less topical issues. The editors evidently eliminated many letters from consideration for publication on the basis of their readability. The writers were painfully aware of their literary deficiencies and often appealed to the editors to correct their solecisms before publishing them. Their apologies were all too apt: countless letters were illegible, and others were all but unreadable for their poor grammar and spelling. Still others employed an earthy language that the editors would have considered too vulgar to print.

Clearly, a large number of letters rejected for publication took too critical a stance in regard to the state's agricultural policy or told stories of village life that conflicted with the official picture of rural conditions.

Because of naïveté, earnestness, unwillingness to conceal the truth, or the psychological need to release pent-up anxieties and emotions, these writers refused to make concessions to official ideology. Aware that their candor could reduce the chances of their letters appearing in print, those who were a bit more sophisticated often employed the obvious device of daring the editors to place their uncomfortable observations before the newspaper's readers. Letters written in this vein sometimes conclude with the reverse-psychological challenge: "But I know you won't do this." The more extreme sorts of these letters express overtly hostile opinions of the regime, its ideology, its leaders, and its rural representatives and mince no words in characterizing them. The authors of these commentaries probably did not entertain any serious notions that their abusive missives would be published and for that reason felt free to say what was on their minds. Such letters rejected the regime's "master narrative" of ever-improving state-peasant relations; putting sharp thoughts down on paper was surely as much a cathartic as a political act.

In the letters peasants express diverse opinions on the questions of the day, and the present collection includes as wide a range of peasant opinions as possible. Letters are taken from the poorer and the better-off, from the highly literate and the minimally schooled. Some are written in matter-of-fact language, others in much more colorful and metaphorical terms. No attempt has been made to purge the letters of "inappropriate" words or expressions, although considerations of space required that some letters be trimmed. The letters touch on all manner of political, economic, social, technical, family, and personal issues. The overarching goal has been to provide as full a picture of rural life in the 1920s as the fund of letters allows. As a rule, from a batch of letters addressing the same or similar themes, letters providing details or concrete illustrations were selected in order to bring particular problems into sharp focus and show how they impinged on the lives of individuals. No conscious attempt has been made to exclude letters expressing particular political or other points of view. Nonetheless, selection is, by definition, a process of exclusion, and other editors would no doubt have made different choices and produced a different book.

In addition to unedited original drafts and typed copies of letters, which have, for the most part, been reproduced in full, this book also makes use of official summaries (*svodki*), produced to provide the political leadership with evidence of popular opinion. Compilers of summaries drew from various sources, including citizens' public declarations, private conversations, gossip, and letters. It was in this usually redacted form that leaders often read citizens' writings. The summaries also contained the contents of secretly printed leaflets, handbills, and appeals that had been surreptitiously distributed among the population. In the 1920s and 1930s a number of institutions—the OGPU/NKVD (the secret police), the Communist Party, the Komsomol, the political departments of state institutions, and newspapers (including *Pravda, Izvestia,* and *Krestianskaia gazeta*)—put together such summaries. During Soviet times the summaries were classified and carried the designations "secret," "highly secret," "confidential," "for official use," and so on.[31] They were typed or mimeographed in limited numbers, distributed to various organs for specific purposes, and intended to be returned after use. However, as often happens in large bureaucracies, stray copies remained in the archives of different institutions. The summaries used in this book, for example, were discovered in the files of the Central Executive Committee (TsIK) of the All-Russian Congress of Soviets and the Commissariat of Agriculture (Narkomzem).

Since these summaries addressed highly sensitive political topics, often contained unverifiable materials, and were composed by subordinates for their superiors, historians have raised a number of legitimate questions regarding their reliability and accuracy. As valuable historical sources, they cannot be dismissed, but variables like the prevailing political priorities, the conflicting agendas of different institutions, and the compilers' assumptions about what their bosses wanted to hear may have influenced decisions to include or exclude certain materials from the reports.[32]

Because this collection is devoted to presenting a broad view of rural life, it is extremely disparate in content. It was not put together to settle any one historiographical question or to advance any particular mode of analysis, so the evidence it presents may lend support to divergent viewpoints. The people of rural Russia did not speak in one voice—that is what we need to remember. Geographical location, economic situation, family relations, gender, personal experience, religious inclination, political sympathies, individual psychology, and an array of other, perhaps unknowable factors contributed to the opinions expressed in these letters.

This is not to say that the editorial voice employed in the following pages is always neutral or dispassionate. The primary purpose of the explanatory text and commentary accompanying the letters is to provide necessary historical context or other information to help make the letters intelligible and accessible even to nonspecialists. Beyond that, however, the text offers analyses and interpretations of the material. These are open to dispute, but even a cursory reading of the letters suggests certain general conclusions.

First and most obvious is that the peasants and the Party often spoke and acted at cross-purposes because of fundamentally different understandings of the revolution and its future course. For the peasants, the revolution was what granted them land. The Bolsheviks, on the other hand, viewed the revolution as the first step on the path to a radical transformation of agricultural organization. Once the state abandoned its attempts to force this change in 1921, the peasants reasonably assumed that state policy would encourage the independent farmer to maximize production. This expectation was frequently thwarted by the Party's rural program. While seeking to guarantee a steady supply of grain and other products to the cities and the army, as well as for export, the Party also sought to inhibit the growth of a class of economically strong farmers. As a result, state policy often appeared irrational and contradictory to the peasants, and they were not averse to saying so.

Second, the Bolsheviks' conception of rural life, especially their division of peasant society into poor, middle, and rich individuals and households and their accompanying assumption that the three strata were naturally and continually in conflict with one another, was highly reductionist, reflecting Bolshevik ideology and preconceptions, and, as a result, tended to ignore the historical, cultural, and social bonds that held peasant society together. The peasants themselves often resented and rejected their categorization or manipulated the Bolsheviks' labels in pursuit of their own ends. Class distinctions and even conflict there certainly were, but as many letters indicate, in peasant eyes class struggle was not the mainspring of rural life, and most peasants who commented on the matter objected to what they saw as an arbitrary division of the inhabitants of their world by ignorant outsiders bent on fomenting strife for their own exploitative ends.

CHAPTER 1

Revolution and War Communism

Forward flies our locomotive.
The Commune is the only stop.
We have no other objective,
Our rifles firmly in our grasp.

—*Our Locomotive,* a civil war song

Few people have ever experienced more turmoil and catastrophe in so brief a time as the inhabitants of the Russian empire in the four years between February 1917 and March 1921. These forty-nine months witnessed the final collapse of the centuries-old tsarist feudal-autocratic system, followed in short order by the establishment of a liberal, then liberal-socialist Provisional Government. Amid unmanageable chaos, the last government quickly fell victim to an adventurous coup that led, ultimately, to the emergence of a one-party Marxist dictatorship. Within months, the new regime extricated the country from a long and bloody world war, but the dictatorial methods it employed in its drive to consolidate power helped precipitate a merciless civil war that inflicted untold suffering over a vast territory. The resulting carnage generated peasant rebellions and military mutinies, armed foreign intervention, hunger and widespread disease, epidemics, crop failure, famine, and near-total economic collapse. Ethno-nationalist struggles broke out across the empire, the social hierarchy was destroyed, along with old-regime elites, and a political drama of the highest order played out on a grand stage. Sober scholarly estimates place the country's total population loss from the fall of 1917 to the end of 1921 at eleven million people.[1] The revolution that ravaged the Russian lands in these years was a revolution in the word's fullest and most ferocious sense.

In fact, the Russian people experienced two revolutions in 1917. The February Revolution, fueled by the world war's endless bloodletting and the hardships it imposed on the home front, released all the social and political resentments that had built up in the years since the "first" Russian revolution in 1905 and in a week's time succeeded in bringing down the 304-year-old Romanov dynasty. In the revolution's wake, bourgeois, proletarian, and peasant Russia united briefly, drawn together by the euphoria that had attended the tsar's abdication. In the capital and revolutionary center, Petrograd, with the consent of workers' and soldiers' councils, or soviets, a Provisional Government consisting primarily of liberal members of Russia's parliament assumed state power. For a moment, the liberal dream of a supra-class solution to Russia's problems seemed at hand. Quickly, however, the fissures that crisscrossed the empire's social structure ruptured. Unfulfilled peasant demands for land, the continued casualties and sacrifices demanded by the Russian war effort, fears of a counterrevolution, and food and supply shortages in the cities—especially in "Red" Petrograd—reignited class animosities. By the fall, the lower classes, no longer trusting the "bourgeois" and moderate-socialist ministers at the head of the Provisional Government to safeguard the revolution, looked to the soviets as the only legitimate and democratic revolutionary authority.[2]

Among the important political parties in the Petrograd soviet, only the Bolsheviks—an audacious faction of Marxist militants committed to worldwide socialist revolution—viewed the deteriorating situation as an opportunity. Since July, when the Provisional Government had all but demolished the Bolshevik Party's central leadership, and even more after the failed counterrevolutionary coup attempt of General Lavr Kornilov in August, the Bolsheviks' popularity had steadily risen among industrial workers and the rank-and-file military. In a historical irony of no small import, this conspiracy-minded organization, headed by intellectuals who prided themselves on being a small, select vanguard of the working class, found itself suddenly transformed into a mass revolutionary party. Seizing on popular support as a mandate, the Bolsheviks' founder-leader, Vladimir Lenin, then hiding from arrest across the border in Finland, ceaselessly prodded his Party comrades to act. Finally, on 25 October, the Petrograd Bolsheviks and their supporters seized key points throughout the city and overthrew the Provisional Government in the name of the soviets and socialism.[3]

For the laboring population that supported the Bolsheviks, the appeal of the Party lay less in its advocacy of international socialist revolution than in its promises to end years of war and months of economic insalubrity. As early as April, Lenin and his colleagues had begun writing out the prescription for Russia's ills in a few simple but attractive slogans: "All power to the soviets!" "Bread, peace, land!" "Workers' control of the factories!" By fall 1917, this program coincided perfectly with popular frustration at the Provisional Government's impotence and the lower classes' understanding of the revolution as a means of recasting their own lives minus the landlords, bosses, and officers who lorded it over them. Braced by rising popular support, the Bolsheviks took power. Now they were expected to deliver on their promises. In the ensuing months, the Bolsheviks and their supporters set out to remake the Russian political, economic, and social landscape ostensibly on behalf of the have-nots who labored in the factories and fields and who were also fighting the Great War in the trenches and on board ship. Their early efforts, however, not only failed to stem the economic and social deterioration that had cleared their path to power but undermined the popular-democratic promise of Soviet power itself.

The new regime's early pronouncements were in keeping with popular expectations. Lenin, in his first public act following the coup, read a peace proclamation to the Second All-Russian Congress of Soviets that called on all warring nations to end hostilities and to renounce territorial annexations and financial indemnities. The proclamation was immediately followed by the Decree on Land abolishing private land ownership and transferring control of farmland to local land committees and soviets pending the convocation of a Constituent Assembly representing the entire nation.[4] Thus, in the matter of a few hours, even before a new government or a legislative process had come into being, the Bolsheviks had tackled two problems that had hamstrung the Provisional Government. In subsequent months, the popular forces set in motion by the October Revolution continued to assault the status quo.

By March 1918, five months after the October coup, the Bolsheviks themselves were facing crises on many fronts. In 1917, unabated economic deterioration had driven the revolution forward as much as any agitator or propagandist. Since taking power, the Bolsheviks were proving no more capable of arresting this decline than their Provisional Government predecessors. Thanks in large measure to the breakdown of transport, city residents faced starvation, and industrial production was in freefall. Peasants, taking advantage of the Bolsheviks' support for land redistribution, were breaking up large, privately owned estates into smaller parcels, thereby undermining commercial agriculture and contributing to food shortages in the cities. Town-country exchange was also seriously disrupted by the decline in industrial production and a worthless ruble; facing a dearth of finished products to buy, low state prices for grain, high retail prices, and currency inflation, peasants had little incentive to bring their produce to market.[5]

The Bolsheviks had concluded an onerous and very unpopular settlement with Germany in fulfillment of their promise to bring peace. The

Treaty of Brest-Litovsk ended Russia's participation in World War I, but only by ceding large, mineral-rich, heavily populated areas of the empire to the Germans. The loss of grain-producing regions, particularly in Ukraine, exacerbated the food crisis. Russians of all stripes—socialists, nationalists, liberals, monarchists, peasants, workers, soldiers—thought the treaty a sellout of the revolution, the motherland, or both. The Bolshevik leadership itself had nearly split apart in the bitter intra-Party debates on the treaty. When the treaty was formalized on 3 March, the Bolsheviks' only coalition partners in the new government—the Left Socialist-Revolutionaries—quit over the issue and began a terror campaign against their former partners. Thus came into being the single-party dictatorship. Civil war loomed on the horizon.

At this point, in spring 1918, the alliance between the Bolshevik Party and the forces of popular revolution also began to fracture. Factory workers, the Bolsheviks' base, were doubly hit by the economic collapse; shortages of fuel and raw materials led to layoffs and plant closings while rising prices and food shortages threatened their very survival. Before seizing power, the Bolsheviks had advocated workers' control over the factories, and this had helped them secure working-class support. In the months that followed, many factory committees had, indeed, established control commissions in fulfillment of that promise. The commissions watched over hostile owners and managers whom they suspected of sabotage and, contrary to the wishes of the Party, took an increasing hand in day-to-day management as the economic crisis deepened, in the hope that they could keep their enterprises solvent and operating. The struggle to stay open gave birth to a competition between factories for resources and orders that contributed to the reigning economic chaos.[6]

Factory committees understood that revolution at the state level would only marginally benefit workers if not accompanied by a revolutionary reconstitution of labor-management relations in enterprises. The committees' desire for local control over production and democratization of factory management, however, placed them at odds with the mainstream Bolshevik view that only the state could effectively organize socialist production. To combat the "anarcho-syndicalist" tendencies of the factory committees, the Bolsheviks created the Supreme Council of the National Economy (Vesenkha) and set about centralizing industrial administration. In some provincial soviet elections held in late spring, workers responded by turning to the Bolsheviks' socialist opponents— the Mensheviks and the Socialist-Revolutionaries (SRs)—in hopes that they would more faithfully implement the October Revolution's economic and democratic promises.[7]

Documents

Social Collapse

The hardships of daily life served as a major source of dissatisfaction with the new regime. Workers, in whose name the Bolsheviks exercised their dictatorship, keenly felt the material deprivations and did not hesitate to express their discontent. One worker who wrote to Lenin called him "friend" because he found the word "comrade" repulsive. Another, clearly in desperate straits, wrote the following: "Comrade Lenin. What have you done to us? We placed our hopes on you, we trusted the governing of the country [to you], and have been left without a crumb of bread on the street. It would be better to have been shot . . ."[8]

In the following anonymous letter to Leon Trotsky and Lenin from November 1918, a Petrograd factory worker attributes his disaffection from the Bolsheviks to the food-supply situation and the actions of the Red Army. The original letter contained numerous spelling errors.

· 1 ·

Letter to L. D. Trotsky and V. I. Lenin from an unknown Petrograd worker on the food-supply situation in the city, November 1918. RGASPI, f. 5, d. 1501, ll. 28–29. Original manuscript.

Comrades Trotsky and Lenin

We, all the proletarians of the Petrograd factories and plants, are quitting you. We are quitting because your government is crushing our strength. Before we experienced it, all of us stood unanimously for Soviet power. But when we saw the ways of Soviet power, then we curse Soviet power and who leads it.

I will explain all the reasons for our cursing:

1) The Red Army steals from us, from the living and the dead, without checking if you have the capital for your family's food or not. They don't check this, just give him what he needs. And if you don't hand it over, then they shoot you on the spot and cook up proof that he didn't recognize Soviet power.

2) They steal from all the passengers on the railroad. Whoever receives these miserable rubles and goes to buy bread for his family comes back without money and without bread. After all, Comrades Trotsky and Lenin, you know yourselves how much bread they are giving per head in Petrograd. A worker can't get his fill this way. Though it's expensive, you have to buy from the speculator, like it or not.

3) Comrades, we ask you as our Soviet leaders to permit all the Petrograd workers free passage to Toropino station in order to buy bread, sugar,

and all provisions.[9] In Petrograd, they're paying fifty-five rubles for four hundred grams.

Comrades, we ask you not to refuse these requests, and we hope that you don't refuse, [and] give us free passage.

Comrades, it should interest you ["us" in the original] that the Germans and the Northern Army feeds us from their sacks.[10] Comrades, when you grant free passage, then in the hungriest cities everything will cost less.

The Germans bring in bread from abroad, but the Red Guard steals.

Comrades, fulfill our request.

This worker begs Lenin and Trotsky to permit the activity known as "bagging" (*meshochnichestvo*) that arose in the months following the October Revolution and continued into the civil war. To compensate for food shortages, traders and other city residents would travel to the countryside, usually by rail, carrying sacks loaded with finished goods or even personal belongings. After trading these items for foodstuffs, they would return home with supplies, which they would either consume themselves or exchange on the black market. The authorities sought to clamp down on this practice, which was not only unregulated, and hence illegal, but so widespread as to tie up transport. As the letter shows, even self-styled conscious proletarians in Red Petrograd recognized the necessity to disregard prohibitions against private trade in order to support their families, no matter what detrimental effect this had on production. So grave was the crisis that workers at the Putilov factory, a center of Bolshevik support, demanded that the entire factory shut down temporarily so that they could go out in search of food.[11]

The archives of the former Soviet Union contain a trove of letters indicating that people attributed the deteriorating economic and social situation directly to the Bolsheviks. Such letters, anonymous more often than not, frequently convey contempt for Lenin and other Bolshevik leaders and for decades remained unavailable to researchers. In the following letter, reproduced almost in its entirety, a young man writes to Lenin in respectful tones, but he does not hide his horror and disgust over the societal breakdown that he has witnessed.

· 2 ·

Letter to Lenin from an unknown twenty-three-year-old man decrying abuses of authority committed in the name of the revolution, 25 November 1918. RGASPI, f. 5, op. 1, d. 1501, ll. 34–38. Original manuscript.

I can't claim the wisdom of a long life. But even though I am only twenty-three, I think that I am able to make sense of and logically understand everything that is happening. Sometimes, seeing the despondent faces that surround me or those twisted in anger, it seems that either some epidemic of madness has everyone in its grip or I'm the one who has lost his mind. Stroll along the streets and you will not see one smiling face. Everyone walks about morosely, dispiritedly. And this at a time when the bright sun of socialism, by all reckoning, should restore in everyone a joy for life. Where, exactly, is the source of this cruel irony of fate hidden? Do not think that I hold You to be the root of this evil. On the contrary, I bow down before You as before a man burning with the faith of his convictions. Where there is sincerity I am prepared to excuse everything. I consider You a man of tremendous erudition, but a man always remains a man—a weak being, easily mistaken. But one thing continues to astonish me: You are truly fenced off from us, the simple, ordinary citizens, by some kind of screen. In the Kremlin—your Parnassus—you are not accessible to us. You do not even appear in print, and only once in a great while is there a note that You spoke at some meeting for workers—only for workers, and what are the other citizens supposed to do? As far as I am aware from my high school classes, workers do not make up a large percentage of the Russian population. If only You could fully see the depth of sorrow, of despair, of individual cruelty, and of tears which now engulf our unfortunate Russia, then You would renounce socialism. It is just this nightmare, it seems to me, that the thick Kremlin walls have hidden from Your sight. Working in the quiet of a study or in a noisy meeting, You are quite distant from this horror.

Everything in the world happens through evolution, and nature cruelly avenges untimely audacity. It seems to me that Your attempt to plant socialism among us, precisely among us in Russia, is indeed audacious, genuinely bold and arrogantly audacious. For decades, under tsarism, they only stupefied and inebriated the unfortunate Russian people with vodka. We still can't air out the lingering effect of its intoxicating fumes. Even the radiant idea of socialism takes the form of drunken fantasies . . . Crudeness and incompetent arbitrariness are now our ruling elements. The Russian is now confronted by his most unattractive side. Having been given a wide opportunity for an orgy of the very lowest passions and the settling of personal accounts, take a look, isn't this opportunity being exploited? Those that earlier foamed at the mouth shouting "Beat the Yids" now, with no less enthusiasm, shout "Beat the bourgeois," and they laugh to prove that they are staunch communists. Their moral outlook is too primitive, and that's

why abrogating any individual's rights, even the most sacred—the right
to life—does not present a dilemma [for them]. After all, any person has
the right to a life under the sun, but in the pages of *Izvestia* and from the
mouths of communist orators the clamor endlessly reverberates, "Death to
the bourgeoisie! Long live the proletariat!" The words "wipe from the face
of the earth, crush, destroy," etc., are always flashing. But for God's sake,
show me the line where the bourgeois ends and the proletarian begins.
There is no such line, and it is impossible to draw one. Given that, what is
to be done with that intermediate group which is related neither to the
bourgeoisie, since the people in this group have no capital or property but
get a wage, and that's how they live, nor to the proletariat, since their out-
ward life is relatively comfortable? These people unwittingly occupy the
spot between the hammer and the anvil . . . You may reproach me, that I
said all this only for the sake of argument, that in reality the only scapegoats
are the big bourgeoisie, but alas, there are facts at hand showing that the
intermediate group suffers equally with the bourgeoisie . . . I think it's crazy
to put almost unlimited authority in the hands of people who work poorly,
who impede the centers, and who, moreover, ignite class hatred, of which
they have a very peculiar understanding. I am not an opponent of the dictatorship of the proletariat, but [it should be] a conscious, organized proletariat . . . Power in the hands of unconscious people or even the slightly
conscious is like a weapon in the hands of the insane.

In conclusion, I should say a few words about myself, personally, so that
you will have some idea about the author of this letter. I am not from the
bourgeois class. I was born and grew up in a family where it was constantly
necessary to think about tomorrow and where individual labor was the sole
source of existence. I am a staunch opponent of any political partisanship
and am equally unsympathetic to both monarchists and anarchists. I believe that partisanship engenders divisiveness and impedes the main task of
humanity—the peaceful construction of life . . .

The idea that Russia was too backward and its people too primitive
for socialism was voiced time and again. The vivid and cogent sentiments expressed here recall the writings of pro-revolutionary intellectuals like Maxim Gorky, who despaired that the revolution's high ideals
had been overtaken by the characteristically base and "uncivilized" passions of the Russian masses, which he attributed, sympathetically, to
their cruel history. Intimate with the psychology of both the intelligentsia and the common people, Gorky wrote knowingly of the fragile link
holding the two groups together: "The Russian intelligentsia is the head,
malignantly swollen with an abundance of foreign thoughts and con-

nected with the torso not by a strong backbone of unified aspirations and aims, but by some barely perceptible and very thin neural thread."

It may, in fact, be that the young man quoted above found inspiration in Gorky's often bitter "Untimely Thoughts," carried in his newspaper *Novaia zhizn* (New Life). The following was written in March 1918.

> Of course, we are conducting an experiment in social revolution, an occupation very pleasing to the maniacs expounding this beautiful idea and very useful to crooks. As is well known, one of the loudest and most heartily welcomed slogans of our peculiarly Russian revolution has been the slogan "Rob the robbers!"
>
> They rob amazingly, artistically . . . They rob and sell churches and war museums, they sell cannon and rifles, they pilfer army warehouses, they rob the palaces of former grand dukes; everything which can be plundered is plundered, everything which can be sold is sold; in Feodosiya, the soldiers even traffic in people . . . This is very "original" and we can be proud—there was nothing similar even in the era of the Great French Revolution.[12]

A woman who had worked for fifteen years in a department of education and considered herself "a proletarian of intellectual labor" expressed similar sentiments in a letter to Lenin written on 19 December 1918: "Russia has never experienced anywhere such chaos, oppression, and illegality . . . It is hardly necessary to drag people into a bright kingdom when they are refusing [to go]. Paradise under the threat of being shot."[13]

Under the dictatorship of the proletariat, members of the educated classes did indeed find themselves "between the anvil and the hammer," as did much of old-regime culture. As Gorky indicates, there was a good deal in the attack on old-regime culture that was driven by ideological fanaticism and a good deal that was out-and-out theft and thuggery. Richard Stites has explored the impulse behind this type of "iconoclasm" in his study of utopianism during the Russian Revolution. "In the Russian as in other revolutions," Stites writes, "the descending fist wielding a hammer against a piece of statuary and the angry manifestoes of poets and anarchists reflected the same underlying desire to smash, demean, profane, mock, neutralize, and ultimately destroy a culture that was perceived as deeply offensive and immoral in order to construct a wholly new one."[14] As understandable as this fury was, such attacks ran the risk of alienating the very people in whose name they were being carried out, especially when the target was a religious one. The following letter from mid-1918, allegedly written by the "comrade of the chairman" of the Orel province soviet, one Alkhimov, describes

such an attack. In these years, the title "comrade of the chairman" usually meant the chairman's deputy, but the letter was probably written by
an angry religious believer or someone from the convent mentioned in
the text. Not surprisingly, given the malevolent and threatening tone, a
recipient wrote on the original document instructions to send copies of
the letter to the secret police (Cheka) and the Commissariats of Internal
Affairs and Justice.

· 3 ·

Threatening letter to Lenin from the "comrade of the chairman" of the Orel
soviet denouncing antireligious activity in Orel province, 28 June 1918. RGASPI, f. 5,
op. 1, d. 1501, ll. 20–21. Original manuscript.

To the Chairman of the Supreme Soviet of People's Commissars
Vladimir Lenin
[from]
the Comrade of the Chairman
of the Orel Soviet

I am notifying you of the events of 24 June in Orel province. Soldiers
from a Red Army unit appeared at the Orel convent to seize the last horse.
The convent houses nine hundred nuns who live by their own labors. They
manufacture shoes and clothing, [sew] gold embroidery, cultivate flowers,
etc. They need the horse to haul firewood, bread, water, etc. When the soldiers showed up without any documents, the nuns sounded the alarm and
gathered up the people. From behind the convent gates they stoned a soldier
to death. Members of the soviet came and quarantined the convent, not allowing anyone into the cathedral for services. The people of the entire province, and especially in Orel, are extremely dissatisfied with Soviet power for
suppressing and mocking religion. They are also disturbed that there are an
awful lot of Poles, Estonians, Latvians, and Jews in the militia, the soviets,
and other organizations. A well-organized and armed group of railroad
workers is preparing a general extermination [*pogolovnoe istreblenie*] of
Jews in Orel for 1 July and the immediate expulsion of Poles, Latvians, and
other nationalities. They have sent secret agents from Orel to Moscow on a
mission to kill Lenin, Trotsky, and Mirbach by whatever means available.[15]
It's plainly understood by everyone that instead of [getting] freedom they
have become the disenfranchised slaves of some kind of impersonal state.
Everyone is choking on this freedom. This freedom is causing everyone to
cry out because it is the freedom of force that favors shameless scum. 90
percent [of the people] expect the tsar [to appear] like manna from heaven,
and [they] blissfully anticipate how they will chop up the Bolsheviks and
hold Lenin and Trotsky in cages, feeding them hay and then feeding them

dog meat. If there is an attempt to register the grain harvest, then all the grain stacks will be burned, since everyone sees this as a humiliation and thoroughly illegal.[16] They consider you, Lenin, an unscrupulous, insolent, and cruel tyrant. In comparison with you, they consider [Tsar] Nicholas [II] an angel, and that mangy Yid Trotsky they consider a traitor. The impertinence of the members of the soviet and the soldiers in the Red Army has everyone in a white-hot rage. The members of the soviet drive through Orel at dizzying speeds, and every day their automobiles crush six or seven people and two or three horses. Rarely does a day go by that an automobile does not smash up a tramcar, maiming three to ten people. Everyone has refused [to undergo] general military training, saying that last year they were forced to kiss the Germans and told to fawn over [them], and now they are being forced to kill their brother little Russians[17] so that they will give us grain and order, but others say that they need to take up arms, get organized, and then slaughter all the Bolsheviks, torture Lenin and Trotsky, and put the tsar [back on the throne]. Comrade Lenin! Before it is too late, get out, stop this rape of the people; otherwise they will chew you and all of us up. You'll be crushed by the masses, and all your resistance and guns will be useless. You must grant the people's wish. Show a drop of humanity. You have inflicted a good bit of torture on the people and drunk almost all [their] blood. The commissars are former clerks and orderlies and are all stupid, crude, and cocky little boys. Everything is teetering. [The people] are preparing to greet the Germans' entry with a peasant procession and bread and salt.[18] Because of the suppression of religion, they would tear all of you to pieces if you happened to fall into their hands. Only the Okhrana's former convict-laborers love you.[19] The people are terribly hungry; they get one hundred grams of oat bread that is more like hog slop—bitter and poisonous. The children are dropping like flies. For the good of the people and their well-being, you must disband the soviets immediately, [and] you and Trotsky should get lost; we've had it—you took over and everything fell apart. The people are hungry, but the members of the soviets, Red Army soldiers, and sailors lead a wild life and gorge themselves on whores. The birthrate has dropped to a minimum. Depravity flourishes. Venereal disease and syphilis are spreading. Heed some good advice—to avoid a slaughter send a telegram now ordering the immediate lifting of the quarantine of the Vvedensky Convent in Orel and the return of the stolen pair of horses and the cart. Otherwise, it will be very bad, and no force will prevent a slaughter.

Lenin! We beseech you—Master of violence, lawlessness, and the desecration of men—and along with you that born defiler of Russia, Trotsky, explain to us exactly who you are. [Are you] the base protégés of [Kaiser] Wilhelm, traitors who are heading by a circular path to monarchism and have already attained [your] goal, or stupid idiots and degenerates who dream of corrupting the state by liquidating culture, and establishing a pastoral ideal, (Diogenes') bliss in a barrel?[20] Whichever it is, spies or degenerates and idiots, either way it is unacceptable and [offers] little joy, but the point is that you are leaving, and the former monarch Wilhelm will give you a pension, etc., but all of us he will crush like mice; you have led us into a quagmire, you ignoble, nasty people. Agents have been sent from Orel to kill you. Be afraid since they are members of the province soviet executive committee and have left here to kill you.

28 June 1918

Just what is fact and what fiction in this letter is impossible to say. Allowing for its exaggerations and hysterical tone, we can take the letter as an expression of outrage at the iconoclasm and arrogance of local representatives of the new regime and the growing hostility in which religious believers held the Soviet state and its leaders. Anti-Semitism and xenophobia, prevalent in society at large, frequently accompanied attacks on the Bolsheviks, as did the charge, often raised in 1917 and resurrected after the signing of the Brest-Litovsk Treaty, that the Bolsheviks were German agents. In this spirit, many letters to Lenin begin with salutations like "Herr Lenin" and "Wilhelm's servant"—referring to Wilhelm II, king of Prussia and kaiser of Germany—before letting forth with a stream of invective. Other messages carry a cynical tone: "A heartfelt thanks for the peace, bread, and liberty, but all the same you will be . . . in good shape."[21]

The belief that Lenin and the other Party leaders remained detached from and indifferent to the daily suffering of the common people is expressed in a letter dating from 25 December 1918.

· 4 ·

Anonymous letter to Lenin expressing hostility toward the Bolsheviks, 25 December 1918. RGASPI, f. 5, op. 1, d. 1500, ll. 31–32. Original manuscript.

Hail, great leader of the Russian proletariat, Comrade Lenin. Sitting in your nest in the Moscow Kremlin surrounded by its crenellated walls, do you have any idea, does your intelligent brow know, how your subject people are getting on?

Do you know what your free people are saying? They are saying, "Lord, how long will you withhold [your] mercy because of our terrible transgressions? Take pity on us, spare us!" They say, "The hour has come. You recall the old regime. We got by well enough. But now they have given us our freedom, and what freedom! They call us free citizens, then they take our last horse, the cow, the last crumb of bread. And for all your sweat and your tears you don't say a word, and if you said anything they would put a bullet in your skull without any questions—there is no one to appeal to." The people attend the village assembly [*skhod*], but they say very little, they bite their tongues. There's freedom for you. Here are my words for you, you bloodthirsty beast. You stole into the ranks of the revolution and didn't call up the Constituent Assembly. You deceived the people. You promised them complete freedom and land. You said, "Down with the prisons, down with executions, down with the soldiers, let there be guaranteed employment." Over and over, you promised there would be mountains of gold everywhere and a heavenly existence, but where is it all? The people sympathized with the revolution. They could breathe easy. They were allowed to assemble, to say whatever they wanted without fear. And then you showed up, the Bloodsucker, and just like that snatched away the people's freedom. And now, under the word "freedom," you act hypocritically; the prisons, instead of being turned into schools, are full of innocent victims. Instead of abolishing executions, you terrorize thousands of people, ruthlessly conducting executions everyday. You shut down industry, bringing hunger to the workers; you stripped the people of their shoes and their clothes; you prohibited free trade; and now the people are left with nothing, no shirts, no boots, no nails, no iron for plows or to cover wheels, no firewood, kerosene—nothing. Remember, you took away the people's freedom and are drinking the last drop of blood that's left after four years of war.

You are mobilizing the troops, but remember, not one mobilized soldier goes voluntarily. A bullet from the mercenary convict Red Army forces him to go. But the hour of reckoning is at hand. The people are secretly preparing their weapons and impatiently await the Whites [counterrevolutionaries], and together with them they will hang all the bloodsuckers of Soviet power.

> Refrain, devil incarnate, all the people curse you.
> Your Red Army volunteers,
> sign up for the sake of a scrap of bread
>
> Down with the pretenders
> long live the president, long live free Russia
> long live America and [President] Wilson

Thanks to the early decrees on peace and land, the army, the Bolsheviks' other pillar of support, had all but disintegrated. Almost on the day of the revolution, exhausted, war-weary peasant-soldiers began to return to their villages in droves to escape death and to participate in land redistribution. The army's collapse aided in the disappearance of the old order and, in theory, foretold the emergence of the new. In more utopian moments, Lenin, in accord with socialist dogma, had called for the abolition of the army and the creation of a "people's militia." As the threat of counterrevolutionary armies became real, however, the Bolsheviks re-created a traditional standing army. Trotsky, the army's chief organizer, successfully lobbied for the reinstatement of former tsarist officers—under strict Party control—to help lead the new force. The Red Army was a highly centralized and—through a commissar system designed to monitor old-regime officers—highly politicized organization. By the end of the civil war, it mobilized five and a half million men, most of whom were peasants.[22]

As for the peasants, to them the October Revolution *was* land redistribution, and their support for the Bolsheviks, if it can be called that, was contingent on its realization.

The October 1917 Land Decree had effected a profound revolution in rural affairs. By its terms, all privately owned land, including landlord, church, crown, and monastery estates, as well as communal and independent peasant farms, was to be confiscated without compensation, nationalized, and pooled in a general land fund. The decree looked forward to the Constituent Assembly as the ultimate disposer of this fund. In the meantime, local land committees and soviets were empowered to distribute land on a temporary basis. All citizens obtained the right to land use; the actual amount of land any household received would be determined by local conditions and need based on the number of laborers and the number of mouths to feed. Significantly, the Land Decree did not express the Bolshevik agrarian program, which favored collective over individual farming. Rather, Lenin had strategically adopted the Socialist-Revolutionary Party program leaving the organizational form that land use would take—individual, communal, collective—to choice. Several months later, a further law, On the Socialization of Land, did refer to the long-term goal of agricultural collectivization, but made no provisions for its implementation and reinforced the terms of the October decree.[23]

In the anarchic spirit of the times, as many commentators have noted, the rural revolution proceeded with little state interference. Peasants and their representatives in the villages and rural district (*volost*) soviets set about seizing and distributing non-peasant lands based on their own notions of justice and the traditions and experience of communal agriculture.[24] By the late spring–early summer of 1918, the seizure of gentry and other non-peasant lands was largely complete. The Soviet government's establishment of a "food dictatorship" in May shattered peasant dreams of autarky, however, and turned muzhik and commissar against one another as the peasants resisted state efforts to extend the rural revolution further in a socialist direction.

The threat of dictatorship and civil war had been implicit in the Bolsheviks' coup from the beginning. They had made a proletarian revolution in an agrarian country overwhelmingly populated by smallholding peasants. In the process, they had alienated the educated classes, who controlled the country's finances and factories and who ran and staffed the state bureaucracy. While their earliest measures in power had acceded to popular demands for land, peace, and workers' control in industry, these were quickly followed by the suppression of other political parties and the establishment of strict controls over the press. In January 1918, a long-awaited, multiparty Constituent Assembly, freely elected by the nation to chart the country's future course, was shut down by the Bolsheviks after just one session. Elections to the Constituent Assembly had been held in mid-November 1917, just three weeks after the Bolsheviks' seizure of power. Turnout had been high, especially in the villages, and in one sense, the results were unambiguous. Socialist parties received the overwhelming majority of the total vote. Peasant support had largely gone to the neo-Populist Socialist-Revolutionary Party, which received the most votes of any party at nearly 40 percent of the total, while the Bolsheviks, whose strength rested among workers and servicemen, placed second with slightly less than a quarter of the vote. The remaining socialist parties collectively tallied an additional 5 to 19 percent.[25]

As a reflection of popular opinion at a critical historical juncture, the Constituent Assembly election results are invaluable. They show that nationwide the majority of the population looked to the parties of the left to consolidate the gains of the revolution. They also show the strengths and limitations of Bolshevik support. In the first half of 1918, another source of Bolshevik strength, the All-Russian Extraordinary Commission for Combating Counter-Revolution and Sabotage (the Cheka), a political police force nominally under state control but in reality an extension of the Party, was progressively accumulating and exercising great authority. In the months and years ahead, the Cheka would undertake a "Red terror" to suppress or eliminate the Bolsheviks' enemies; a multifarious category that included not only counterrevolutionary "whites," representatives of the old regime, and liberal politicians but intellectuals, anarchists, non-Bolshevik socialists, traders, and peasants who refused to hand over their grain to the state. From a very early date, Soviet power, understood as a bottom-up, broad-based socialist democracy representing all shades of left-wing opinion, was quickly displaced by Bolshevik power.[26]

Contrary to the ex post facto myth of the omnipotent, monolithic party, factionalism and dissension had marked Bolshevism since its origins early in the century, and voices objecting to the latest course of events had also been raised within the Party's ruling councils. Grigory Zinoviev and Lev Kamenev, two members of Lenin's inner circle, had consistently opposed Bolshevik unilateralism, first on the eve of the October Revolution, when they broke ranks by publicly exposing their party's coup preparations, and afterward, when they fought for a socialist coalition government. In each case, finding themselves in the minority, they accepted the dictates of Party policy. Not so the Russian people. Opposition to the Bolsheviks' actions was expressed not just by political opponents but by workers, trade unions, and white-collar professionals. Bank and government employees had greeted the Bolshevik coup by going on strike. A very serious and embarrassing crisis developed when the executive committee of the railroad workers' union (Vikzhel) threatened to paralyze the country with a strike if the Bolsheviks did not agree to form a coalition government.[27]

Economic failure fueled political disenchantment, deepening the crisis still further. The economic collapse did not begin suddenly in October 1917 with the Bolshevik overthrow of the Provisional Government and the proclamation of Soviet power. Indeed, the Provisional Government had proven tragically incapable of securing food supplies to the cities or arresting drastically declining rates of productivity that brought shortages and unemployment. Even earlier, the failure of the tsarist ministries to feed the cities, especially Petrograd, had been one of the immediate causes of the February Revolution. Under the Bolsheviks, however, the crisis worsened exponentially. Few governments have had to contend with such a simultaneous and complete collapse of their technological and productive legacy as that which beset the fledgling Soviet republic in the years 1918–1921. Even a cursory glance at a few key indicators conveys the scope of the economic cataclysm. By 1921, gross industrial production was only 31 percent and agricultural production 60 percent of what it had been before World War I. From a high of twenty-nine million tons in 1913, coal production stood at nine million by 1921. Electrical generation measured in kilowatt hours was down 75 percent. Oil stood at 41 percent, cement at 2 percent, and pig iron and steel, at 3 and 5 percent, respectively, of prewar output levels.[28] The foundations of modern life in the Russian empire lay in ruins.

Major cities, the centers of modernity, felt the brunt of this disaster. The twenty-three cities in Russia with more than fifty thousand inhabitants lost, on average, 25 percent of their population between 1917 and

1921. Thanks to the epidemics of typhus, cholera, scarlet fever, diphtheria, and smallpox, among other diseases, Moscow's death rate nearly doubled in the two years after the revolution. Even more than disease, flight drastically shrank the number of urban residents. As civilization collapsed around them, townspeople escaped to the countryside in search of food, a movement facilitated by the urban workforce's rural roots. Foreign emigration and mobilizations into the new Red Army also contributed to this contraction. Although it may be an exaggeration to speak of widespread "deurbanization" in these years, the figures are stunning. Between 1917 and 1920, Moscow lost approximately one million people—close to half its population; Petrograd lost 850,000 in 1918 alone. According to Daniel R. Brower, in the Russian republic between 1917 and 1920, northern towns (excluding the two capitals) averaged a 24 percent population loss; those closer to the food supplies in the south and east lost 14 percent.[29]

Although the countryside held the promise of food, it could not guarantee complete shelter from the storm for those who sought refuge there. Russian rural life, never an idyll, could be positively hellish in these years. Contending armies of Reds, Whites, and nonaligned "Greens" fought the civil war in the vast expanses of provincial Russia and Ukraine, bringing death, destruction, and retribution. To alleviate food shortages in the cities and towns, the Bolsheviks initiated grain seizures in spring 1918. Armed detachments were dispatched to the villages with instructions to appropriate, by force where necessary, the "surplus" grain that peasants were allegedly withholding from the state. To facilitate grain collection, Party organizers also created poor-peasant committees (*kombeds*) designed to split the peasantry along class lines. Whether war communism grew out of the Soviet state's desperate economic situation or the Bolsheviks' maximalist and utopian proclivities is a question that continues to engage historians.[30] There is no doubt, however, that for the mass of the population war communism brought inconceivable privations and traumas that remained embedded in popular memory.

War Communism

The "slide into war communism," to use Alec Nove's imagery, began on 9 May 1918, when the Soviet government established a "food dictatorship" to be exercised by the People's Commissariat of Food Supply (Narkomprod). And food, its collection and distribution, remained at the heart of war communism until the Bolsheviks replaced forcible grain requisitions with a "tax-in-kind" that signaled the start of a new economic arrangement in March 1921. Despite the name, which was applied after the fact, the establishment of the food dictatorship before large-scale combat operations had begun strongly suggests that military considerations were not chiefly responsible for the severity associated with "war" (*voennyi*) communism.[31]

Making the case for the food dictatorship to the workers of Petrograd in May 1918, Lenin portrayed the economic crisis as the latest manifestation of the class struggle between bourgeois and proletarian Russia. Urban traders and rural "kulaks," he asserted, were profiteering in grain to starve the revolution. He could not have drawn the battle lines more sharply:

> Either the conscious advanced workers triumph, uniting the poor peasants around themselves, establishing iron order, a mercilessly severe rule, a genuine dictatorship of the proletariat, then force the kulak to submit, and institute the proper distribution of bread and fuel on a national scale.
>
> Or the bourgeoisie, with the help of the kulaks and with the indirect support of the spineless and muddle-headed (the anarchists and the Left Socialist-Revolutionaries) will overthrow Soviet power and install a Russo-German or a Russo-Japanese Kornilov, who will reward the people with a sixteen-hour workday, a scrap of bread per week, mass shootings of workers, and torture in dungeons, as in Finland and Ukraine.
>
> Either—or.
>
> There is no middle ground.[32]

This taut and inseparable interplay of economic necessity, ideological ardor, and desperation generated and sustained war communism, making it seem both obligatory and desirable, a duality that Bolsheviks later subsumed under the rubric "heroic." Key here is Lenin's emphasis on compulsion leading to submission as the only way to prevent the counter-revolutionary apocalypse.

War communism's distinguishing feature was its use of coercive, not economic, tools to extract food from the countryside. To bolster the existing state monopoly on grain, individual peasant households were made to account for their grain holdings. Any surplus beyond consumption and seed requirements could be taken by the state at fixed prices and paid for with inflated currency. Food brigades (*prodotriady*) composed of factory workers and déclassé elements—often urbanites who knew food shortages firsthand—served as the dictatorship's shock troops enforcing the collection of the surplus. Interdictory detachments (*zagraditel'nye otriatdy*) enforced the ban on trade by patrolling roads, rail lines, and waterways for contraband grain being illicitly transported between country and town.

The Party also tried to incite the village poor against their better-off neighbors. In June, based on a crude, highly politicized urban sociology that considered poor peasants the equivalent of village proletarians, Party functionaries organized the poor-peasant committees to locate and seize the grain reserves of richer peasants. In fact, rigid class divisions were an ideological fantasy, and poor peasants were more than likely to rally around their fellow villagers in resistance to the outside invaders. Large segments of the kombed membership, particularly its leading elements, did not come from the working peasantry. Ignorant of village life and agricultural practices, the committees confiscated seed and food reserves together with alleged surpluses and were not above selling their spoils for their own profit. They indiscriminately targeted middle peasants along with kulaks, and their actions frequently degenerated into out-right theft of property and possessions. By November, as peasant resis-tance to the kombeds and their violent tactics rose, the Party leadership called for their disbandment.[33]

Under the new procurement policy (*prodrazverstka*), promulgated in January 1919, attempts to determine individual peasant surpluses were abandoned. Now, regional authorities established district delivery quo-tas and left it to local officials to determine the share to be fulfilled by an individual village. The village population was then held collectively re-sponsible for meeting its assigned quota. Responsibility for enforcing collections fell on local soviets. To dampen peasant resistance, the center toned down class-war rhetoric against kulaks in favor of a policy that sought to placate or "neutralize" the middle peasants. To this end, central authorities raised official prices and promised the revival of a commod-ity exchange between town and country. But again, inflation and the lack of finished goods severely restricted the success of such economic measures.[34]

The state's inability to satisfy urban food requirements throughout the war-communism period guaranteed that the black market would flourish. Major city markets like Moscow's "Sukharevka" became virtu-ally open centers of this nominally illicit trade. In 1918 and 1919, ac-cording to William Chase, Muscovites purchased half their food at such markets, but the high prices charged there took 90 percent of a worker's total food budget. Official measures to halt "speculation" proved ineffec-tive, however, and by the fall of 1918, in a concession to reality, bagging was allowed, but with a limit of twenty-five kilograms of food.[35]

The following letter to Lenin was discovered in the Commissariat of Food Supply files. Written in March 1920 by an unknown woman living in the Ukrainian city of Sumy, it describes just how blatantly the black market operated and how unintelligible the Bolsheviks' hostility to free trade appeared even to someone who professed to support Soviet power. Noteworthy, too, is the woman's equating of profit making with speculation.

· 5 ·

Letter to Lenin from O.M., a resident of Sumy, Ukraine, on the prohibition of free trade and its effects on the supply situation, 25 March 1920. RGAE, f. 1943, op. 1, d. 693, ll. 26–27. Original manuscript.

Respected Comrade Lenin!

Bowing before your brilliant intellect and deeds, I ask that you explain to me, a little person, what is the point of prohibiting unregulated trade when there are bazaars? It seems to me, if you are going to prohibit unregulated trade, then there should not be any bazaars. In my view here's why: if unregulated trade is prohibited, it means that all items, without exception, should be available in the state stores with a ration book. But there is nothing in the state stores. And, if by chance, they do get something, one has to abandon one's house to the whims of fate and get in line at four o'clock in the morning. This crisis is felt especially hard in the city of Sumy, where there is absolutely nothing. But at the bazaars, for insanely high prices, there is everything. The merchants charge anything they want for their wares in violation of the prohibition on unregulated trade. The poor proletarian toiler walks through the bazaar hungry, stands, looks, and then goes to the state store on the off chance that there will be something to get there. But alas, there is nothing there, and he goes home with a heavy heart, where he faces his hungry and cold family of six or seven souls—his wife and children. But the rich man doesn't go to the state store to stand in line; he goes right to the bazaar and buys everything. What's it to him? He has money and he still speculates. At home his family is happy and glad because they are stuffed. Respected Comrade Lenin, what kind of fight for the proletariat is it when toilers die from hunger, cold, and typhus? What is the poor proletarian to do? Either he honorably supports Soviet power, or he speculates, [which is] what Soviet power is fighting against. But he must speculate because he wants to eat. Of course, he has to buy at one price and sell at another—that is, as they say, skin another poor soul who also has to eat. I understand that unregulated trade is not allowed in Soviet Russia, but perhaps this could be postponed because the people still haven't adapted. But when they catch on to what Soviet power means, then unregulated trade could be outlawed. It seems to me that if there were no bazaars, then there would be more goods in the state stores, and here's why: the peasant would make faster deliveries, and if he wouldn't [do this] voluntarily, then it's all right to make seizures and requisitions. But now he is better off delivering [his produce] to the bazaar and naming his price. As soon as he hears that there will be a requisition, he hauls everything off to the bazaar; otherwise he'll be [caught with] everything at home. In my view, if there are bazaars, then there is unregulated trade. The prohibition on unregulated trade is only written on paper, and this is to the benefit of the traders who'll get a higher price. But if unregulated trade is allowed, then there will be more goods at lower prices, and the people won't go hungry. Or else the state stores have to be improved: they should be ordered to work better and not let things slide as they do now. The secondhand market is another matter. Life would be impossible without it because someone will buy the things that you bring to the bazaar and [there] you can sell your last crumbs from a hunk of bread—since our husbands' wages are still enough to buy bread—but you need to buy other things, too. You'll have to excuse me, and I sincerely apologize that at this instant I am taking you away from business and am asking for an explanation, I personally, as a proletarian, and this affects me greatly, and, I think, others like me. I beg you not to refuse my request and to provide an answer in the *Sumy gazette*. I am not giving you my exact address because on receiving your answer I will write you about something more that you don't know about and that would be of interest for you to know, and then I will send you my address.

With sincere respect for you
O.M.

Securing the food supply by extracting as much grain from the peasants as possible was the sine qua non of war communism. The administrative system that emerged to carry out this policy combined centralization (with all its bureaucratic inflexibility) with emergency measures that often devolved into arbitrariness, an unstable mix that placed Moscow at loggerheads with local Party and soviet officials. When directives and threatening telegrams failed to achieve their purpose, central authorities dispatched plenipotentiaries vested with extraordinary powers to the provinces to cut through red tape and obstructionism and to ensure that local officials carried out instructions. Resentment on the part of local officials at these intrusions into their bailiwicks by individuals who either were ignorant of local conditions or behaved in a bullying, highhanded manner could reach white-hot intensity.

No better illustration of war-communist methods and the accretion of power by central institutions can be found than the People's Commissariat of Food Supply, Narkomprod. The successor to the Provisional Government's Ministry of Food Supply, Narkomprod moved quickly to extend its hold over food-supply administration following the October

Revolution. As the food crisis worsened, its authority over all other food collection and distribution agencies substantially increased. With the establishment of the "food dictatorship" in May 1918, Narkomprod became the chief executor of a strategy summed up by Lenin as "centralization of the food supply, unification of the proletariat, and organization of the rural poor." To this end, the government granted Narkomprod emergency powers that gave its commissar, A. D. Tsiurupa, near-dictatorial authority in relation to food and allowed him to subordinate or ride roughshod over other, especially local, institutions.[36]

The May decree spoke of grain in the hands of wealthy peasants and called on "all laborers and propertyless peasants to immediately join together for a merciless struggle with the kulaks." In this way, the collection of grain became transformed into a class war, and Narkomprod acquired the license to employ armed force against those who would oppose the expropriation of their produce. Its food brigades were often supplemented by troops from the Cheka's Internal Security Force of the Republic (VOKhR). Still, finding grain remained difficult. On 3 August 1919, a food brigade working along the Volga River in Lenin's native Simbirsk province reported: "Our work is going very poorly. Peasant grain deliveries are bad. We've picked over the district they sent us to five times. There is no grain. The peasants are extremely hostile. Our agitation is useless. We'll have to use our guns."[37]

On 7 March 1920, Vladimir Meshcheriakov, the chairman of the Novgorod province soviet executive committee, sent a lengthy and acrimonious report to Narkomprod collegium member A. P. Smirnov. In it he details the negative effects of food requisitioning in the province and fumes at the arrogant, ill-informed central administrators supervising local affairs. Other, similarly overburdened local officials no doubt shared Meshcheriakov's anger and frustration. The following excerpts are taken from that report.

· 6 ·

Excerpts from a report by Vladimir Meshcheriakov, chairman of the Novgorod province soviet executive committee, sent to A. P. Smirnov, Narkomprod collegium member, on the food situation in the province and Narkomprod's relations with local officials, 7 March 1920. RGAE, f. 1943, op. 1, d. 693, ll. 73–74. Typewritten copy.

———

[. . .] 6. You have been completely misinformed that we are prohibiting work with the [food] brigades. Before, the Provincial Food-Supply Committee [Gubprodkom] let the brigades in for any reason and without any over-sight. But these brigades completely failed to meet their obligations. Wherever they went, the villages rebelled against Soviet power. Crudeness, illegal de-mands for provisions for their own use, confiscation when faced with a re-fusal [to hand over] livestock, and its ostentatious slaughter and consumption on the spot undermined any faith that the peasants may have had that the requisitioning of provisions was, and would be, done for the good of the workers and the army. There have been frequent cases of out-and-out plun-der (an accordion, a ring, dresses, etc.).

The province is starving. Large numbers of peasants are eating moss and other trash. Since the fall they have stocked up on bark, grass, moss, etc. This is the third year of famine in the countryside. A mood has taken hold such as can only be found in a famished countryside.

To make matters worse, until summer–fall, the center paid no attention and sent only one responsible comrade from Piter [Petrograd] for Party work. Party committees and organizations did not begin to adopt the VIII congress's line on the middle peasant until the fall.[38] Until the summer, the attitude to-ward the muzhik was fierce, that of the poor-peasant committees. As a re-sult, last spring was marked by violent uprisings that were put down cruelly.

All that was left for me was to continue to mollify the peasantry and to insist on cautious relations with the countryside per the [Party] program. Judging by several congresses, it seems to me that we have attained the [de-sired] result.[39] The cowed and broken peasant is beginning to stir. He is be-ginning to speak out and lodge complaints. He is coming to understand that he also has rights, that the era of the ferocious "gun-toting commissar" is passing.

By purging the Party—getting rid of alien, parasitic, and plundering elements—we are becoming stronger. We are trying to strengthen Soviet power all over the countryside. Now, more than anything else, the biggest obstacles are continued famine and the lack of goods.

In a word, the legacy of this is that we need a sensitive line toward the peasant of famine-stricken provinces. Even before, there was almost no ru-ral bourgeoisie here—the Cheka had completely wiped it out.[40] The middle peasant is starving. But the work of the reckless VOKhR unquestionably drove the peasants to food riots. I know that these considerations are for-eign to Narkomprod and are unacceptable to the psychology of its officials and that I risk being misunderstood, but duty obliges me to present the ac-tual situation, which you in your offices don't know.

There was only one way out—to place the work of the brigades under the supervision of the Party. But Comrade Tsiurupa needlessly, relying on secondhand accounts, and without deigning to ask us what was the matter, rushed off a telegram saying, "Illegal. I countermand." The Party has the right of supervision over the brigades and is even obliged to monitor their work. Comrade Tsiurupa himself has repeated this over and over, and there is nothing illegal about it.

7. "You asked for more men, they were sent to you, and then you refused them." Again, you have incorrect, imprecise information. Once more, I have to express my regret that the officials at the center, such as yourself, find it unnecessary to inquire here first to find out what is happening, and then write telegrams.

"We," that is, the Provincial Party Committee [Gubkom] and the Executive Committee [Gubispolkom], requested real forces—that is, responsible administrators returning from the former front. We didn't ask for VOKhR [troops], [and] we don't know what to do with the ones we already have. The hay department and the Chrezkomzagsen [Extraordinary Committee for the Collection of Hay] requested these forces, and the Piter comrades correctly thought to send someone in advance *to find out* what we need, and as it turned out, they didn't *send* us any "forces."[41]

8. Even before [receiving] your telegrams, with their "brigade" recipes, the Gubprodkom was taking all the measures necessary in order to utilize the time remaining. Through the Gubkom and the Gubispolkom, they put forward a resolution on "hay week" (15–22) using printed and oral agitation. For this week, the counties were instructed to throw everything that could be taken from fuel [collection] into the countryside for hay [collection]. With the village soviets, special hay meetings were organized in the districts for carrying out the assigned requisitions. The provincial labor-conscription department [Gubtrudpovinnost] had already instructed the district labor committees [Ukomtrudy] to ease up on the timber obligation in order to make it easier for the peasants to haul the hay.[42] When necessary, we even allowed the [food] brigades [to take part]. Even a bonus of salt for hay was announced.

These are the latest plans for [our] current work, and it is too early to say how they are being carried out. In view of Narkomprod's usual lack of faith in local officials, I myself will not begin to evaluate our work but will offer the words of the plenipotentiary of Chrezkomzagsen—a person who has no reason to relate well to us (following the orders of Narkomprod, we exclude him from the missions). This is what this stranger, this man who does not belong to our apparatus, telegraphed to Chrezkomzagsen about our work: "Gubprodkom together with the hay department took the most energetic measures to pump out the hay."

9. "The collection of oil-producing seeds proceeds poorly." Again, [we see] the cardinal sin of Narkomprod: You don't know what your corresponding department does or says and to no purpose whatsoever tie up the line with meters of absurd telegrams. Just recently, the province food-supply commissar, Bazanov, made a special trip to Narkomprod to clear up the confusion it had created with this dispatch. He explained that as long as they don't send [anything] in exchange for these seeds (as it says in the corresponding decrees), then he can't conjure up oil from nowhere. This is so clear that even you in the department agreed. They reassured the commissar and promised to send the oil. They admitted that it is impossible to change

the seeds to oil without any oil [to begin with]. And now you come down again with a categorical reprimand, even though the matter has already been clarified and settled, without bothering to find out the real story.

10. And that's how it always is. They promised to carry out orders, but they don't hand the grain over to us. They promised to send seeds, but they don't send [them]. They promised to supply oil, but they don't supply [it]. They promised that the supply trains would arrive in accord with strict rules and instructions, but not one train came according to the instructions. Narkomprod tried to uncouple (and did uncouple) each one. The supply train underwent unusual ordeals. They took it, and it required a tremendous effort in the face of Narkomprod's screwup to get a little something to a starving province.

We are keenly aware of the circumstances in which Narkomprod works, how difficult its job is, the obstacles it faces at every turn, and we would not have brought up all these and other sins of Narkomprod if the latter had given as much consideration to the circumstances and obstacles of our work, if it had troubled itself to evaluate this situation, if it had observed the most fundamental principle: first get the lowdown from the source, and then judge the matter and write threatening telegrams, etc.

11. Take some time to look through your telegrams: "I reprimand," "I demand," "All discussions are through," "If it is not taken care of, I will personally drag the Gubprodkom and Gubispolkom to court." This is the full "bouquet" of phraseology from the Narkomprod lexicon.

Long ago, local administrators came to the dismal conclusion that, apart from idle talk, Narkomprod has little help to offer. We are quite accustomed to Narkomprod's slighting of local officials—it simply doesn't want to take them into consideration. And we have long known that Comrade Smirnov, in particular, is an extremely crude individual . . . And we all have pictures in our minds of the shameful condition in which the [All-Russian] Central Executive Committee [VTsIK] inspection found Narkomprod. And when several responsible leaders of Narkomprod, without any compunction whatsoever, stray from the truth, it is incumbent on us to respectfully remind them that the Party congress and the VTsIK session are not far off. And we consider it our comradely duty to warn those commissar-minded [*zakomissarivaiushchikhsia*] soviet generals: be more careful, comrades. We will discuss our work and the breakdown of Narkomprod at that court with which you so often threaten us. You can answer these arguments at the Party and soviet congresses.

Vl. Meshcheriakov

In theory, grain requisitioning meant a direct exchange of food between town and country. In exchange for the "surplus" removed by the food brigades, items like cloth, salt, sugar, and kerosene should have been delivered to villages for distribution, especially among the poor. Throughout the civil war, however, these and other goods were in extremely short supply in the countryside. In the following letter, dated 6 April 1920, an unknown peasant from Glazov county, Viatka province, provides Lenin with an account of product exchange. The original letter was written in pencil without punctuation marks and contained many grammatical errors.

· 7 ·

Letter to Lenin from an unknown peasant in Viatka province detailing abuses committed during grain requisitioning, 6 April 1920. RGAE, f. 1943, op. 1, d. 693, ll. 32–33. Original manuscript.

Moscow, to Comrade Lenin.

Deeply respected comrade, I apologize for taking up a few minutes of your time. I want to apprise [you] of how things stand in the countryside in relation to provisions. Maybe I am frightfully mistaken, but to me, as well as to the peasant, there is much that seems abnormal. It is precisely about these abnormalities that I want to inform you. I can provide you with several extremely noteworthy instances that conflict with instructions from the center. As far as I understand order no. 88, which deals with the food norms for the agricultural population, it reserves two hundred kilos of flour and sixteen kilos of groats, as well as the necessary amount of seed for each person, during the levy. But something else entirely is going on when the assessment is carried out. They don't take into account any norms and they don't allow for the seed—they take everything. In Glazov county, I have observed many cases where they cleaned out the oats down to the last grain. What further can we expect from the countryside at the present time? Whoever pays the assessment of eight hundred to sixteen hundred kilos [of grain] should get the [food] norm set on 29 February by the Glazov county food-supply committee [*uprodkom*] at six and one-half kilos per month per head. How should the peasant view the instructions of the central authorities now? Are all the orders simply on paper? Things work completely contrary [to the instructions], and it's the same with the seeds they want to provide from Urzhum county; they haven't managed to provide them yet, and after a certain time they won't be needed. And what would happen if they should provide them on time? The organs of supply would obviously have nothing to do. At first, they took [grain] and tormented the peasant. They conveyed [the grain] to the collection point, then winnowed it in a simple winnowing machine and say that it is sorted. They take it from the peasant for forty-one rubles but turn it in for one hundred rubles, and the oats are already going bad, smelling a little moldy. No, Comrade Lenin, this is not a policy! You must, must pay the most serious attention to labor productivity in the countryside and give the peasant a human standard of living. They take everything from the countryside, and what do they give in return?—nothing! [Each peasant gets] a meter and a half of a kind of shoddy calico, but the factory worker and the city dweller [get] more. The peasant has no use for your rags! Give him a plow, a sickle, a scythe, give him iron—everything he needs for agricultural equipment. If any one of you found yourselves in the situation of the peasant of Viatka province, you would no doubt complain.[43] This unfortunate peasant carries the whole world on his shoulders, like a dumb animal. Everything we do could be done faster and more easily if all the comrades in authority tried to avoid any mistakes. A lot of mistakes are made unintentionally. This is excusable. But there are those who know well enough that it's a mistake, but they do it anyway. I request that you give serious attention to this letter!

A peasant.

Economic collapse, shortages, the breakdown of order, civil war, and war-communist policies pushed the people to the brink. Even ardent supporters of Soviet power felt compelled to point out the inequities and dangers in the existing situation. The Red sailor, Ya. Lachugin, who spent his leave in his native village, Shue, in the central Russian province of Ivanovo-Voznesensk, wrote a letter to Lenin on 2 April 1920.

· 8 ·

Letter to Lenin from the sailor Ya. Lachugin describing the food-supply situation in his native village, Shue, Ivanovo-Voznesensk province, 2 April 1920. RGAE, f. 1943, op. 1, d. 693, ll. 34–35. Original manuscript.

Comrade Lenin

Owing to domestic circumstances, I find myself at home on a seven-day pass. I have heard, seen, and am convinced that to a man the peasants in our county all oppose the government. Because of the famine's effects, this holds for the workers, too. The communists are moving to the right. With every hour discontent grows, and, not surprisingly, there are all sorts of shenanigans going on in soviet institutions and all because of the famine, for a crumb of bread. This past March each worker received two kilos of flour and three boxes of matches and nothing more. Can one really survive and not croak on two kilos? But the market prices are hellishly high— sixteen kilos of flour [costs] 15,000 rubles; potatoes, 1,500 rubles per portion— meanwhile, workers are making 800 rubles, and from that a number of petty deductions are taken. At the moment, the factories are idle. It's impossible to transport [provisions] on the railroads; it's forbidden. What is there to do? A hungry man is capable of anything. Everyone's new slogan is "Any system you like as long as there is bread!" Dark forces use this for their own ends. The people are not against the system (although it needs to be improved), but against the bureaucrat-reactionaries who, one after the other, have snuck into soviet institutions and try any way they can to wreck and slow things down. The food-supply situation in particular [serves] as the main weapon against Soviet power, and all the while he [the reactionary] shouts from the corner: "There is your Soviet republic! It was better before, under Nicholas!" The dark mass understands little of this—their outlook on life is too narrow. Corresponding measures must be taken. Famine doesn't wait. Otherwise, it could take an undesirable anti-Soviet form.

Sailor YA. LACHUGIN

By the time of this letter, the civil war had decidedly tipped in the Bolsheviks' favor, but this did not lead to a moderation of war-communist policies or practices. Although the country's new leadership was aware of the people's continued sufferings and the consequent erosion of support for government policies even among steadfast supporters like Lachugin, it took another year of countrywide hardship before Lenin and his colleagues came to the same unhappy conclusions as this loyal seaman and changed course.

To the popular masses who actively supported the October Revolution or saw their salvation in it, the revolution promised to deliver them from the domination of the old state and its elites. Institutions of direct democracy like the soviets and the committees of factory workers, soldiers, and peasants allowed the lower classes to exercise an unprecedented degree of control over their work and their daily lives. But the sort of independence envisioned by each of these groups became untenable in the face of economic and military exigencies and the Bolsheviks' brand of state building. Between 1918 and 1920, the Bolsheviks instituted a strict regime, a dictatorship of coercion in order to prosecute the civil war while keeping the cities fed and the factories running. Survival demanded taking harsh measures. That many in the Party also believed these measures were carrying them nearer to the socialist future, however, says a great many unhappy things about the revolutionary millenarianism that fired communist thinking in these years. Just as important, it testifies to a fundamental cleavage between the people's understanding of revolutionary liberation and that of the revolutionaries who now exercised state power.

No cleavage could have been any greater than that separating the Bolsheviks and the peasants who were 80 percent of the population. As Marxists, that is, advocates of working-class revolution, the Bolsheviks had little empathy for the peasantry. In Party eyes the peasant class, for all its size, was a medieval remainder, a sociological anachronism, and, as such, an obstacle to progress. Still, after the October Revolution the peasants had been allowed—by the force of circumstances—all the independence they could have desired. Within five months, however, the Bolsheviks had embarked on a policy of forcible grain seizures that struck directly at that independence and at the peasants' understanding of their revolutionary achievement. For the next decade, the shifts and strains in the relationship between these two forces—Party and peasantry—would be a central factor in determining the character of the Soviet republic. In the chapters that follow, the voices of the peasants are given pride of place. They have much to say regarding their relationship with the new authority and on the many issues of the day. At times they speak with a voice of reason, at times of incomprehension, anger, and desperation. Whatever the mood, their voices deserve a hearing if we are to come to any understanding of how the people—the *narod*—experienced the Russian Revolution and its aftermath.

CHAPTER 2

The Old Village and the
New Economic Policy

We must, in accord with our worldview, our decades-long revolutionary experience, and the lessons of our revolution, put matters plainly: the interests of these two classes are different—the small farmer does not want what the worker wants.

—V. I. Lenin, March 1921

By late 1920, Red victory in the civil war was certain, but at a tremendous economic and human cost. Citizens safely beyond the reach of the fighting as well as those in the war zones had endured extraordinary privations. The winding down of the war, therefore, raised popular expectations that the harsh war-communist regimen would relax. No such change was immediately forthcoming, however, for Party leaders invoked economic collapse to justify the perpetuation of extreme measures. Grain requisitions continued, penalties for infractions of labor discipline were stiffened, trade unions came under increased state control and were redefined as organizers of production, and Leon Trotsky championed the "militarization of labor" through the creation of "labor armies." In response to these measures, the deepening economic crisis, and the continued suppression of democratic practices, large segments of the lower classes turned to active opposition against the new regime. Though socially, ethnically, and geographically diverse, the protesters shared a desire to see the Bolsheviks' political monopoly ended and the ideal of soviet democracy realized.

Throughout the fighting between the Reds and the Whites, the peasants had been waging regional wars in defense of their recently won freedoms and against forced grain requisitioning. These rebellions were widespread, numerous, and of varied scale, involving from a few dozen to tens of thousands of peasant participants drawn from every economic stratum; the resistance did not come just from kulaks, as official Bolshevik accounts claimed. Many Red Army soldiers and, in more than a few cases, entire units deserted and joined the peasants in their fight. Between 1919 and 1920, large-scale peasant rebellions erupted along the Volga River and in the North Caucasus, Belorussia, Central Asia, and Western Siberia. In Ukraine, under the leadership of the charismatic anarchist Nestor Makhno—himself a poor peasant from Hulyai-Pole in Yekaterinoslav province—the peasant "Insurgent Army," which at times numbered as many as thirty thousand, conducted a guerilla war against its former Bolshevik allies into the summer of 1921. In August 1920, peasant resistance to forcible grain requisitions ignited a major rebellion in Tambov province. For nearly a year, tens of thousands of armed men under the command of the former Socialist-Revolutionary A. S. Antonov made war on Soviet authority with the support of the local peasantry. Antonov's movement controlled much of the region until it was crushed by fifty thousand troops led by the future Red Army marshal Mikhail Tukhachevsky in June 1921.[1]

Although smaller in scale than the Makhno and Antonov movements, the mutiny of Baltic Fleet sailors stationed at the Kronstadt naval base in the Gulf of Finland in March 1921 was an especially demoralizing blow for the new regime. The Kronstadt sailors had provided the Bolsheviks with a reliable base of support in 1917. Now, after witnessing the effects of three years of the Bolsheviks' economic and political dictatorship, and with industrial strikes under way in nearby Petrograd, the sailors called for the immediate implementation of the popular socialist democracy heralded by the October Revolution. Among their specific demands, they included freedom of speech, freedom of the press, and freedom of assembly for workers, peasants, and parties of the left, free trade in grain, and the right of peasants to use their land as they saw fit. For sixteen days, beginning on 1 March, they defended their island fortress before succumbing to a combined Red Army–Cheka onslaught led by the ubiquitous Tukhachevsky across the frozen gulf.[2]

Under these pressures and with food still in short supply, the Party leadership at last changed course. In March 1921, as the Kronstadt rebellion was unfolding, the Tenth Party Congress voted to replace compulsory grain requisitioning with a tax levied as a percentage of peasant output. The turn to noncoercive exchange signaled the end of war communism and the start of the New Economic Policy (NEP)—a series of ad hoc measures intended to stimulate peasant productivity and encourage voluntary grain marketing. Peasant farmers would now be free to dispose of their surpluses as they saw fit once they met their obligation to the state through the payment of a "tax-in-kind" (*natural'nyi nalog*). Allowing private exchange meant legalizing trade and tolerating the commercial activity that trade would engender, but, as Lenin noted, under war communism mistakes had been made and things had "gone too far." Now it was time to accept that Russia's cities could be fed and its shattered productive plant repaired only by granting peasants the right of "free exchange."[3]

Free trade implied that the peasants would be fairly compensated for their grain and other products. The confiscatory policies of war communism had arisen, in large part, precisely because industry could not provide the manufactured goods that peasants desired. If, in the future, urban-rural trade were to be maintained and peaceful relations with the peasants sustained, then industry had to satisfy peasant demand for finished goods. As Lenin wrote after the congress, this "is the only kind of food policy that corresponds to the tasks of the proletariat and can strengthen the foundations of socialism and lead to its complete victory."[4] On this point, Party policy and Soviet reality collided. All Bolsheviks agreed that only large-scale industry could provide the economic basis required for socialism; the sooner the Bolsheviks constructed a heavy industrial base, the sooner socialism would be realized. Now, however, Lenin was suggesting that Soviet industry gear itself toward satisfying peasant demand for light industrial products, which meant that heavy industry and the promise of socialism it carried would have to be put off. Delaying the construction of "machine-building" industries also meant delaying the state's ability to provide the means to mechanize agriculture. Tractors, reapers, combines, and other modern machinery were the carrots that the Bolsheviks intended to offer peasants to induce them to abandon their small, independent holdings and join collective and state farms. Absent the new equipment, the superiority of socialized agriculture remained purely theoretical, and the peasants' attachment to individual farming would not be overcome.

The change from war communism to the NEP, then, had profound implications for Party doctrine and the economic development of the country. By reintroducing bazaars and other aspects of the market, Lenin and the Bolsheviks were accepting that, given Russia's small-scale, private agricultural base, there could be no direct transition to socialism. On the contrary, it now seemed certain that the transition from capitalism to socialism would take a long time. Lenin himself spoke in terms of "generations" and "decades." During this period, state and private economic sectors would have to coexist. The state would exercise a monopoly over banking, foreign trade, and large-scale industry—the so-called commanding heights of the economy. In the meantime, agriculture would largely remain in private hands and be organized on the basis of the peasant farm. The Bolsheviks fully expected that as the economy grew and industry recovered, the socialist sector, by virtue of its size, productivity, and efficiency, would outperform and eventually incorporate the private one. Yet they were also mindful that the NEP provided the means for a capitalist revival in both the city and the village. Thus, although a slogan of the day confidently asserted that "NEP Russia" would become "socialist Russia," the outcome of the competition between the two sectors—the *kto-kogo?* (who will triumph?) question—remained uncertain.

Abandoning coercive war-communist methods, however, meant that the socialist-capitalist struggle that lay at the heart of the NEP was to be conducted by economic means. In this sense, the NEP was a call for social peace, an effort at repairing the broken relations between urban, proletarian Russia and the 80 percent of the population that labored on the land. Indeed, as soon became apparent, the NEP was premised on making concessions to the peasants even when it hurt workers' living standards. Consequently between 1921 and 1925 peasant interests made steady gains. The Land Code adopted at the end of 1922 secured in law the prerogatives granted by the October 1917 Land Decree. The code upheld land nationalization and the prohibition on buying and selling land, but it also preserved the peasants' freedom to choose the form of land tenure under which they worked their allotments, and guaranteed those who so desired the right to separate from the commune. Moreover, contrary to the 1917 decree, the 1922 code allowed for the leasing of land. Similarly, the code also allowed for the use of hired labor, though under strict regulation.[5]

Likewise, Soviet tax policy favored the peasants. To encourage peasants to plant more, the total amount of the food tax was announced each year before the spring sowing, thereby assuring peasants that they would retain control over their surplus production. The tax amount was also set considerably lower than previous requisitions. For the fiscal year 1921/1922 the total tax was approximately 180 million *puds* less than the previous year's quota (240 million versus 423 million puds; one pud is 16.38 kilograms). In March 1922, the government simplified assessments by replacing the various food taxes, which numbered between thirteen and eighteen, depending on the region, with a "unified food tax" computed on the basis of a pud of rye. At this time, progressive changes to the tax code were also made that favored smallholders and families with members in the military. The Twelfth Communist Party Congress, held in March 1923, sought to improve the peasants' situation still further by allowing some of the food tax to be paid in cash and by amalgamating all the peasants' tax obligations—food, household, cartage, and local—into a single agricultural tax. The congress also advocated an overall reduction in peasant taxation and an increase in exports to raise grain prices. In January 1924, the tax-in-kind was replaced entirely by cash payments.[6]

All these measures were intended to encourage peasants to expand the area they placed under cultivation for, just as under the tsars, the rate of Soviet industrialization depended on the amount of surplus capital that could be extracted from the countryside in the form of taxes and exportable foodstuffs. In fact, the Soviet state was even more dependent on the countryside as a source of investment capital than its tsarist predecessor had been, since the revolution had led to a dramatic reduction in foreign investments in Russia. Thus, industrialization, which the Bolsheviks considered the economic prerequisite of socialism, was, as Nikolai Bukharin would note, indeed being pulled forward by the peasant oxcart. And herein lay the dilemma that occupied so much of the Party leadership's attention during the 1920s. If agriculture, by providing the state with taxes as well as industrial and export crops, remained the nation's largest single source of investment capital, then its continued organization along traditional lines utilizing antiquated methods meant that the country's industrial transformation would materialize very slowly.

The pro-peasant policies that the Party leadership pursued through the mid-1920s arose naturally from the NEP's market orientation. In this, they resembled the agrarian reforms of a generation before that had sought to undermine the commune and create a class of independent smallholders. Like the proponents of the earlier "Stolypin reforms," the Bolsheviks aimed at eroding the commune and its equalizing mechanisms and replacing the peasants' inefficient agricultural practices with up-to-date techniques. For many Party members, however, this strategy remained highly controversial, not least because, like the earlier reforms, it was expected to encourage social differentiation and strengthen the strong, capitalist-minded producer at the expense of the weaker subsistence farmer, on whose shoulders the Bolsheviks hoped to build socialized agriculture. Nevertheless, for the time being, the Central Committee majority succeeded in deflecting this criticism and, in 1925, provided the peasants with a new series of concessions that favored the better-off, more enterprising peasant still further.[7]

The food shortages, economic collapse, and popular unrest that resulted from war communism had forced the Bolsheviks into this rapprochement with the peasants. In the long run, however, the truce was built on a fundamental contradiction. Stated baldly, the advance to socialism as the Bolsheviks understood it could not succeed without a thorough transformation of the world the peasant inhabited and the agriculture the peasant practiced. The contradiction was inherent in the Bolshevik project of building socialism in a peasant country and reflected the existence of two, interdependent, yet very distinct Russias: one urban, industrial, and proto-socialist; the other rural, parochial, and fragmented. Necessity had forced a compromise between the two, but to successfully achieve Bolshevik aims, the former had to overcome and ultimately integrate the latter into a single, comprehensive, and large-scale socialist whole.[8]

For this reason, the Bolsheviks' acceptance of small-scale, independent agriculture (and the market) was highly conditional. The fact remained, however, that attempts during the civil war to implant agricultural collectives in the Russian countryside had met with little success. By the end of 1918, only an insignificant fraction (0.15 percent) of the peasantry had collectivized. In the Moscow region, even the poorest peasants—those without a horse or their own agricultural equipment—refused to join collectives. By the end of 1920, some 10,500 collective farms of various types had been established in the country, but organizers had induced only 131,000 peasant households to join them. Moreover, the vast majority of collective farms had been created, not through the amalgamation of individual peasant farms, but out of recently seized monastery and landlord estates. Overall, in terms of total land area and peasant households, collective farming accounted for slightly more than a paltry 0.5 percent. According to Lenin, the peasantry had laughed at inexperienced organizers' clumsy efforts to substitute "socialized" for traditional forms of work and association. Vasily Afanasiev Biakov, one of our letter writers, saw what these Bolshevik "dreamers" (Lenin's term) apparently could not. His observations led him to the sage conclusion that the sort of person prepared to live in the *kommuna*—the most highly socialized type of collective farm—had yet to be born (Document 21). Nevertheless, the rational, scientific, and productive virtues of large-scale, mechanized agriculture retained their hold on the Bolsheviks' Marxist imaginations. By consolidating the peasantry's small, scattered holdings into larger units, socialization promised to end the muzhiks' isolated existence and to increase urban (that is, proletarian) Russia's influence on them. In this way comprehensive and rational order would ultimately supplant the old, illiterate, disjointed, and superstitious muzhik culture. As Marxists, the Bolsheviks viewed this integration of city and village as a prerequisite for a truly planned economy and as the means for delivering the countryside from ignorance and backwardness. So the long-term task the Party set itself remained what it had been before the NEP: "to establish a more perfect system of agriculture, a communist system, which will be competent to deliver our rural population from the barbaric waste of energy that occurs in the extant system of dwarf agriculture."[9]

This visionary note sounds all the more fantastic given what the rural revolution had actually wrought. The old regime's collapse had strengthened rather than weakened peasant Russia's "muzhik" character by fortifying the peasants' devotion to traditional organizational forms and practices centered on the repartitional commune (*obshchina*), the village assembly (*skhod*), and the household (*dvor*). Throughout the NEP years, the communes greatly outnumbered rural soviets and exceeded them in genuine authority. Stated programmatically, peasant aspirations were quite simple: land to the tillers and as much aid with as little interference as possible from the state. Peasants focused their attention not on socialization but on equalizing holdings. To this end, between 1917 and 1921, to justly apportion recently confiscated estate and church lands, the peasantry engaged in what Moshe Lewin calls a "frenzy" of land redistribution. Another of our correspondents, Ya. Shepelyov, probably expressed the thinking of most peasants when he explicitly defined "socialization" as equalizing the amount of land given to those who actually worked it (Document 10).[10]

The peasants' proclivity for the small, traditional, and local was but one of the many differences in outlook that told against conciliation with the new regime. To most of the urbanized Bolshevik Party cadres the muzhiks were benighted at best and petty bourgeois at worst and would sell out their proletarian allies for a few kopeks. In cadres' eyes, the village was a bastion of ignorance, a fly- and priest-infested class battleground where avaricious kulaks lived off the toil of impoverished peasant farmers and pauperized landless laborers. To the Western-oriented Party elite as to previous generations of city-dwelling intellectuals, the peasants' world was an alien Russia that few of them understood, that many of them, no doubt, found discomforting, and that most of them were dedicated to eradicating.

As is evident in several letters below, peasants could agree with the Party's negative characterization on at least one point: the village was backward, and most of its members were ignorant. In their letters, for all their mistrust of outside authority, villagers were also not averse to slapping the "kulak" label on a neighbor and exposing his or her machinations if it suited their purposes. Where the Party's class language and ideological prejudices coincided with the peasants' material interests, peasants were quick to utilize the new, politically charged terminology for their own ends.[11] This was but one way the peasants endeavored to influence state behavior in their favor. In the end, what the peasants expected from the state was the means to improve rural conditions within the existing village structure and notions of justice: where former gentry or wealthy peasants acted underhandedly or in cahoots with the local authorities to gain economic or political advantages, now, peasant letter-writers call for investigations to set things right; when a legislative act they find useful and just is subverted, they raise a hue and cry. Like their fathers and grandfathers, most Soviet peasants wanted minimal contact with the state and its agents; they preferred self-regulation and were prepared to police themselves according to their own rugged customary law. But where their personal interests, concepts of fairness, or understanding of revolutionary achievement were threatened, they called on the state to step in. As one offended party writes: "It wouldn't hurt if the higher organs took a peek into this little corner" (Document 14). That, after all, was often the whole point of writing a letter.

Apart from more land, what the peasants wanted most from the state was the means to promote and improve agricultural production and the situation of their *khoziaistvo* (farm; economy). Here their requests and suggestions could be very specific. The previously mentioned Biakov, no defender of the traditional village, demotes the government's Promethean plan to electrify the countryside to "last on the list" of things to do, at least until cement can be supplied in sufficient quantities to satisfy the peasants' need for cattle-shed foundations. Much like agricultural leaders in Moscow, who considered rural electrification premature, Biakov put his priorities elsewhere, and he willingly eschewed the light bulb for the kerosene lamp.[12] Several writers admonish the new regime for failing to do as much as its predecessor in providing seed, livestock, and agronomic advice. For these peasants, the soviets failed to measure up to the prerevolutionary *zemstvos* as promoters of agricultural development. Others simply wish to be accorded the rights and dignity they associate with being a laboring citizen in a socialist state.

These shortcomings notwithstanding, the peasant proprietor could view the change in official policy from war communism to the NEP as a step in the right direction, the rightful outcome of the revolution that had destroyed the old regime, broken the back of its landholding gentry, and had finally provided land to those who cultivated it. In fact, the switch to the food tax in 1921 was so dramatic a change that in the months following its announcement peasants in many locales did not trust the Bolsheviks to put it into effect. Nevertheless, police reports from across the country indicated that peasants, overall, received the news positively and expected the changes to address the food and goods shortages. That April, the Kaluga province Cheka reported that news of the legalization of private trade had significantly improved the peasants' mood and that the population there now considered Lenin a defender of peasant interests. After the publication of the decree on trade, the price of bread in the province fell by two to three times.[13]

Still, the peasants' initial response to the NEP was not uniformly enthusiastic and often ran along economic lines. Ravaged by crop failure, the poorest peasants in Samara province equated the reopening of the bazaars with speculation and worried that their situation would become even worse with the return of the market. Among middle and poor peasants who had fulfilled their grain-collection norms in Stavropol province, the state's failure to supply basic necessities in return as promised was a source of discontent and caused them to wonder if there was any significant difference between requisitioning and the new tax. The continued imposition of cartage and other burdensome, unpaid labor obligations gave rise to this view in other places as well. Into the fall, reports from Orel province contrasted the satisfaction of the well-off at the new turn with the worsening mood of the poor, who could not afford the prices charged at the markets or in cooperative stores and who had to pay dearly to rent plow horses. Families of Red Army soldiers living in the province were said to be in particularly dire straits.[14]

For many millions, as agonizing and murderous as the previous years had been, the first year of the NEP turned out to be even more horrific. Crop failure and famine occurred in Bashkiria, Kazakhstan, Western Siberia, Southern Ukraine, along the Don River, and in the Kama River basin. Poor weather, consecutive years of war and destruction, and requisitioning brigades' practice of commandeering seed and reserves along with "surpluses" combined to undermine agricultural productivity in these areas. Hit worst of all were the Lower and Middle regions of the Volga. Here, crop failure in 1920 was followed by killing frosts, then drought, which all but annihilated the next season's planting as well as the current season's. Since 1918, grain collections in the region had been particularly aggressive and thorough, continuing even into 1921 and leaving peasants with little or no food in reserve. When calamity struck, they were defenseless. By the end of the summer, thousands of villages were experiencing catastrophic food shortages or were in the grip of famine. Scurvy, dysentery, typhus, and cholera took many victims. Instances of cannibalism were widely reported. Maxim Gorky and other prominent figures organized a private relief effort, and foreign aid, particularly American, poured into the country despite Bolshevik obstructionism. In total, hunger and disease claimed five million lives.[15]

Famine and widespread food shortages in 1921 indicate how close to the brink the countryside had been pushed. Peasants in the affected areas held the new regime directly responsible. Local Chekas reported instances of peasants seizing and redistributing grain, sometimes at fixed prices, from grain-collection points. Peasants were also deeply disturbed by the frequent lack of adequate seed supplies to complete the spring sowing. Many officials were simply and openly corrupt. For others, war-communist methods had become customary and were not easily abandoned. The dire food situation, and the resulting pressures on local officials to collect grain, only reinforced the command-administrative approach. As a result, the first year of the NEP remained a desperate time despite its promise of relief, and most peasants continued to view Party and state organizations with suspicion and hostility.[16]

Thus, even as commissar and muzhik moved toward their newfound accommodation, the peasants were constantly reminded that their interests and those of the regime were far from identical. In fact, as the documents in this chapter indicate, much of Bolshevik agrarian policy simply mystified the peasants. Implicitly or explicitly, those who troubled to write to newspaper editors and government officials could not understand why those in power would not do all they could to facilitate peasant farmers' productivity. After all, when they were feeding the cities, the industrial workforce, and the army, did it make sense to deny peasants the same civil rights granted to other laborers? Why would the authorities want to encourage the breakup of large, successful family farms into smaller units that would surely be unable to support themselves and would doom their members to lives of poverty? Most puzzling perhaps, why did Soviet policy favor the village ne'er-do-wells and idlers while punishing the able, hardworking, and ambitious?

Peasant certitude that the state should spare no effort to ensure the individual farmer's success could not be squared with the Party leadership's fear that the individual farmer's success, by strengthening private property, would indefinitely delay (and perhaps doom) socialist construction. As a result, the NEP compact between regime and peasantry was tenuous and fragile and liable to fracture under duress.

Documents

Land Reform

Although early attempts to inflame rural class warfare via the poor-peasant committees (*kombeds*) and to establish collective and state farms failed in the main, these measures did undermine peasant confidence in the permanence of their revolutionary gains. The conclusion of the civil war and the change to the New Economic Policy therefore necessitated a reappraisal of the agrarian situation with an eye toward stabilizing peasant landholding and resolving questions of land redistribution. The terrible famine that struck many grain-growing regions in 1921 added urgency to these questions. The reappraisal culminated in the passage of the Land Code at the end of 1922.

Responsibility for drafting the new agrarian legislation fell to the People's Commissariat of Agriculture (Narkomzem) of the Russian Soviet Federated Socialist Republic (RSFSR). Like many government institutions, during the 1920s the Narkomzem leadership consisted of communists and noncommunist specialists (many of whom had belonged to political parties that opposed the Bolsheviks), and each group sought to put its stamp on the final draft. As early as fall 1921, their various proposals were carried in the press and initiated broad discussion of agricultural policy in both the center and the provinces.[17]

Other organizations had a hand in the legislation, and their deliberations also garnered press coverage and stimulated discussion among the peasants. On 2 December, 299 delegates representing local land departments (*zemotdels*) gathered at an All-Russian Land Congress. Two weeks later the Eleventh Communist Party Conference met and established general agrarian policy guidelines. These formed the basis of the Ninth All-Russian Congress of Soviets resolution on agriculture. In accordance with the October 1917 Land Decree, the resolution affirmed both the nationalization of land and the right of choice in the form of land utilization. Contrary to the 1917 decree, the resolution allowed for some renting of land and hiring of labor. To end further parceling of land, it also sought to limit excessive redistribution.[18]

Throughout all these proceedings, peasant attention focused sharply on questions relating to land reorganization (*zemleustroistvo*). This was especially the case following the First All-Russian Congress on Land Reorganization, held in February 1922. In a March 1922 letter to the newspaper *Bednota,* the peasant V. Platonov, from Olkhova district in Cherepovets province, expresses his concern that without extensive land reorganization and redistribution, poor peasants would be denied sufficient land to support themselves.

· 9 ·

Letter to *Bednota* from the peasant V. Platonov regarding land reorganization, 27 March 1922. RGAE f. 478, op. 7, d. 564, ll. 47–48. Original manuscript.

I most humbly ask you, citizen editor of the highly esteemed newspaper *Bednota,* to accept my manuscript and to spare some small space in your newspaper for my article. I would request that you print it word for word to prove, once again, that your newspaper truly is *Bednota*—standing on guard for the interests of the poorest of the laboring peasantry. As an honorarium, send me a copy of the newspaper in which my article will appear.

Once again by the broken washtub[19]

It fell to me to be a participant at the All-Russian Congress of Soviets held at the end of last December 1921. [I] even attended [sessions of] the commission preparing the draft law on the land question, which was passed unanimously by the congress, where not a word was spoken about the recognition of any sort of property rights in land. When I gave my reports to the local peasantry, as I was obliged to do, they were quite satisfied that the congress had devoted so much attention to the restoration of agriculture. Only a tiny contingent of property owners, the kulak element, was unhappy.

But within a month and a half, another congress gathered in Moscow under the clangorous title the Congress of Land Regulators. [It] slumbers in Moscow under Comrade Mesiatsev's[20] lullaby and submits a resolution on the recognition of private ownership of land that does not propose to carry out any radical land reorganization but rather to reinforce [the rights] of those who [already] own [land] and not to subject [land held by] peasant allotment deeds to any redistribution.[21]

Comrades, if up to now the peasantry believed in the revolution, I say squarely that it was in anticipation of that peasant dream—to destroy the antediluvian confusion that existed under the old capitalist system: insufficient land, landlessness, strip farming. Up to now we at least had hopes, but now all our hopes have crumbled: everyone keeps what he now has, and [land parcels] will be assigned to those settlements that are now cultivating them. This, comrades, is not an answer, nor is it a law, because no one can be sure that today or tomorrow his land won't be taken from him. Why will they take it away[?] Because the fact is, they have every legitimate right to, because it wasn't really allotted or sold [to the peasant]. But [if this happens, then] one [peasant] will end up with eight *desiatinas* [8.7 hectares] [to support] four people, and another a half desiatina for six people. How I would like to speak personally with Comrade Mesiatsev and demonstrate that your venture is not necessary. Take a look at how the peasants walk with scythes and pitchforks for five to eight kilometers through two or three villages to mow, and how the peasants of those villages through which they walk look at them, and they mow right up to the house itself and the local peasants have nowhere to let out a hen. Comrades, I never sound [the alarm] for nothing, but when it rings out, then I will prove whose interest is being served—the kulaks or the proletarian International [Comintern]. For a long time before the revolution I had already fought for the existence of the land rights of the laboring peasantry. I attended eleven district, eight provincial, and two All-Russian Congresses. From this revolutionary work I've gone gray, and that is why I am saying how stupid this resolution is. All the honorable officials of Soviet power will be convinced this summer that I have spoken the truth. With the recognition of the right [to have] land deeds, all those big and tiny leeches on the laboring peasantry will stop at nothing to try and number themselves among the holy assembly[22] of the laboring peasants and to get back the land taken from them. That will start a new struggle, because the small landholder and the landless peasantry will not give up their just land rights for a mess of pottage. This summer all the agricultural offices of the Soviet republic, beginning with the district land department and ending with Narkomzem, will be daily swamped by carloads of every possible petty litigation, complaint, and appeal. And I am absolutely sure that the enormous Narkomzem building will be unable to contain the land spiders from throughout Soviet Russia. They will hound Comrade Yakovenko[23] from his office. If he manages to read my article, I am quite sure, as I write it, that he will remember and say more than once that Comrade Platonov was right.

Further, I would ask Comrade Sosnovsky[24]—he remembers Platonov, we were on the commission at the Ninth Congress together—to ask Comrade Trotsky at some point if he would take up this subject. If it is confirmed that millions of Red Army soldiers will be left without land and if somewhere it will again be necessary to smoke out the bourgeois like at Kronstadt, then before you get to say anything, no one will believe that you are defending freedom and land. You won't instill that much courage just by [singing] the "Internationale." But I believe that our leaders will consider the situation and will stop before they chop off the last branch they are sitting on. They will revoke this stupid resolution and will once again choose the path of radical land reorganization for the elimination of land shortage, landlessness, and inaccessible land.

V. PLATONOV, delegate to the Ninth All-Russian Congress of Soviets

---

The completed Fundamental Law on the Laborers' Utilization of Land was passed at the Third All-Russian Central Executive Committee (VTsIK) session in May 1922. Later, the law was included almost without change in the Land Code introduced on 1 December. As the letters show, many peasants were far from satisfied by this attempt to mediate between socialist principles and the reality of private farming. The code put to rest any expectations some peasants may still have entertained that the revolution amounted to a "black repartition"[25] and that the land had become theirs. The code reaffirmed land nationalization, declaring all land "the property of the worker-peasant state" while also eliminating private-property rights to minerals, water resources, and forests. The purchase, sale, and bequeathal of land were prohibited.

To stabilize land utilization and halt the continued division of allotments into ever-smaller, less productive parcels, the Land Code also limited general land redistribution to three full crop rotations. Where the three-field system was in use, this meant a minimum of nine years between redistributions. Rational though it seemed, this provision struck directly at the peasants' attachment to just land distribution, and they insisted on maintaining a system that reflected demographic change. Demobilized soldiers and peasants who had left for the city but had now returned to the countryside were, along with their relatives, expecting their allotment of land from the commune (obshchina), and the Land Code notwithstanding, the commune sought ways to accommodate them. In the peasants' eyes, the revolution's success or failure, and their support for it, depended in large measure on the degree to which it made possible the closely-held ideal of land equalization. This is evident in a letter by Ya. Shepelyov, a former Red Army soldier from Borkov-Shubatskie village, Olkhova district, Cherepovets province.

· 10 ·

Letter to *Bednota* from the Red Army veteran Ya. Shepelyov calling for the equalization of land holdings, 7 March 1922. RGAE f. 478, op. 7, d. 564, ll. 188–189. Original manuscript.

What is the October Revolution giving the poor peasantry? If someone already has a lot, then they will give him even more, but he who has little has this little bit taken from him—that was the law in prerevolutionary days. When the October Revolution flared up, all the poorest populations of town and country, like a drowning man [grabbing] at a straw, took up the defense of this just cause in order to win their freedom and to receive land. The Council of People's Commissars placed all their hopes for a victorious revolution on the mass of poor peasants. Everywhere assemblies and various sorts of rallies were held at which orators spoke, addressing those poor peasants whose help had successfully kept Soviet power in their [poor peasants'] hands. The people's blood has flowed for four years, mostly [the blood] of the poor peasants, who had to endure so much cold and hunger. The mass of Red Army soldiers did not return from the front; they laid down their lives for freedom. Those who did return to their villages and hamlets as cripples found their farms in even worse condition than they were before. Returning to their native nests and seeing the ruins, these eagle-warriors for land and liberty did not lose heart, hoping that, having been victorious at the front, they could now engage in peaceful construction. Most of all, the poorest peasantry waited for socialization [to come] like the biblical manna from heaven. Instead of manna, the revolution would give the poor peasant wounds. At the Congress of Land Regulators, the majority in the congress, led by Mesiatsev and Professor Rudin,[26] negated equalization and stood for immediately allotting to all settlements and communes the amount of land that they are now using. In the opinion of Comrades Mesiatsev and Rudin, by means of this land usage it will be possible to quickly deliver the country from the economic ruin in which we now find ourselves; moreover, it will be possible to quickly calm all those who do not want and fear such a redistribution (the rich and non-poor, of course). To abandon socialization is to distort the revolution. The revolution should establish equality, not shed blood. It is to no purpose to fight your way into the Kremlin only to toss the have-nots overboard. For the Comrades Mesiatsevs it is not enough just to return the land to the gentry; the factories and plants [should also be returned] to their former owners, and [Tsar] Nicholas put back on the throne, so that they, at any rate, will also be pacified. I am amazed that the officials at the center support this. Specialists[27] do not want to work but only to receive a salary, or they already belong to the bourgeois class. The war is over, [so] now we can boldly get down to peaceful construction. When the housewife gets up in the morning, lights the oven, and prepares a meal, a lot of dust gathers about her, but it is as if she does not notice it. When she is done and everything is ready, then she takes out a broom and sweeps up all the trash and throws it all away. So it is for the leader of the revolution. Comrade Lenin can, I think, try and clear the Kremlin of trash and give the poor peasant not only wounds but liberty and land, which we will breathe like the air. [. . .]

YA. SHEPELYOV

However much the peasants may have desired just and equitable land reform, the situation arising from the war and the revolution greatly increased the opportunities for larceny. In a letter from 23 June 1924, the village correspondent I. Filonov (who wrote under the pseudonym Svistunov) from Voronezh province writes of the fraud and land grabbing that occurred in his village: "For our village there are no decrees or land codes. The allotting of land began in 1921, and the schemers helped themselves to an extra desiatina, the best, and a few swindlers even managed to get three to five desiatinas of better land closer to the village. This much is obvious, but secretly they took as much as they wanted. And so that such a big field should not, by some fluke, catch the eye of an observant peasant, they cut it up into three or four fields. A lot of this land has been tracked down, but the cheats still hold a lot in their sharp claws. This is the situation that the demobilized Red soldiers are coming home to, and after the famine, widows and orphans are returning from all over and still have not found their allotments."[28]

By concentrating land in a few hands, this sort of banditry outraged peasant sensibilities. But peasants themselves were not unaware of the drawbacks arising from their own equalizing practices; frequent repartitions made rational land management impossible. They not only preserved but increased inefficient strip farming, with its far-off fields, subdivided plots, and tortuous configurations. In the name of equality, peasants were given an abundance of small parcels of land in different fields. At times, the number of such parcels belonging to one peasant could reach twenty, thirty, forty, or even more. In October 1924, *Krestianskaia gazeta* received an article written by the peasant P. Yerastov from Kostroma province that he entitled "The Ossified Parasite." "Imagine for yourself," he writes, "the peasant's one-and-a-half soul allotment[29] that has nine desiatinas in all, of which three desiatinas fall to common pastureland, and up to one desiatina is just no good, and the remaining five-and-a-half to six desiatinas are meadowland and under cultivation, and so it happens that this one-and-a-half soul allotment of 5.5 desiatinas of arable and meadow land numbers up to eighty strips, and sometimes more. And the field has, in addition, up to fifty boundary strips, and in three fields there are one hundred fifty. And so, peasant, here is the parasite on your farm, up to eighty individual strips on one allotment, and how many times a summer do you get to each one? From this it's clear how much time and work you are killing in vain, to no purpose."[30]

Only the state had the resources and means to improve these irrational and unproductive features of Russian agriculture. Between 1919 and 1921, the state had undertaken land reorganization largely to promote collective farming, but after 1922 the land given over to collectives in this way declined into insignificance. During the NEP years the state redirected its land reorganization efforts toward improving the efficiency and productivity of Soviet agriculture generally. In addition to resolving land disputes between villages and communes, the state now placed emphasis on replacing the antiquated three-field system with improved multifield rotational systems, consolidating strips, and breaking large communes into smaller ones.[31]

Demands for more land and access to common lands and forests are familiar to those who study the prerevolutionary peasantry. In the following letter, G. Yerofeev, a Red Army veteran living in Pasma village, Kostroma province, points an accusing finger at the Soviet state for its failure to act in accord with peasant interests.

· 11 ·

Letter to the People's Commissariat of Agriculture from the Red Army veteran
G. Yerofeev calling on the state to establish land-use norms and to provide
financial assistance to peasants, 19 March 1922. RGAE f. 478, op. 7, d. 564, l. 258.
Original manuscript.

"We need to think about this"

For four and a half years the Russian worker and [the Russian] peasant have
been carrying a terrible but glorious burden on their shoulders: the struggle
for the great slogans of socialism, freedom and equality, and for their country's property, for all its riches, which rightfully belong to them. He has
borne many physical and spiritual tortures, many deprivations. The best
comrades gave their lives. Many returned unable to work and remain in
poor health. And their relatives, who stayed at home, were not disinterested
in this struggle. They gave of their labor, of their last bread, and they themselves ate substitutes. But they bore everything with dignity, trusting in the
future.

But what is actually happening[?] A lot of words, a lot of talk. Articles
are written in the newspapers about improving agriculture, but they are all
only phrases on paper. It turns out that Soviet power is probably no better
than the tsar. In a word, the impending decree on land utilization once more
states that all allotted and deeded land is again transferred into the hands of
the peasantry. This is good, but has Soviet power really forgotten that very
little [land] was actually allotted to the peasantry on 19 February 1861, and
that the allotments they got were not the best?[32] [Since then,] many [peas-
ants] have been born, and many, who had earlier been thrown off their
landholding by tsarist despotism, have arrived from the cities. Are they really [such] stepchildren of the republic that it doesn't speak about an increase in the land and of an increase, in general, of all the norms [i.e.,
amounts] denied in the earlier allotments, [that it] has not ordered a precise
determination of the percentage of land ownership by the peasantry? This is
why disorders will break out between settlements, since one will have more
land and another less. That is why it is necessary to determine a definite
norm for land in use by the peasantry of each province based on the revision soul,[33] and to divide the rural population according to categories, and
to give [land] to each wishing to work, and to carve [this land] out of the
state [land] fund, and then there will not be hostility between peasants, because everyone will have the same[-size holdings]; only let Soviet power
prove itself kinder and not just return deeded land but parcel out [land]
from its own fund.

Second, the peasantry is excited by the slogan "All the wealth of the republic to the people," but in fact it's not like this; by the will of Soviet power,
the peasant has become a thief. That is, [to make up] for the seven years of

war, the ruined peasantry demands improvements, the creation of the new. He also needs fuel for heat, which is why he demands wood. But here the peasant is at a dead end and does not know how to get out from the situation that has cropped up. For all the lumber to build a home and even for firewood, the peasant must pay money, and at quite an exorbitant rate, too, but here is the problem: where is he going to get the money for the payment to the state? He doesn't have land with forest—it has all been nationalized. To equip a house one needs, at the very least, 800,000,000 rubles for firewood and then has to pay 260,000 rubles per *sazhen* [approximately two meters]—really, is this conceivable?[34] Where will he find that kind of money? In addition, he still gives away a lot for nothing. And this dictates that he must steal lumber for the repair of his farm (and this in the RSFSR!!!)—by right the timber belongs to him. Give him some indication where he can get such capital. But Soviet power authorizes nothing in this regard. If, as it says, the collapse is serious and money is needed, then let it spend less; that is, get rid of the 100,000 forestry officials since they are useless and receive a ration for no reason. Let Soviet power act by the slogan "To each according to his needs." As you know, if the peasant requires some material for his holding but cannot pay, he will pilfer it, that is, make off [with it] secretly, and that is why any need for forestry officials is moot. As a result, [it will be possible] to reduce the large expenditures for their upkeep—a large gain in the opinion of the peasants. Most of all, he will not have to take what he needs. Repressive measures regarding payment for materials will call forth indignation from the population.

[. . .] I am a Red Army man who returned from the front on 20 December 1921. On the basis of the VTsIK decree on the supply of needed timber material, I placed an appeal with my land department, as a soldier, for a free allowance of one hundred trees to build a house. I filled in a questionnaire and sent it to the forestry official, he in turn sent it to the provincial forestry [department], and now three months [have passed] and there has been no answer. I do not know who is at fault; either some grease is needed, or the VTsIK decree exists only on paper.

But time is passing and is being lost in idleness; where can this be fixed?

I request some indication of where [I] still need to turn in order to receive lumber, or is this just a cry in the wilderness? As for the payment of money, I have received nothing for two and one-half years.

I request an answer.

———————————

Whatever regulations issued from the center, local considerations greatly determined how land reorganization actually proceeded.

· 12 ·

Letter to *Bednota* from an anonymous Red Army veteran on land redistribution in
a Moscow province village, 7 March 1922. RGAE f. 478, op. 7, d. 564, l. 21.
Original manuscript.

The Land Question in the Village

Two months after the decision of the All-Russian Land Congress, the countryside is beginning to stir. The other day I attended a village assembly in
Tiazhino, Bronnitsy county. The question under discussion was quite topical: the redivision of the land according to the laws that were published after the congress.

Every face in the packed schoolhouse was serious. It was obvious that
this question greatly interested the assembled peasants! They listened calmly
to the instructions from Narkomzem and then promptly forgot them as
they resolved to decide the question in their own way.

No land at all would be given to anyone who resided "outside" because
there was little enough for those of us living in the village. Let everyone
from a different village who had entered a house through marriage go and
cultivate the land of their own village. At that, the voice of someone or
other's wife rang out from a corner. "She may make use [of land] here," the
chairman explained.

Further on, they began to consider each family separately, and here, under
the pressure of clamoring kulaks, they made a few departures from the recently passed legislation.

They decided to give land to the peasant I. F. Makarov, who all his life
has served and lived with his entire family in Moscow. But to those peasants
who live in the house of their wife [*sic*] and work the land they gave no allotment. Exactly where the legality or even the common sense is [in this] is
hard to understand.

There are even more interesting goings on in this village. The land is divided according to a head count; that is, the more mouths there are to feed,
the more land [the household receives]. A year ago, several resourceful muzhiks declared that they would soon marry off their sons, and therefore land
should be given for their future wives. The peasants were all honorable:
why not grant their request? So to this day they own excess land, but their
sons still haven't married.

In a word, there is arbitrariness at every step. I think that freedom of
choice in the form of land utilization for each settlement is one thing, but
freedom of arbitrariness, to give to one and take everything from another,
this is too extreme. Those progressive peasants whom the village despises for
their innovations should not be placed under the thumb of any local kulak-
boss. It is incumbent on the Bronnitsy land department to look in on the

village, if only occasionally, or else the scavenging crows there will build
themselves a sturdy nest, and no number of decrees will touch them.

"Yours"

However crucial to peasant survival, land reform often exacerbated
social divisions within the village. In June 1924, the head of the OGPU,
F. E. Dzerzhinsky, informed the Communist Party's leading body, the
Politburo, that poor and middle peasants who had been deprived of part
of their land during the famine years supported reorganization but that
kulaks who had the best land and higher norms "speak against the land
reorganization policy everywhere."[35] The following letter from P. I. Dukh,
a peasant from Arpachin village in the Cherkass region, Don province,
reveals how important land reform was for the poorer peasants and illustrates their ability to close ranks in the face of opposition from the
better-off.

· 13 ·

Letter to *Krestianskaia gazeta* from the Don province peasant P. I. Dukh detailing
the conflict between rich and poor peasants over land reorganization, 17 March
1924. RGAE, f. 396, op. 2, d. 16, l. 330. Original manuscript.

How we are carrying out land reorganization.

The citizens of our village, the poor and middle peasantry, led by the soviet
chairman, Comrade Manatskov, have tried for a full year already to carry
out land reorganization. They have repeatedly elected representatives, but,
as is always the case, the more prosperous [peasants] end up as representatives, and rather than try to carry out [reform], they try to stop [it]. But the
poor, seeing all the tricks of the kulaks, threw out all the elections and selected representatives from among themselves who immediately went to the
regional land administration and concluded an agreement for six thousand
desiatinas of land. On their return, they gave a report of their work at a
general assembly held on 17 March. Hearing the report, the poor were satisfied and rejoiced that at last they had succeeded in obtaining an allotment
of land. The former kulaks and their allies spitefully hissed: "Go ahead be
happy, be happy. How will you pay off [the debt]? You yourselves will wear
the horse collar." But the poor paid no attention to this and attended to their
own business. And then, when the questions had ended, the leader of the
kulaks of the entire village showed up late for the assembly—the well-known

kulak Ignat Kovshikov. And without any idea of what had gone on in his absence, but sensing that his voracious appetite was nearing the end [of being satisfied], he began to shout: "What are you doing? They have taken our last pair of pants. We need to keep things as they are for another year or two—a year, anyway. Let's see what the future brings. We have no money. How will we pay for the allotments?" The poor became agitated. Lebed, the chairman of the assembly, started to ask Kovshikov to stop, but he would have none of it. Kovshikov became enraged and took the bit between his teeth and bolted. His kulak friends and their stooges gathered around him. Seeing the kulaks' behavior, the poor categorically demanded Kovshikov's arrest and removal from the assembly. The chairman of the assembly was powerless to prevent the intensifying storm, but the soviet chairman, Manatskov, helped him calm the commune, and then, continuing his report, the representative for land reform, the poor peasant and former humble peasant[36] Vasilenko added: "Citizens, we all ant to carry out land reform, and we will carry it out. And if all the well-off, especially the wealthy Kovshikov, don't want to and point out that we lack the means to pay for the reform, then we will pay, and I myself will pay Kovshikov's share. Only don't let them break up the assembly and block [our work]." Vasilenko did not have to say anymore, since the arch kulak left, and yelled at Vasilenko: "Ruffian, cheat! Who are you to give us orders?" Seeing how their elected representative had been personally insulted at a public assembly, the poor swelled up, rose in full force, and threateningly began yelling, demanding the arrest and expulsion of Kovshikov. The storm worsened, and this time the soviet chairman Manatskov could not soothe [them]. The affair would end in fisticuffs. After it all, the kulaks, seeing the resolve of the poor, quieted down, and the assembly on land reform adjourned. Now they are already delivering money to the representatives, and in the not-too-distant future we will have land reform.

P. I. DUKH

In 1926, the head of the People's Commissariat of Worker-Peasant Inspection (Rabkrin) reported that more than one-half of the complaints his commissariat received from peasants concerned the work of local land departments (*zemorgans*), especially the conduct of land reorganization. In Perm region, for example, each land regulator was responsible, on average, for 1,200 to 2,300 farms and approximately 162,000 to 769,000 hectares of arable land. By the end of 1927, only 11,500 surveyors and other personnel were engaged in land reorganization work. When the land regulator did undertake work in a village, individual interests often superseded those of the whole. The peasant A. D. Usachev wrote that in his village, Ternovka, land-reorganizing work had been going on for three years but had yet to finish. The rich—those who could bribe the land regulator—had, however, seen to their needs. According to his account, a group of well-off peasants drank thirteen buckets of homebrew with the surveyor "and [gave him] chickens, butter, and eggs, no charge," and they reaped the benefits. The poor peasants, even though they had organized an association called the Path of Communism, did not have the means to offer the land regulator any "vittles" and were excluded from the process.[37]

The inspectorate's report cited above contains an account that testifies to the extreme necessity, in certain cases, for land reorganization: "In one province the remnants of so-called *barshchina* exist to this day—the division of a village into groups according to their [previous] ownership [as serfs] by different gentry. These groups continue to have different norms of land per head, which gives rise to countless misunderstandings. As a result of their organization into separate communes [*zemelnye obshchiny*] within the village, they even graze their cattle separately, and when several herds arrive in the village simultaneously, it incites all manner of arguments. This feudal residue—the partition of one village according to the old barshchina—is fully preserved to the present day. The peasants are requesting help to amalgamate, but the land departments are doing nothing."[38]

Landlords

Many peasant letters touch on their relations with former gentry landlords (*pomeshchiks*), indicating that a number had weathered the revolutionary storm. As a result of the rural revolution, about fifty million desiatinas of gentry land had passed into peasant hands. Yet individual representatives of this class could still be found tending remnants of their property into the late 1920s. In the letters, peasants freely expressed their feelings of distress. S. V. Kuznetsov, the head of a village reading hut in Kaluga province, wrote in 1924, "A landlord lives among us. He has forty desiatinas of land. The village land encircles his land. And he disturbs us a lot because in the summer he puts fences up everywhere, and there is not even a good well. There is a law to give him a place in the rear [of the village] and [for us] to take and divide [his land], since he troubles us so."[39]

Peasant hostility toward the gentry ran deep and was often exacerbated by the practices some gentry engaged in after the revolution. That is the subject of a letter from the peasant F. K. Yevseev.

· 14 ·

Letter to *Krestianskaia gazeta* from the peasant F. K. Yevseev on the commercial activities of former landlords, 1 August 1924. RGAE, f. 396, op. 2, d. 16, l. 38. Original manuscript.

I sincerely request that you place the following [letter] in your pages. In the village of Kodushkino, Finiaevo district, Skopin county, Riazan province, there is a certain former landlord, Klavdia Matveevna Klykova, who has a garden (two and one-half desiatinas) and a kitchen garden (one and one-half desiatinas) in her possession at the present time. In one year, this garden yields no less than fifteen hundred puds of apples. This garden is not on the books, and neither is the kitchen garden. The landlady who owns the garden is pretty old. She has two daughters, the entire family consists of three people, [and] they don't work but stuff their pockets. This has gone on from the start of the revolution, that is, from 1917. Consequently, in 1924, the committee of mutual aid repeatedly petitioned the county land department to take this garden for its own, that is, for the poorest of the population. But everything has been unsuccessful. Why unsuccessful? Because this landlady has a partner in this garden, a neighboring landowner or a well-off kulak, who, thanks to his shiftiness and cleverness, has managed to bribe the county land department not to see. It is not enough that this chiseler has a garden of his own no smaller than the one already mentioned, but he was still grabbing [land] in another place. This chiseler's name is Matvei Matveich Golovin. By spreading out his tentacles and stuffing his pocket, in one year he managed to pick up six horses and various sorts of cattle. And thanks to (frauds)[40] or an underhanded trick, he exploits his garden and also benefits from the garden of the above-mentioned landowner and calmly stuffs his pocket. Consequently, what comes of this? These same former landlords, former kulaks, have not died yet but are still alive, sucking dry the blood and sweat of the poor. They hire workers, then go riding around on *tarantasses* and calmly spin their webs.[41]

Where is the government, what is it looking at, whom does it exist for, the poor or the rich? In the countryside there are still dark [ignorant] people; these things are beyond their understanding. They put their faith in the village, the district, and the county administrations, but the administration is only [a place] for furniture. The only good thing is that there is moonshine in the countryside, even if just enough for a swig. I will give one more example: at the time of the mowing they sold one-third of the meadow, and the money they made on this the peasants squandered on moonshine, they played songs, and nothing came of it. But money is needed to repair the school and for the fire brigade. Everyone says: even if my life depended on it, I couldn't find a kopek.

Written by F. K. YEVSEEV

It wouldn't hurt if the higher organs took a peek into this little corner.

A large number of letters sent to newspapers, government institutions, and officials were denunciations. Greed, envy, or personal hostility often lay behind the politically sanctioned class discourse. Whether Yevseev was grinding a personal ax here is difficult to say. His indignation that those whom Soviet power considered opponents—in this case, a former landowner and a well-off peasant/kulak—were prospering with the support of the local administration at the expense of the village appears genuine. In any case, the letter does provide evidence that rural dwellers were learning to frame their charges in terms that would catch the eye of authority and, in so doing, were acting as de facto agents of the state.

State policy and the 1922 Land Code restricted but did not entirely prohibit former gentry landlords from practicing agriculture. According to the code, gentry could receive allotments comparable to those of other cultivators, but not in the same place where their former estates had been located. Again, the law and reality were often at odds, as explained in this excerpted letter.

· 15 ·

Letter to *Bednota* from the peasant N. I. Bogomolov detailing the conflicts between peasants and former landlords, 14 May 1922. RGAE, f. 478, op. 7, d. 564, l. 246. Typewritten copy.

The peasants of Seslavl village, Merenishchensk district, Kozyolsk county, [Kaluga province], were extremely interested in the work of the All-Russian Congress on Land Reorganization. On hearing of the draft law on land the peasants immediately became cheerful, their backs straightened, their eyes sparkled, and during the reading of the decree, exclamations of "There you are, Soviet power has not forgotten us ... thank you comrades!" were heard. [...]

After the reading of the decree, the peasants of Seslavl village went to their landlord, Ivanov, and announced that he, Ivanov, on the basis of the law did not have the right to utilize more land than the norm, and that excess [land] was subject to alienation to the commune. They drove his female worker, who had plowed up an area of land which had served as common pasture for livestock, from the field. Great, then, was the peasants' surprise when late in the evening on the second day, i.e., 25 April, the surveyor,

Comrade Larin, arrived from the city of Kozyolsk with the head of the Eighth Division of the county militia, Rodyn, his assistant, and the militia man Grishinsky, from the Merenishchensk district militia. He [Larin] parceled out and staked off from the village's common pasture a section of land with a garden for the landlord, Ivanov, without asking the consent of the citizens. To their protestations, the peasants received the response: "We don't recognize you . . . make a revolution, etc. . . ." The citizens' faces immediately became wan and pale. [. . .]

On 28 April, the workers hired by the landlord, Ivanov, began to dig a ditch from the village along the border of [his] parcel. Seeing this work, the women (all the men were in the field) began to fill in the excavated ditch. Just then, two Red Army soldiers jumped out from the landlord's house (it later came out that they were agents of the Kozyolsk criminal investigation department, Neustroev and Silin, the acting head of the Kozyolsk county militia) and behind them, smiling and wringing his hands, the landlord himself. One of them, Neustroev, brandishing his revolver, headed toward the women and cried out in a shrill voice, "Don't move . . . hands up!" The women began to wail from fear, and several started crying. Running toward the noise, the menfolk, indignant at what was happening, drove the workers from their chores and demanded an explanation from the criminal investigation agent Neustroev. Neustroev, probably thinking that the citizens wanted to make an attempt on his life, mounted his horse and fled to the city of Kozyolsk. That evening, however, the entire garrison of Kozyolsk arrived in the village to put down the mutinous peasants. The brigade of soldiers and militia interrogated the citizens but failed to find anything illegal and left, but not before promising the citizens that they would try to bring the entire matter to the attention of the center.

After the foregoing incident, on 2 May, the head of the Kaluga province agricultural subdepartment [and] the chairman of the Kozyolsk County Executive Committee, Novikov, arrived in Seslavl. They conducted an investigation of the disputed garden and parcel of land given over to the landlord and acknowledged that the allotment was incorrect and that the land that had served as a common pasture should be transferred to communal use, but that the disputed garden of four desiatinas, having belonged to the landlord, Ivanov, N. I., before the revolution, remained, by decree, [the property] of the former owner until the further notice of Narkomzem.

The citizens of Seslavl promise the Soviet republic that if Ivanov's garden was turned over to their use, the funds earned from its cultivation [would be used] to open a school.

———————

In addition to illustrating the peasants' ability to join together in a collective action to defend their interests—in this case, common land— this account also reveals the influence that some former landlords could still bring to bear with local officials nearly five years after the revolution. To address such situations, the government passed the resolution "On Depriving Former Landlords of the Right to Cultivate Land and to Reside on Farms That Had Belonged to Them before the October Revolution" in March 1925. That June, it also issued the instruction "On the Order of Expulsion of Former Landlords and the Liquidation of Their Property Holdings." To implement these decisions, special commissions were created at the local level. By the end of 1927, as many as 10,756 former gentry still living on their estates had been registered on the territory of the RSFSR. Of these, 4,112 were evicted, with the result that 43,706 hectares of additional land passed into peasant use.[42] In practice, however, evictions often followed their own logic, as shown in this excerpt from a secret 1926 report to the RSFSR People's Commissariat of Agriculture.

· 16 ·

Secret report to the Narkomzem RSFSR on peasant efforts to evict former landlords, 20 November 1926. RGAE, f. 478, op. 1, d. 1970, l. 177. Typewritten copy.

———————

Former landlords resided in the region of Shtevets village, Nizhny-Terebuzh district, Shchigry county, [Kursk province]: the family of General Ragozin (shot in 1918 as an active participant in the shooting of Leningrad workers), Kurlov, Goncharov-Dombrovsky, Shatilov, Bogdanov, Mesh-cheersky, and Poliakov. At the Shtevets peasant assembly, on 11 January of this year, with 560 people in attendance, a resolution was proposed "to liquidate the lords" and take [their] gardens and houses and transfer them to the committee of mutual aid. On 10 and 17 May of this year the question of the eviction of the landlords was again debated at the general assembly of Shtevets village, and again resolutions on evicting the former landlords were put forward, but in spite of this, the landlords were not evicted. On 1 November, at a general assembly of village activists, the question of evicting the landlords came up again, particularly in regard to the former landowning woman Kurlova, whose house they wanted to use for cultural enlightenment work, and a resolution was put forth on taking the house under the protection of the commune and transferring it to cultural-enlightenment [work]. The red tape over the eviction induced several peasants to organize an attack on the houses of the landlords on the evening of 2 November. They smashed the windows with stones, damaging some things in the process. Other [items] they broke up and divided among themselves. On the night of 4 November, the house of the landlord Shatilov burned down to the foundation, but up to this time, the question of the eviction of the former landlords of Shtevets has not been settled in its final form. At night, groups of peasants hacked the wooden ornamentation off the manors of former landlords. Among themselves, the peasants spoke of evicting the landlords on their own, without the participation of the authorities.

The Cheremisinovo VIK ordered the Korandakovo village soviet to clarify the relationship of the population to those landlords subject to eviction. The village soviet member Apalkov raised the question of an addition to the proposed list of landlords subject to eviction. An atmosphere of panic was created among the evicted. They tried mightily to find people powerful enough to support them in the assembly, and in the village an almost general drunkenness began. One time, those who had been evicted brought a large amount of moonshine, enough to treat almost the entire population, to the house of citizen G. Grinev, and they all drank together. The peasant Poliansky carried moonshine around in goblets. Another group of evicted [people] brought moonshine to citizen M. Apalkov's, where there was also mass drunkenness. Soviet member N. Shakov, being a daily inebriate, went to the homesteads of the evicted and proposed that they treat some of the peasants [to liquor] to gain their support in the assembly. The main initiator of the additional evictions, village soviet member Apalkov, together with village soviet member N. Apalkov, showed up at the houses of citizens slated for eviction, in particular at citizen Pisarev's, and suggested that they toast their support for the evicted at the coming assembly. On the question of eviction, a resolution was put forward "to leave evicted [persons] in the commune, in order to give [them] the approved sentence." Leaving the assembly, some peasants said, "They sold out for booze."

During land reorganization the new village leadership sometimes also took advantage of their position to the detriment of the poor and less astute, as explained in this letter from Donetsk province.

· 17 ·

Letter to *Krestianskaia gazeta* from Donetsk province peasants on irregularities in land distribution, 5 May 1924. RGAE, f. 396, op. 2, d. 18, ll. 270–271.
Original manuscript.

Red Landlords!

Comrade editor, we earnestly request [that you] publish our letter in *Krestianskaia gazeta*.

We, the commune of Slavianoserbsk village, Donetsk province, Lugansk county, have yet to see Soviet power for ourselves, and all the citizens are asking one another, "When will we get soviet construction?" Somehow or other, we began to get *Krestianskaia gazeta* with the supplement on the law code on land. When [we] read how Soviet power promises well-being for the peasants and the poor, and how it gives advantages and full rights to all citizens of the USSR, we were horrified. The citizens of Slavianoserbsk village have not yet gotten such full rights. Why? Because our village is located in the sticks, and the people are unsophisticated and intimidated. The administration, exploiting the ignorance of the peasantry, holds it in a bridle and does not allow the peasant to speak or contradict [it]. If someone says anything, then [they are] immediately [placed] under arrest: "You are a counterrevolutionary." The people just skulk about, you only talk to yourself, and at the assembly the administration on its own, in the name of the citizens, makes decisions and does things arbitrarily in its own interest. We offer an example related to the allotment of the peasants' lands. Last year, 1923, the head of the land department, Comrade Zinchenko, together with the representative from the commune, Comrade Medin, allotted hayfields to the peasants in the following way: to the rich peasant they gave as much as he could cultivate, and [they] left the poor peasant without hay. Now, in the fall, they allotted the cultivated land as follows: all the rich elements got land behind the village as a group, then the KNS [Committee of Indigent Peasants][43] took the land in the center for itself and left patches, hillside, rocky land, and ravines for the commune. Is this justice, is this support for the poor peasant? The commune was left with one [option], to divide the remaining land into six sections and, where necessary, give a sazhen per head, but the clique and the KNS took two desiatinas per head, and all in one location. They forced the commune to take the garbage, and these six sections were spread out over a distance of eight to ten kilometers [from the village], and the poor peasant reluctantly refuses it because it is impossible [to cultivate], but the law says to give the poor peasant the nearest land so that it would be possible to cultivate it. Besides this, the KNS took for its own [use] all [the other] important [land]: the forest and kitchen-garden land. The KNS commission sold the forestland. As a result, each citizen had to buy timber to rebuild his holding at the going rate. For example, six to seven *vershoks* [25–30 centimeters] of oak, cost one ruble, but they jacked it up to three to four rubles. Customers refused [to pay], and the commission held [on to it] for themselves, and then, when everything was arranged, they listed construction [high-grade] lumber as firewood and sold it themselves for their price. Comrades Terentieva, Tsymbal, Yanenko, and others did this. New landlords have appeared who are selling off precious state property. We also have a collective, named in honor of Artemis,[44] occupying 132 desiatinas. Last year at an assembly they ordered all citizens to plow the land—allegedly social fund [land]. The peasants plowed. They were forcibly driven, coerced as under barshchina, to sow, to weed, to mow, to cart the grain. But the threshing they [the wealthy peasants] did themselves, and then divided [the grain] among us. Isn't this barshchina? The poor peasant knows only that he is working. For whom? For the big shot [*na bugaia*], as they say. It's time to destroy this barshchina. This year the same KNS is grabbing anything it wants; they let their cattle graze in the forest, but under no circumstances can the rest of the citizens. Isn't this barshchina? [If] all the party members and the KNS are allowed [to do this], then why are the rest forbidden to? Before, everyone understood that this is the master's, but now it is social [property], but only a well-known bunch has the right to use it because they have power. They say, "We won out, and we have the full right to do with it what we will, as we need, and don't you point to our legal codes. We are the law." You may ask, "How did they win out? Where were the rest of the citizens? Really, this bunch won?" Where were all our sons then? In the Red Army. And what did we do? We gave up everything for the struggle with the bourgeoisie, we gave up our sons, our livestock, our possessions, our grain; everything that we had, they tore to the ground. This means that everyone who remained and didn't take off with the bourgeoisie won out. And now every peasant should be free to restore his ruined economy, but they won't let us. They only squeeze the food tax out of us. The KNS grabbed up the best land, at two desiatinas a soul, and gave about half of it to the powerful peasant, and the poor peasant was left without land. It is awful for us poor because the administration works in cooperation with the well-off element. [. . .]

THE SLAVIANOSERBSK COMMUNE

Improving Agriculture

The NEP obliged Party leaders to explore ways of raising productivity within the framework of small-scale peasant agriculture. In this regard, cooperation appeared particularly attractive. Agricultural cooperatives had existed in Russia since the mid-nineteenth century and had experienced a spurt of growth following the 1905 Revolution. During the First World War, when the urgent need to keep the military supplied led to a general relaxation of government prohibitions on private activity, the cooperative movement took off. By January 1917, peasant membership in various types of cooperatives stood at well over ten million. Although cooperatives had declined drastically during the civil war, Lenin recognized their suitability to the free-trade conditions of the NEP and had spoken of them positively at the Tenth Party Congress. Admitting that in the short run they would strengthen capitalist relations in the countryside, Lenin envisioned cooperatives as transitional organizations capable of fostering the association of millions of people. "The cooperative policy," he proclaimed, "if successful, will result in raising the small economy and in facilitating its transition, within an indefinite period, to large-scale production on the basis of voluntary association." Under Lenin's influence, strong supporters of the NEP, like Nikolai Bukharin, came to view cooperation as *the* peaceful path to socialized agriculture. For this reason, Party members placed special emphasis on the formation of production cooperatives. But, just as before the revolution, this proved to be the least popular form of association among the peasants, who were much more inclined to join consumer, marketing, craft, and other specialized cooperatives. Between 1924 and 1927, for example, membership in village consumer cooperatives came to encompass more than one-third of all peasant farms, growing from just under 3 million to 9.8 million members. As their critics maintained, production cooperatives, because of their market orientation, were more beneficial to the better-off peasants engaged in commercial agriculture and of less utility to the poorer and middle strata, whose interests the Party hoped to promote. As a result, Party members often criticized cooperatives for their susceptibility to kulak domination even as they accepted their utility.[45]

Given the peasantry's meager financial resources, state aid was required to establish a widespread and effective cooperative movement. Beginning in 1922, the government endeavored to make credit more easily available, especially to poorer peasant proprietors. This was intended not only to provide the means to improve agricultural holdings but also to undermine the economic leverage enjoyed by better-off peasants, who continued to play their traditional role as village moneylenders. In 1923 the State Bank made twenty million gold rubles available to help fund agricultural-credit societies, and local agencies provided an additional twelve million. A Central Agricultural Bank integrating all local agricultural banks and credit societies began operations with assets of twenty million rubles in July 1924. A special fund to aid poor peasants—in particular, to purchase animals and equipment—was created in spring 1926. Nevertheless, to the Party's dismay, lending agencies generally found it safer to advance funds to better-off peasants, who were more likely to repay debts.[46]

Beginning in 1922, important general indicators such as total sown area and area cultivated in grain showed annual increases. The Party's "face to the countryside" policy of the mid-1920s further stimulated agricultural production. This was only part of the picture, however. Each year the Kremlin leadership nervously awaited the reports of provincial Party organizations and collection agencies as to the size of the harvest and the amount of grain reaching the market. Drought in the grain-growing regions in 1924, for example, dashed expectations of a third good harvest in a row, forcing a halt to grain exports—which had also been expected to rise significantly over the previous year—and necessitating instead the importation of 540,000 tons of grain. The resulting rise in prices benefited the well-to-do peasants; poor and middle peasants, with no surpluses to fall back on, became consumers themselves, having to buy grain at the inflated prices.[47]

For all the concessions granted under the NEP and for all the advances made, benefits did not fall uniformly across the peasant class. This point is made plain in a letter from the village correspondent F. Romanovsky, of Zadore village, Shchuche district, Smolensk province, written on the occasion of the ten-year anniversary of the Bolshevik Revolution. (In January 1918, the Bolsheviks decided to abandon the Julian [old-style] calendar in favor of the Gregorian [new-style] calendar in use in the West. The new calendar was thirteen days ahead of the old one. With the change the anniversary of the October Revolution shifted to 7 November.) Romanovsky's appeal is noteworthy for its emulation of the by-then-familiar official style employed in the Party's public pronouncements.

· 18 ·

Letter to *Krestianskaia gazeta* from the village correspondent F. Romanovsky on the achievements and shortcomings of Soviet agricultural policy, 18 November 1927. RGAE, f. 396, op. 5, d. 4, ll. 636–638. Original manuscript.

Long live the union of workers and peasants of the whole world and our Soviet government, and heartfelt greetings to the leaders of the people and the revolution, and on the ten-year anniversary of the October Revolution! Comrades, Soviet power has existed for ten years, and still there are many, many memories of tsarism, about how the worker-peasants lived under the yoke of capital. Even now they often recall that for firewood [you had to] go and cart [something]. For a log, you have to haul manure all day. For whom? For the landlord. But nowadays every peasant feels free, and life has gotten very good. Before the revolution our village had fifteen wooden plows and three of iron, eight wooden harrows and seven with iron teeth. But now our village feels as if it has come to life. It has nine iron plows and fourteen iron-toothed harrows, and they are transferring everything to a crop rotation system. Until the war, our village had not sown a single *funt* [0.4 kilogram] of clover—now we have sown five puds of pure clover. Life has improved for the entire peasantry. There are many pure-bred stallions—four—and two young pure-bred bulls, four heavy wagons [*drogi*] with iron wheels, two *lineikas,* and one *drozhki.*[48] We had none of these [things] in our village until now. They used to plow with a wooden plow, a *sokha,* they also harrowed with a wooden harrow, and they traveled in wagons with wooden wheels. That is how our village is coming to life. Only there are often land disputes. One side wants [to redistribute] by the head, and one part doesn't. It would be desirable to find out if a farmstead [*khutorskoe polzovanie*] can be divided by the head. [. . .]

Comrades! I feel free and know that Soviet power will not forsake us, because it loves the workers and peasants who are striving for life. And, therefore, comrades, I ask you, as your worker-village correspondent, to help me in my life and intervene in my situation, as I am myself a person in poor health, an invalid of the civil war, with poor use of my right leg. There are five souls in my family, and we all live thanks to my wife's efforts. She plows, sows, and harvests. Sometimes she beats me—why exactly? I am unable to either plow or harrow, I occupy the place of the wife. In the summertime, I heat up the oven, milk the cow, and feed the family while my wife plows and harrows. Then I serve as the social chairman of the committee for mutual aid, which requires a lot of work for the commune, and she argues with me—she says, don't go, refuse, you are a sick man. But, comrades, I feel that this is what Soviet power is about, and the aspirations it has [to make] a life for the workers and peasants, and that's why I was its defender. And wherever it was necessary, we endured tremendous cold and hunger, as well as being on the front and [living in] fear. And we feel that the worker-peasant government liberated us, and we feel that now we do not live under the yoke of capital, and each feels that I [*sic*] am a free citizen, and no one has to go to work for the landlord to get firewood or haul manure all day on a horse for a log. Soviet power freed the peasants from this. Once a year it sets aside a week for clearing the forest tracts, and ten days to chop and haul firewood and deadwood without charge. It has helped us learn, it teaches everyone from the big to the small. It has established an agricultural credit association for the poorest of the population to improve their farm. The poorest of the population get a tax reduction. And Soviet power has put all the land in the working hands of the peasants, which is all now in our hands. But, dear comrade leaders of the people and the revolution, this is what I am writing you: Soviet decrees proclaim how things should be, but our local leaders are governing the localities differently. Of course, I can give you specific facts about how things are being run here. [. . .][49]

As is evident from Romanovsky's letter, modest gains in technology were being made, if only slowly. During the 1920s, there were signs of development and modernization in the countryside. Through the separation of some peasants from the commune the foundations of the obshchina were weakened. Enterprising peasant-separators sought to consolidate their allotted land into independent farms known as *otrubs* and *khutors*. (In an otrub the field is separate from communal land, but the peasant retains a house and a garden plot in the village. In a khutor the peasant resides on a farmstead.) The peasants also discussed the transition to intensive forms of agriculture and multifield crop-rotation systems. Tradition, however, was not easily overcome. The peasant M. Kusakin from Kamenka village, Tula province, wrote to the People's Commissariat of Agriculture on 11 May 1922 about the difficulties the "separator" faced: "It is nice to work an individual farm, but not everyone is capable of resettling, of moving buildings, digging a well, damming or digging a pond. The detached farm is not comfortable. There is nowhere for cattle to graze. It is necessary to say that the peasant is somewhat afraid of all this."[50]

In the following letter, written on the occasion of the October Revolution's ten-year anniversary, S. A. Shlapakov describes the difficulties that separators and religious believers encountered. He writes from Pskov province, in the northwestern RSFSR, where peasants attempted separation more frequently than elsewhere.

· 19 ·

Letter to *Krestianskaia gazeta* from the peasant S. A. Shlapakov explaining the obstacles faced by religious believers and commune separators, 11 November 1927. RGAE, f. 396, op. 5, d. 4, ll. 516–519. Original manuscript.

Long live ten-year old October [Revolution]! Long live our revolution! Long live our Soviet power, which for ten years has maintained three positions [*pozitsii*]: The first position is to protect the revolution because the revolution has many enemies. The second position is the rooting out of the wolves in sheep's clothing. The third position is the publication of new laws. In agricultural and economic life these positions are very difficult because the capitalist enemies surround our Union from all sides. We have to be ready in case of an invasion to repulse the enemies so that human blood will not flow and human labor, which in time of war perishes like food and materials, will not be needlessly wasted. For the end of the war, hail to the Bolsheviks who in 1917 at the congress in Vitebsk tried to stop the war.[51]

Now I implore you to give me an answer [to a question] that greatly disturbs me, since faith or religion should be free. In Izochensk district, in Karatai parish [*pogost*], land reorganization was carried out for a farmstead [*na khutora*], and a fund of about thirty desiatinas from the state's hundreds of desiatinas has been apportioned; [the land] is rented from the executive committee, and the tenant has enclosed everything, so we, a group of believers, who come from ten to twelve kilometers [away] to attend church, have nowhere to put people. So we request that this reserve state fund be transferred to us, a group of believers, for which we agree to pay the state tax, since it is very necessary to us. We have to build a house for the priest, the sexton, and the elders. But they left us only one guardhouse. They sold every last church building that had been acquired through church funds. The clergy have nowhere to live.

I am signing as a member of a group of believers, STEPAN ANDREEV SHLAPAKOV.

I search for the truth but can't find it anywhere.

I have appealed to the newspaper *The Pskov Plowman* [*Pskovskii pakhar*] and to the provincial land regulation [department], and as for the county land regulation [department, I've gone back and forth so many times] that my legs ache and I've worn a rut in the road just to get my working norm of land in one place and to set up a model farm, but it's all been for naught. This has been my wish since I was very young, and I've lived sixty-six years. But I haven't gotten it because I have always lived in the village, and I live there now, in a village of twenty-one households. Since the village won't agree to give me [land], there is no way [to get it]. Why do I want it? Because I am a cultivator [*kulturnik*] and practitioner [*praktik*] of all types of agriculture, and I am a gardener. I want to show people what I can do. Why can't I show them? Because my land is scattered over three kilometers in sixty strips of arable land, hay land [not clear in original]—I can't even figure it out—so I carried on, I worked, and my labor gets me nowhere.

How our land regulation works.

[. . .] In issue no. 7 (155) of *The Pskov Plowman* for 11 February 1927, they write that 158,000 rubles have been given to the poor for land regulation, and in the peasants' newspapers they write, "Help the poor," but there is no help anywhere. The rich don't want the individual farmsteads; because he [the farmer] settles anywhere he likes, [they'd prefer] to keep the poor like birds, all in one nest. In the field he [the rich man] takes any strip he wants and doesn't want to pay the land regulation [department], saying, "I will pay when the surveyor comes." Tell him that you only get what you pay for, and then he'll find [the money] to pay for the whole village. The land regulation [department] is misguided: whenever there is a lord's estate it gives out two desiatinas a head, and where there is open or worthless land it also gives out two desiatinas a head. Some end up with good [land] and others with poor, and now in our area some work seven desiatinas a head [and some] five or four or two or one desiatina a head. Some live well and some badly. Some have land that goes to waste while those who want [land] to cultivate have nothing. In our area life is very difficult for the laboring, grain-growing peasant. It takes many years to fertilize our land, and after you fertilize [it], there's a redistribution and you're stuck with wasteland [*pustosh*]. It's the same with hay mowing. I'll give you an example: In our village, for forty years, the waste land was separated out into one field. Some put their strips to use, and others let them grow wild. This year there was a division. They gave three strips to the laboring [peasants], those who used them, but the lazy kept the cleared [land], and the laborers got all the bushes and have to break their backs all over again. Our dear leader Vladimir Ilich [Lenin] issued decrees that all the land should go to the laboring people, and small, remote fields should be removed [from cultivation], but with us just the opposite happened; they used the smaller, distant strips. But this doesn't work in our area, because it is very hilly and swampy. But our swamps are small, they can be turned into excellent hayfields, but in the village this is impossible; those who want to do it aren't allowed to. So they go to waste, useless.

In our area, the communes got hold of the [gentry's] estates after the revolution and broke them all up. Some made them into strips that they distributed among themselves, some broke them up into individual farmsteads, but some hadn't lived in the commune at all, but only joined the commune in 1920 because they could grab more land from within the commune and produce a bigger yield, because at that time commune members got all kinds of aid. [. . .] I cannot hold with my orthodox faith and not pray for Vladimir Ilich Lenin. His memory is forever. He was a leader and a fighter for the life of the poor people. He did not spare his life. He was surrounded by enemies—open and secret—who posed as his friends, but in reality were his enemies. I regret that he died so soon and that after his death his plans were not put into effect. But he explained everything that is good for the life of the laboring folk, who should not forget Vladimir Ilich Lenin.

This letter presents the reader with an intriguing discursive strategy. Shlapkov, a religious believer, opens his plea for state funds for the local church with a conventional homage to Soviet power. He goes on to denounce officials who are obstructing his attempt to establish an individual farmstead apart from the village. And he concludes by promising to pray in memory of Lenin, who is portrayed as a martyr for the common people, a saint. The letter is all the more remarkable for having been written in late 1927, during the Party's turn against "rural capitalism" and at a time of resurgent atheist militancy.

I. Chugin, from Prudovka village, Voskresensk district, Krasnye Baki county, Nizhny Novgorod province, also illustrates the travails of the enterprising peasant in this excerpt from his letter of 15 September 1924. (For the bulk of Chugin's letter, see Document 25.)

· 20 ·

Excerpt from a letter to *Krestianskaia gazeta* from the peasant I. Chugin on the village commune's opposition to individual separators, 15 September 1924. RGAE, f. 396, op. 2, d. 18, l. 368. Original manuscript.

———

[. . .] How the peasant moves to improved farming methods when the village commune obstructs him!

For citizen Smirnov, communal disorder had become unbearable. Crops were often damaged because unattended cattle [escaped] the broken-down enclosures of lazy owners. Pigs destroyed meadows. Owing to the commune's carelessness, grazing cattle went hungry because pastureland was not maintained. Working days were needlessly lost as a result of frequent commune assemblies. And people rob each other of stocks of firewood and lumber and [take] many other scarce [items] from the commune that are supplied to the citizens to improve agriculture. So he decided to pull out from the commune to a farmstead in order to establish a model farm and to escape communal practices. But the village commune greeted Smirnov's appeal with hostility and refused to allow the apportionment of an individual parcel. It would be worthwhile for the village commune to allot [the land] to the applicant and observe the farmsteader's new life and conduct, and if it turns out well, others will follow his example.

———

Opposition from the commune discouraged many would-be separators from setting out on their own. Those who persisted and managed to establish a farmstead anyway faced endless official hostility and bureaucratic inertia. In a letter from November 1927, V. A. Biakov, from Seleznyov district in Viatka province, lists the many obstacles confronting the enterprising separator.

· 21 ·

Letter to *Krestianskaia gazeta* from the peasant V. A. Biakov on the shortcomings of Soviet agricultural practices and policies, 23 November 1927. RGAE, f. 396, op. 5, d. 30, ch. 1, ll. 82–83. Original manuscript.

———

To the editorial board of *Krestianskaia gazeta*.

I am taking the liberty of writing to the editor and his colleagues that it is high time to change the writing motif of the peasants' newspaper and *Be Your Own Agronomist*. I receive the one and the other. From the very beginning the writing has been one and the same. I receive the peasants' newspaper, and they always print the same thing: that the old regime was worse, i.e., under tsarism, but that Soviet power is better. But when you look at everything in detail, then in my opinion I would say that if one wanted to live better, there was more help for improving agriculture under tsarism. Before, for example, the zemstvo gave clover seed without charge, and that was that, whether for the entire village or for an individual. They also gave seed for forests. There is a planted forest growing in one village from seed given by the zemstvo. The forest already has produced [trees] twenty-six centimeters thick at the base. Before, they gave grants free of charge for [creating] ponds and they gave a lot for [putting up] fireproof buildings. For example, the zemstvo built us a clay-brick house with a non-repayable loan of fifty rubles. The zemstvo supplied the craftsman. They gave loans of one hundred rubles [payable] over five years. The tile was given in installments. And in the village we had a bull of the Kholmogorsk breed, and the young bulls [in] the district were respectable. As I remember it, and I am forty-three, every year they brought in state foals. Before, they gave long-term credit for brick houses and in general for all types of agricultural improvements, for the cultivation of meadows and other tracts of land. Whoever was not a dunce could receive more than a little help under tsarism, certainly more than now. Now, one can only get verbal help, as much as you want, and from the newspaper too, but go into a credit association, for example, and ask for help and say that such and such is written in the newspaper, and they'll answer, "Who knows what's printed [there], but we don't have permission." That's that, and off you go.

I received an individual farmstead from Soviet power and have worked it since 1923. I have not received one kopek, and where haven't I been: in the province and in the county, not to mention the district, and all my appeals were refused! Once an agronomist purposely fetched me. They were giving out loans for the cultivation of meadows, and he knew that I needed the loan. I submitted an application to the county agronomist, and he believed that they would approve [it]. I got back the answer, "Not granted to individuals." I radically disagree. Is it possible that Soviet power would be any the worse if I established a meadow? I would use it and not send the revenue across the border. But to kommunas they give unlimited loans. I see this for myself, personally. And so, Soviet power has sons and stepsons. Before receiving that farmstead, I myself inspected everything for thirty-two kilometers around and took my wife along so that there would not be any conflict [between us]. We looked at kommunas, *artels,* and farmsteads. The best way to live is on a farmstead. Let them all criticize; whoever criticizes is stupid. The people [able to live and work in] the kommunas have yet to be born. I have seen it myself—their careless attitude toward property, [their] lazy work habits, and [there are] many other examples. The kommuna will be good when the people live as one, but such people have not been born yet. The farmstead is better because no one interferes, and you never have to wait for anyone. I would say that maybe Soviet power helps those close at hand, that is, in Moscow and Leningrad provinces. As far as our locale goes, [Soviet power does] very little. Sure, where there were landlords, there is a benefit [to Soviet power]—they [the peasants] got the land. We had no landowners in Viatka province. So the center is far off, and there is nowhere to turn. I would request that Soviet power turn more of its attention to the peasants as the foundation of the state. Don't just build electrical power stations. I believe it is necessary to attend to the building of cement factories if possible, to make [cement] cheaper and [so it is possible] to give out [cement] in installments and [to provide it to] craftsmen free of charge. First, build up-to-date cattle sheds according to the latest science—i.e., heated and indestructible. Our buildings stand for five or ten years, then sink into the ground. This is a useless waste of lumber! The huts we throw up will never last more than twenty years! The walls are put up any old way, and the floors are bare wood and not covered. In this kind of hut so much extra firewood is needlessly wasted! If you want to show your love for the peasants, then first of all put up buildings; it is the most important subject for the peasants. The second subject is to bury the three-field system. The third subject is mechanization. Animal rearing has increased and has improved. As for electricity, it is last on the list. Vladimir Ilich Lenin's words or bequest is good: the village needs electricity. But electricity will be necessary only when the economy is in order. Kerosene will do [for now]. Electricity doesn't very much interest me.

I will definitely write and will not be silent in the least. Let them do what they want to me. That's why so many dislike Soviet power and disapprove of it. It's often heard that a large segment in the cities and in the factories curse the communists because they get the best positions and their apartments are better, [because] they get better food and more of it. I have never lived in cities, [but] in the villages among the peasants, they stick a communist into elective office somewhere or in service somewhere. Everywhere doors are open to them, but the people all revile Soviet power. The truth is that there would be a lot of pride if everything would go according to the Bolshevik's program. Then only a fool would begin to curse at Soviet power, only a merchant. Many of our villages went over to the multifield rotation, [but] more than one-half have returned to the three-field system. Why . . . ? Because of livestock: there is no pasture, [so] they hurry back to the old [ways]. I'll write more. [Tsar] Nicholas gave freedom, then took it back.[52] More than likely, this is the truth. Soviet power published a Land Code. According to the code, one may choose any [form of] land tenure and get [land] on the basis of the code, period. But now they don't do it. [. . .] For example, neither a khutor nor an otrub gets a thing. I will give you an example. We seven household heads got individual farmsteads in 1923. Now, in 1927, they gave us a slice of a hayfield, 26.8 hectares, and we asked for permission to distribute it among ourselves all by mutual consent. The county land department refused our request. We appealed to the province, and there they rejected the complaint, and now we don't know where we can get satisfaction. We all are in agreement, and there is no rivalry. This is like Nicholas's freedom: you may, but you can't. Now there is absolutely no improvement. The only additions we get to the old land are former common forests. Now they only give out meadowland or a forest of local importance. I will praise Soviet power only because I received a farmstead, and there is a second [reason] that I can write down, freedom, even though my words have no effect whatsoever. Cement is the first thing a farm needs—under the foundation of buildings, for the chopping block in the ice house, the well, and the pits for root vegetables. This would save a lot of lumber. I am not saying that all buildings should use cement. It is time the editorial board lowered prices for newspapers and magazines. Compared to the old days a newspaper costs three times more. Before, the book *Be Your Own Agronomist* cost two kopeks, but now it costs twelve. I subscribed to the peasant gazette through the Zuev postal department of the Perm railroad for a year. Since 1 October 1927, the editorial board has promised to send a calendar with [issue] number 42. Not only [have I not received] the calendar, they didn't send issue 43 at all.

Written by VASILY AFANASIEV BIAKOV

This letter vividly expresses the concerns of the independent-minded, enterprising peasant. Separating from the commune was risky, and the separator often incurred the enmity of fellow villagers. Biakov clearly understood the risks and was up to the challenge, but resolute separators like Biakov were not the norm. For most peasants, communal traditions and mentalities served as a brake on individual undertakings.

The highly idiosyncratic letter of P. A. Burkov from Krutenko village, also located in Viatka province, stands in sharp contrast to Biakov's letter with its prosaic imagery. Yet in his own way he, too, advocates change in peasant agricultural practices. This letter was written "from the earth," in a clumsy hand without any punctuation and a large number of errors; here, it is somewhat tidied up.

· 22 ·

Letter to Krestianskaia gazeta from the peasant P. A. Burkov calling for technological advances in agriculture, 3 July 1929. RGAE, f. 396, op. 7, d. 14, ll. 74–75. Original manuscript.

A Peasant's Discourse

We peasants are called the rural masters. My opinion, my point of view: we are not deserving of this title, but deserve only [to be seen] as three orphans. The first orphan: the master. That's why he cannot run his farm according to new and correct [methods]. The second orphan: his cattle. He feeds them poorly, and the cattle cost him and do not nourish the master. Why? Because they [the peasants] live by the old ways without feed grass and do not allow culture on their farms and don't keep written records. The third orphan: the land. It is waiting [to see] if its master will soon turn to a new, more cultured life. Only then will all of us orphans be happy. The cattle on his farm will eat clover and will give lots of milk and dairy products. Then the master will marvel—that very same cow began to give twice as much milk from [eating] clover, and the cattle will be plump and happy, and when the cattle fatten up, then the manure will be rich and give nourishment to the earth, and the earth will provide abundance [*rodit khorosho*], and the master will be happy, and there will be no orphanage on his holding. Then he will see the new life, and his cattle will appear like a machine that converts feed into dairy products and will lighten all the muzhik's burdens. He will not think of the old three-field system and the pathetic sokha which did not plow the earth, but the muzhik will think only of those machines that lighten the labor of horses and men and unite the mass of humanity. He will dream not of the old life but of the coming bright happiness. I request that the editorial board of *Krestianskaia gazeta* place my discourse in an issue of *Krestianskaia gazeta*.

PAVEL ALEKSANDROVICH BURKOV

Soviet agriculture was highly labor intensive; only a small percentage of peasant budgets was expended on materials like tools, seed, and fertilizer. From an early date, the Soviet state had committed itself to agricultural modernization and mechanization, and during the 1920s some progress was made. On the eve of collectivization, the number of horses—which had been greatly depleted by World War I and civil war— had nearly recovered to prewar levels. Modern tools and horse-drawn machinery were also becoming more common in the village. This did not mitigate the need for labor: nearly 75 percent of the 1928 spring sowing had to be done by hand, and tractors had plowed less than 1 percent of the area sown.[53]

Social Divisions

The richer, more secure peasants might have been expected to support the introduction of innovative methods and techniques. But as V. Arkhipov, from the village of Churashevo, Voskresensk district, Chuvash autonomous region, explains, selfish interests often undermined agricultural modernization. The letter has been edited owing to the inordinate number of repetitions.

· 23 ·

Letter to *Krestianskaia gazeta* from the peasant V. Arkhipov on the
conflicts between poor and rich peasants, 9 January 1924. RGAE, f. 396, op. 2,
d. 16, ll. 106–107. Original manuscript.

The Kulaks—Opponents of Multifield Rotation.

Most highly esteemed comrade editor! Could a place not be found for my
humble and ungrammatically composed article in the pages of your respected
Krestianskaia gazeta?

Why are we dependent on the rich kulaks? Just like the old imperialists,
the rural kulak-bloodsuckers[54] even now hold the premier places at village
assemblies and gatherings. They dictate to the village poor whatever notions happen to pop into their kulak heads. The reasons for the rule of the
rich over the poor were and are one and the same, but they are only rightly
understood only by him who himself lives poorly and personally experiences the bitter insult of kulak rule. I, who am writing these lines, am truly
poor and often fall into despair when my voice, and the voices of those who
are also poor like me, remain, at the assemblies, like a voice crying in the
wilderness. They often insult us for our boldness, calling us ragamuffins, the
peddlers of Russia, etc. Right then and there, you, the poor peasant, get a
shiver down your spine, and it's as if you would tear the kulak to pieces
with your teeth, but they hold you back, these poor circumstances, which
often force you to appeal to that rich man who has insulted you. As a result
we, the poor peasants, are trading our superiority for a mess of pottage.

It frequently happens that the poor peasant not only lacks grain but also
lacks any agricultural tools, and he must borrow them from his rich neighbor. The poor peasant has no horse; for this he must turn to this same rich
neighbor, and horsepower is necessary on a farm. The poor do not even
have a cow. Although we poor are used to meager portions without any
dairy, sometimes it happens that even the poor peasant needs milk; little
children get sick and ask for milk. It's well known that we rarely have money
in our pockets, and whether you want to or not, you have to go to a rich
neighbor for a glass of milk. The poor do not have baths, and even though
he is used to the dirty life and to bugs feeding on his blood, there are times
when he needs a bath. Again, the poor peasant goes to the rich neighbor so
that the latter will allow a poor family to wash up in his bath.

I have only selected several examples here that force the poor peasant to
bury his consciousness and his reason in the trash and to give himself up to
the rich bloodsuckers. But this occurs in rural life almost everywhere. For
example, I will take the reelection to the village soviets. I proposed the candidacy of my poor comrade, Ivanov, but my rich neighbor, from whom I
borrow the agricultural tools I lack, puts forward the candidacy of a rich
comrade whom he favors, Petrov. Preventing the registration of candidates

is, of course, to their benefit. Any voter has the right to point out the good and bad sides of a proposed candidate. Here again our rich neighbors always get the upper hand because they try to show the best sides of their candidate and try to discredit and toss out our poor candidates. Here the rich get the upper hand, not because their candidate is more fit for the job than our poor candidate, but because it is not possible to speak against our rich neighbors and godfathers for reasons of conscience, because otherwise our neighbors not only would not lend us the implements necessary for agriculture but would drive us from their houses with a shower of assorted insulting words.

We the poor encounter a similar cabal not only at reelections but in all social dealings. Now, when so many have begun to write and speak of the transition to multifield rotation, we, the friends of multifield rotation from among the poor must, more than before, contend with kulak oppression because one rarely finds a supporter of multifield rotation among them. Why are the rich uninterested in multifield rotation? It is clear that the majority of the village rich are traders, and they don't care about multifield rotation. Many of the village rich are the same people who could be found at home during the imperialist and civil wars, and those of them who served in the army took shelter in lucrative positions and robbed the national purse. During the imperialist and civil wars several land repartitions were carried out. Then, of course, the kulaks didn't miss a trick. They took advantage of our absence, lorded it over the old ones, the women, and the young and got their hands on all the best spots. With the transition to multifield rotation the kulaks are loath to part with their spoils. We the majority, the poor peasants, are poor because we squandered our health and our young and most productive years at the front and in captivity. Many of us, even in these difficult years, did not fritter the time away but tried to observe and learn how the Germans, the French, and others conduct their agriculture. In spite of all this, our knowledge and efforts are wasted because our attempt to change over to multifield rotation is stymied by the kulaks. Only the Soviet government, as the defender of the poor and oppressed, can get us out of this predicament. Only it can pave a strong and true road for an exit from this oppressive situation by establishing cheap agricultural credit. Only then will we tear ourselves away from the hands of the kulaks and our weighty words to them—Get your hands from our heads! You are not our guardians but brakes. The Soviet government and the Communist Party are taking guardianship over us. Long live cheap agricultural credit and its organizers!

Peasant V. Arkhipov

As starkly as any other source, this letter describes the economic, social, and even political divisions that existed within the village, and the feelings of hostility and humiliation that they could engender. In addition to denouncing kulaks in traditional terms, as exploiters of the poor, Arkhipov employs the new regime's own terminology in order to denounce the well-off as opponents of agricultural modernization and, hence, the state's best interests. In this way, he appeals to the state and the Party to be true to their principles and to defend the poor from the depredations of the rich.

The actual degree of social differentiation occurring within the village at this time was a hotly debated topic in the Party's leading institutions. Beginning in 1925, the left opposition charged that the Party's "face to the countryside" policies had gone too far in fostering the growth of the kulaks, and as a result, their influence in many areas of rural life—most critically in the area of grain collections—had become negative and dangerous. In response, spokesmen for the Central Committee majority, like Bukharin and A. I. Rykov, denied that differentiation posed a threat even if it existed. Muddying the waters still further was the lack of agreement among contemporary scholars investigating the subject who disputed the size of the kulak stratum and even the criteria to use in determining who should be classified as a kulak, not to mention the role—positive or negative—the kulak played in village life.[55]

The Siberian peasant A. Kalinin, from Stolobovskoe village in Novonikolaev province, calls attention to important regional differences in rural life, offering a different view on the matter.

· 24 ·

Letter to *Krestianskaia gazeta* from the Siberian peasant A. Kalinin explaining the difficulties in defining a kulak, 22 May 1924. RGAE, f. 396, op. 2, d. 16, ll. 159–160. Original manuscript.

An Important Question

In Russia the peasants are discussing an important question: Who, in the village, is considered a kulak and who is not? So, for example, the peasants of Vladimir province express their opinions in the newspaper *Bednota*. These [opinions] are being read throughout all Russia. These thoughts even reach us here in Siberia. Then, having read [them], each peasant ponders who in the village he regards as a kulak, and he also expresses [his opinion] in the newspaper. This does not prevent us Siberians from mulling over this question in the newspapers for ourselves, because for us in the countryside, particularly in Siberia, if you work like an ox and acquire, let us say, three cows and three horses, you often find yourself [identified] as a kulak, but your next-door neighbor has [only] one horse and a pair of cows but loves to speculate, and they are ashamed to call him a kulak: "He," they say, "still hasn't gotten out from under his poverty—how is he a kulak!?" In Siberia, you probably will encounter peasants who have ten horses and as many cows; they are often called kulaks. In fact these kulaks work like crazy. In the summer, because of work, they don't sleep at night. And then you will see the type that considers peasant work degrading[56] and not for him, and this peasant devotes all his energy to some other pursuit, trading fish in the bazaars, or hauling salt and selling it, or [hauling] tar. In general, he hits the trade sector, and the smell of the earth and mown hay are not for him. There are people whose heads can't "figure out" which will bring a bigger profit, tar or salt. Just look at it: Which of these is a kulak, he who honestly labors with his family on the land or he that considers fieldwork a dirty business?

For us Siberians this is an important question because in Russia our middle peasant would be considered a wealthy peasant because here with one horse the plowing is poor, but in Russia one horse feeds a whole family with some left over.

I am writing this article in order to force Russian[57] correspondents to address this question, and then the peasants, who have yet to correspond. It is an important[58] question, which, in my opinion, deserves to be broadly discussed. So, comrade peasants, everyone who is writing and has yet to write to a newspaper, tell us what you think, who, in your opinion is the village kulak. We must sketch him out so that he can be quickly identified, but now we do not know whom to regard as a kulak. I will be very grateful to those who respond to my call.

A. KALININ

The 1918 RSFSR Constitution withheld the vote from certain social groups. In general, the disenfranchised (*lishentsy*) included those who hired labor, had sources of non-labor income (interest, rents, etc.), engaged in trade, were members of the clergy, or had worked for the tsarist police. The 1924 constitution eased these restrictions somewhat by returning the vote to individuals falling in the first three groups if they now lived by their own labor. Despite the center's efforts to define the disenfranchised, however, local officials continued to apply their own standards to decide who should vote and participate in public life.[59] In the following letter from September 1924, Ivan Chugin, a disenfranchised village handicraftsman, spells out the policy's negative effects. In a section of his letter not reproduced here, Chugin explains that he had engaged in trade ("sold his wares") before the revolution, but that he ceased the practice after the revolution because "trade died off."

· 25 ·

Excerpt from a letter to *Krestianskaia gazeta* from the peasant I. Chugin on the deprivation of civil rights, 15 September 1924. RGAE, f. 396, op. 2, d. 18, ll. 368–369. Original manuscript.

We really have to think about this.

There is a mass of citizens who are very insulted by Soviet power, and the insult is undeserved and is harmful to the general peasant mass. This should be taken into consideration, and maybe the government will find a way to correct [it] and to restore justice. And if this is done, then I am convinced that these citizens will bring about great benefits to the overall peasant cause. This is who they are.

Before the revolution, the overwhelming masses of laboring peasants, those having five to ten desiatinas of land that they cultivated and harvested, figured that the harvest from the land that the entire family had worked all summer would fail to support their families—not simply over the course of the year but by midwinter. And this was the case every year and is so even now. So those who had no savings or the trust of others or [were] backward in development and [could not support themselves] would seek a supplement to their agricultural earnings as laborers in lumber production or hauling freight and building materials, and that's how they supported their economy. It cannot be said that these [people] are badly treated by Soviet power.

But others from the peasant masses had savings (accumulated over the decades), or they were trusted, or were more quick-witted, smart, and developed than the first [group], and found a way to augment their agriculture, that is, to supplement their earnings. Some [engaged] in petty trade and handicrafts, others sold a cow or something from their landholding, or they took out a loan, and they somehow scraped together twenty-five to fifty rubles in advance, and they organized a five- to ten-person artel of laborers; that is they became construction-work contractors and did this work on the side, leaving only women and old men at home to carry out agriculture. After working seven to ten weeks—some successfully and others at a loss (as they used to say: he just worked to get a cow)—they returned to their homes. These laborer-peasants—that is, petty traders, handicraftsmen (working without hired labor, not selling their products from home to a middleman but selling them at bazaars), and petty seasonal (not permanent) side-workers— are definitely badly treated by Soviet power and are thrown overboard as worthless elements. The insult to these peasants would be tolerable if, after the revolution, they had continued to carry on trade or had been contractors. But if they have been engaged solely in agriculture since the revolution, even if [that work is] supplemented by either carting or handicraft production, and [they are] honest citizens, intelligent and experienced, and some of them before the revolution were organizers of many different cooperatives and agricultural improvements and were productive public figures in cooperatives and other organizations, and if, under Soviet power, being in service to it, they render significant services recognized by the county executive committee and military revolutionary staff (if the newspaper would like, it is possible to identify these people and their services in detail), and their economic position is either middle or poor, then for these peasants, this is a great insult! And it's also a great [insult] that these people—that is, former traders who have quick wits, are fast thinking, and have initiative (as one great scholar has put it), who are living among other laboring peasants, observing different kinds of deficiencies and imperfections in administrations, in instructions, and in abuses [*sic*], especially in cooperation—are deprived of the possibility of intervening or taking part in eliminating the shortcomings or abuses by officials. When at any criticism or exposure of abuses the faulty elements (who have civil rights) say to them: "According to article 65, letters 'a' and 'c' of the constitution, you are deprived of [civil] rights, and therefore you should move away or shut up."[60] And then all that will remain for the citizens who witness this evil and disorder is to watch or even take part in covering up the losses brought about by this evil or by dishonest, inexperienced elements.

Really, isn't this insulting! Is it right [to insult] these people when the community of laboring peasants itself wants to entrust them with some business or other, when it admits that they are honest and experienced citizens in this or that matter? Then can't they bring real benefits to society and the state if they would be pulled back on board—that is, from the ranks of the worthless elements—and added to the flock of the electable?!

Let us reserve letters "a" and "c" of article 65 of the constitution for the following people:

1) for traders, both former and current;
2) for traders not engaged in agriculture;
3) for contractors, large and full-time, who are not engaged in agriculture.

But it is not necessary to apply the constitution:

1) to former traders for whom trade served as a supplemental income to agriculture but who no longer engaged in trade after the revolution;
2) to present-day traders whose [trade] supports their agriculture but whose agricultural output cannot meet their needs;
3) to handicraftsmen and traders, both former and current, who do not exploit the labor of others;
4) to contractors, both former and current, who may be considered small-time and irregular and who themselves work the same as

workers and whose practice serves as a necessary support to their agriculture and is carried out by them or their family without employing workers;

5) to traders or contractors who were or are honest citizens, not reprimanded by society for depraved acts, and whom the community of laboring peasants wishes to call or entrust with some matter or leadership;

6) to those former traders or contractors who have rendered some kind of service for Soviet power, whether or not in Soviet service.

———————

Disfranchisement was vulnerable to the same complexities and capriciousness as the task of defining who was a kulak. As Chugin's letter indicates, peasant economic activity and social relations formed such a tangle that no formal legislative act or abstract definition of class could ever hope to unravel them completely.

During the civil war, peasants ceased their prerevolutionary practice of migrating to cities and towns as seasonal laborers (*otkhodniks*), but the flow resumed as the economy revived in the 1920s. By decade's end, more than four million peasants, about half the prerevolutionary total, were supplementing their earnings in this fashion. To migrate, peasants needed to obtain an internal passport. This could be an involved procedure, as Chugin shows in the following vignette from the life of an otkhodnik that he includes in his letter.

· 26 ·

Excerpt from a letter to *Krestianskaia gazeta* from the peasant I. Chugin illustrating the bureaucratic obstacles to obtaining an internal passport, 15 September 1924. RGAE, f. 396, op. 2, d. 18, l. 386. Original manuscript.

———————

"Have you traveled far, Uncle Ivan?"

"Yes, to Voskresensk, to the district executive committee."

"What for?"

"[I] intended to work in Nizhny Novgorod, so I needed a passport. The village soviet gave [me] this certification to get the seal, so I'm going to the district executive committee, and from the district executive committee I'll go to the village of Krasnye Baki, to the militia that issues passports."

And our Uncle Ivan punctuated his answer with a vulgar expression toward Soviet power and after the last words added:

"Just look what they've done! To get a passport I have to lose two days of work, go 15 kilometers to Voskresensk just for a stamp and 40 kilometers from Voskresensk to Krasnye Baki, 55 kilometers in all, then back 25 kilometers from Baki—[that's] a total of 80 kilometers I have to travel just to get a passport, on top of which I have to go another 120 kilometers to Nizhny Novgorod to get to the worksite. So I'll end up going 200 kilometers on foot."

After this, I asked myself a question, "What sort of worker has enough patience not to curse when he is confronted by such difficulties and loses two days to get a passport?" Before, this worker as well as others received passports at the district executive committee at a distance of only a kilometer. Is this normal?

Doesn't it make sense to go back to the earlier convenience so as not to force workers to walk 80 kilometers for a passport and lose two working days? Losing two days is also a loss for the USSR.

An eyewitness, I. CHUGIN.

Peasant society divided along many lines. In an April 1927 letter, I. Danilenkov, of Tolpek village in Smolensk province, expresses the hostility that peasants harbored toward those who avoided agricultural labor by either letting out their land or leaving the village to find nonagricultural work.

· 27 ·

Letter to *Krestianskaia gazeta* from the peasant I. Danilenkov criticizing state support for peasants who do not practice agriculture but rent out their land and engage in off-farm work, 1 April 1927. RGAE, f. 396s, op. 5, d. 30, ch. 1, l. 297. Original manuscript.

On the occasion of the All-Union Congress of Soviets, what needs to be corrected[?] To distinguish sharply between the poor peasant and the loafer.

I don't know! Maybe this question has already been discussed and a resolution on it has been proposed, but I have not yet found a resolution, neither in the newspapers nor in any pamphlets, and therefore if such a resolution exists on this problem, it has certainly not gotten down to [us at] the bottom. In the village soviets, in the committees of peasant social mutual aid, there is no sharp line between the loafing poor peasant and the working poor peasant. Freeing both from the rural tax, giving aid—legal and material, in the form of loans, etc.—this in general incurs losses and especially angers the middle-peasant segment of the peasantry. I will give several cases in local life. I will take two identical households, only in the second household there are two and a half times more family members. The first household, the Kovalyovs, formerly had two cows and two horses, as well as small livestock, all the buildings of an average peasant, eight desiatinas of land, and a family of five souls. With further developments the household decreased, and by 1917 it was reduced to only one hut, four souls, and eight desiatinas of land. In 1918–1919, as a poor peasant, this household gets a cow, which they will sell after two or three years. Twice the KKOV [Peasant Committee of Social Mutual Aid][61] provided seeds for sowing and arranged a group plowing for this household's benefit; all the same, a fish doesn't pick up a chisel, and as before, they continue to rent land to their neighbors. Moving ahead to 1925, [as a result of] land reorganization, for five souls they receive about six desiatinas of land, including a one-desiatina farmstead. They continued to rent out their land while they themselves often went off to do work on the side. This winter he went to the mines, along with others, to make big money.[62] He comes back after a little while, critical of all the miners, to save his health.[63] And with this, roughly speaking, it would be possible to live, the resources are enough to exist on: the return from land rent, from the sale of part of the property, as well as the work on the side. But here is the root of the evil! In this fiscal period, they're getting a rent payment and other [income]—breaking all records—so that their table is never without wheat bread, *barankas*,[64] there is no dinner without a glass of moonshine or a card game for money, etc., etc., and [they enjoy] all manner of delicacies and diversions.

I will dwell briefly on the second farm, which is trying to raise its standard of living. The Kondrashovs' farm is not as rich as the first farm. Since every minute of life is devoted to raising their economic level, there is no extra energy for work on the side. As a result, it happens that four households [came] from this one household, each ten times richer than the first. Improvement and development came about not by exploiting hired labor but by self-exploitation, and that's how they developed their household. It is possible to present many more such examples, and the All-Union Congress must turn its attention to this and distinguish the lazy loafer from the poor working peasant so that the loafer doesn't ride along on the state account but gets down to work, to labor for the advancement of agriculture.

IVAN DANILENKOV

Apart from calling on the state to give its support where that support is due, this letter reveals another more or less open secret of peasant life—the widespread practice of leasing land. The law allowed temporary leasing in cases where labor power or inventory to adequately work the land was lacking, or when necessary to help in recovery from a natural disaster. In general, legislation prohibited the leasing of land to increase income not derived by one's own labor. However, peasants, and not just "kulaks," found numerous ways around the regulations and leased land to supplement their earnings.

Taxes

The state attempted to regulate social relations in the village by means of progressive taxes intended to strengthen poor- and middle-peasant farmers while restraining the growth of the wealthy. The tax policy often failed to satisfy any of these groups, however. This is discussed in the letter of one Narusbek, a resident of Liutovko in Kharkov province. The section of the letter where the author requests to be appointed a village correspondent has been omitted.

· 28 ·

Letter to *Krestianskaia gazeta* from the peasant Narusbek describing the negative effects of Soviet tax policy, 15 September 1924. RGAE, f. 396, op. 2, d. 18, l. 375. Original manuscript.

As we workers and peasants are aware, we won Soviet power and intended for our lives to improve. But we still haven't seen it, and no one from the center hears our heavy sighs, [hears] how our peasants cry in the distant villages: they can only sob into their hands. I live in the countryside, and with nothing better to do, I passed through a certain village and heard what the peasants are saying and how they are getting along under the worker-peasant state. I heard a peasant crying, and I asked him what the matter was, why he was crying, what was wrong. And he answers: "How can we not cry, my little dove, when this year's harvest is so utterly bad and doesn't even yield seed, and now they are coming to get the tax. Where in the world will I get it when I have one young cow and a family of four souls and four desiatinas of land, and for taxes, they record the cow as one desiatina, and we have nothing to eat, but they say, "Hand over the first half of the tax, or else the percentage will increase.'" And it is the same in another and in a third village, and everyone sighs as one. This year the peasant's life is truly sad, and it is even worse in Akhtyrka region, Siniansk district, [Kharkov province], where there is no harvest at all. If only someone would plead for the peasants, but there is no one. There is no one who will open his mouth on behalf of the peasants. Everyone is afraid of something or other. The peasant needs help from somewhere or to be permitted to stay in arrears until after the next harvest, but the local authorities say: "It's not up to us. They send us [orders] from the center, and we have to get the tax. No one can remain delinquent."

NARUSBEK

The following report from the village correspondent F. T. Laptiev from Rossosh district, Voronezh province, expresses the middle peasants' resentment of high taxes and the favored treatment they believed poor peasants enjoyed. The first point, however, takes up a problem that existed in the 1920s along the border between Russia and Ukraine. During the civil war, anti-Bolshevik and bandit groups operated in these regions. Later, to the consternation of the population, the local administration was often drawn from ethnic Russians, outsiders, referred to by the author as *moskaly,* a common Ukrainian term of derision for Russians derived from the word "Moscow."

· 29 ·

Letter to *Krestianskaia gazeta* from the village correspondent F. T. Laptiev on the inequities of the agricultural tax, 4 December 1924. RGAE, f. 396, op. 2, d. 18, ll. 442–443. Original manuscript.

1) I am reporting to the editorial board of *Krestianskaia gazeta* that the peasants have a good relationship with Soviet power. There are no [bandit] gangs,[65] and everything is going smoothly. But they [the peasants] are gravitating to Ukraine and hate the moskaly because the top is all moskaly, or, to put it simply, all the bosses are moskaly. The question is why. They [the peasants] still have not withdrawn to Ukraine, but it will happen soon.

2) They are offended that horses are assessed at two desiatinas a horse, and cows at one desiatina, and for a third [category of animals] there's a tax, so one must sell off the livestock.

3) The single tax reduction was done incorrectly. The reduction was given to those who never paid and never considered paying. And those who always paid were pressed for 75 percent immediately, and later for the rest that was due. They had to sell that which should not be [sold]. If such support for the middle-peasant economy continues, then whenever they dream a dream it would be that dreamt by the Egyptian pharaoh: seven lean cows and seven fat, and the lean ones ate the fat ones and did not gain weight. [...] But for us middle peasants, it is difficult to support the state, and soon even we, the middle peasants, will have to abandon our farm and live like poor peasants. If Soviet power would remit the tax, who, in this strange time, would help Soviet power more[?] The middle peasant gave [his] grain and [his] wagon and [his] horse and [his] meat and went [to the front] carrying his rifle, and now they pile all sorts of burdens on him. Now, it's

more expensive to maintain livestock than oneself. You have to pay for it and to feed [it] from that land you use to feed yourself, and you have to pay an incredible [amount]. Now there are no kulaks, only toilers.

4) I would propose that the tax never be increased and that it be set once and for all at no more than one ruble or two rubles per desiatina. Or else we'll have overpaid [the tax] with livestock for the land we have.

5) I request that you do not lump us into groups, or else the poor will walk about in galoshes, the middle peasants in tattered boots, and the communist in jodhpurs. But, some get a salary of one hundred rubles and do not want to pay the tax.

I request an answer.

Household demographics, no less than taxes, also weighed heavily on the peasant mind. The following letter, sent from Saratov province, outlines the difficulties that large peasant families encountered.

· 30 ·

Letter to *Krestianskaia gazeta* from the "weak peasant" M. S. Shcherbakov on the burdens that taxes placed on large households, 21 April 1924. RGAE, f. 396, op. 2, d. 18, ll. 289–290. Original manuscript.

A request from Mikhail Stepanovich Shcherbakov, a weak peasant.

I am a citizen of Bobylevka, Romanov district, Balashov county, Saratov province. I have a family of fifteen souls. My property consists of a house, a shed, a barn. For livestock, I have one horse, one cow, and nothing else. I have a land allotment of one and one-half desiatinas for each soul: arable and pasture land. I ask you for an answer, if you have the decrees; if not, then inquire at VTsIK so that it will give an explanation. I don't have the strength to work this land with only one horse. I pay the single agricultural food tax in full, but I don't have enough grain to feed my family; as a result, my family is extremely hungry. It is a sad thing just to look at my family. We are three brothers. We considered dividing [our landholding], but then we would be in a completely destitute situation. We would split this holding up into three parts, then none of us would have anything. Our communist Red party calls us to the socialized [forms of] work—artels, collectives—but I asked my brothers not to split up but for us all to live together. I ask you earnestly to give an answer in writing and to carry it in *Krestianskaia gazeta*. Let VTsIK create the possibility of living in a large family with such a small number of livestock. If we divide our holdings, then no one will have anything, but the food tax won't be paid, because we'll have nothing. I have observed many, many cases in our village [where] the family is big and livestock are few. They suffer from having to pay the single agricultural tax. They divided up their holdings, and for them the tax is reduced completely. It staggers me that it is impossible for a large family to live. Isn't it possible for VTsIK to make the appropriate decree.

[...] I beg your indulgence, comrade editor, my head is so distracted from need that it is difficult for me to correct [my prose]. My request of you is that when you go to put out the newspaper, that you people, who know what is written in this letter, make selections and compose a short article [from it]. We receive *Krestianskaia gazeta*. But I ask earnestly, do not refuse my request.

M. S. SHCHERBAKOV

Not all large families found themselves in such a miserable position. In fact, a 1927 Central Statistical Administration (TsSU) survey found that rich and kulak families tended to be larger than those of poor and middle peasants. But, the economic well-being of any family depended on the ratio of mouths-to-feed to able workers. The decision to maintain one large household went against the trend in the mid-1920s, a period marked by partitions and a diminution in average household size. As a result, the number of peasant households in the USSR increased every year between 1923 and 1927, with the total peaking in the latter year at approximately twenty-five million.[66]

Nature also took its toll. Drought, early frosts, hailstorms, and other natural phenomena could lead to a sharp reversal in a family's economic condition. M. S. Bida, a villager from Odessa province, describes the effects of one such misfortune in a letter from November 1927.

· 31 ·

Letter to *Krestianskaia gazeta* from the peasant M. S. Bida describing the
effects of a hailstorm on his economic situation, 12 November 1927.
RGAE, f. 396, op. 5, d. 4, ll. 169–170. Original manuscript.

A Notification

I wish to inform *Krestianskaia gazeta* and its employees how our village celebrated the ten-year anniversary of the October Revolution, and how our village of Gniliakovo conducted [the celebration] in particular, and what the peasants are saying about this day, and what occurred on the eve of this holiday. Our village has 600–odd households. But at the rally there were schoolchildren, of course, all the [Young] Pioneers and Komsomolists, but from among the peasants [there were only] about 60–70 men and women—100–150 [people in all] and no more. The rally began with a report by speakers from Odessa on the achievements of Soviet power after ten years. But, hearing this, the peasants quietly say: "Soviet power has done all right, but we're not so well off." I should write that our village of Gniliakovo suffered terribly from the natural disaster of drought; and, worst of all, on the evening of 11 June, hail came down so hard and this region got [so much] that absolutely nothing in the ground remained alive. Our old folks could not remember such hail. Just on the very eve of these holidays, by order of the finance inspector and the chairman of the village soviet and representatives of the village, a list of property to be seized for the food tax was published. But they took whatever they could get their hands on—a samovar, a mirror, a sewing machine, curtains, cornices—in general, whatever they saw. So, as a result, no one came to the celebration. How is it possible that Soviet power can allow such violence and such tears to occur among us? This is not an achievement: it is an undermining of Soviet power, it is a hundred times worse than the old tsarist yoke. I would ask *Krestianskaia gazeta* to bring this to the attention of the necessary people so that they completely write off this food tax for 1928 without interest and without a fuss, not because the peasants don't want [to pay] but because it is quite unbearable. They have [already] paid what they could. If [they intend] to pump the entire food tax from the peasant, then it will be necessary to ruin his economic situation and weaken him. However, the Soviet government [should] not intentionally ruin the peasant's economic situation but, on the contrary, [should] raise it, but here the local authorities utterly and intentionally ruin [the peasant's economic situation]. Speaking for myself, I have [to pay] a food tax of sixty rubles, but I paid only ten rubles and then sank into grief. They took my samovar, mirror, and sewing machine and will not return [them] until I pay the remaining fifty rubles. But I do not have the fifty rubles at home, not even fifty kopeks, and there is nowhere to get it, there is no work to do on the side and no grain to sell, nor was there any. The hail knocked down everything. I have four desiatinas and three-quarters [of a desiatina], respectively, of sown and unsown fallow land because the credit society gave seed for sowing to support and not to ruin the economy. For me to come up with fifty rubles, I would have to sell a horse or a cow or some agricultural machine. But this is wrong, this is not an improvement but a devastation of [my] economic situation. We are seven souls in my family—me, my wife, and five children souls; the eldest son, twenty-two years old, is an invalid who was stricken at four years of age in the last scarlet fever [outbreak]. His entire left side is paralyzed, and the remaining children are unfit for agricultural labor. I have tried everything so that they would not apply the food tax to my invalid son: nothing helps. Can you help me, *Krestianskaia gazeta?*

I have a pair of twenty-year-old horses, one pedigreed cow, and two pedigreed calves. There is a mower, a seed machine, a plow, harrows, and my entire landholding. If I were to sell one of these and pay the food tax, then this would not be an improvement but a devastation of [my] economic situation. It took ten years to assemble this inventory. I would ask *Krestianskaia gazeta* to petition wherever necessary on my behalf. and on behalf of those peasants who have paid a portion but cannot pay the rest, to remit all the food tax, and if it is not remitted, then to write off the upcoming [tax] for 1928. Then our peasants and I would joyfully remember the ten-year anniversary of Soviet power, and each peasant would remember the proclamation of forgiveness and not of ruin.

Be a defender, *Krestianskaia gazeta;* defend us who suffer from natural disasters and hail.

MIKHAIL SAFRONOVICH BIDA

Unable to defend themselves against the arbitrary actions of local officials, this final sentence takes literally *Krestianskaia gazeta's* proclamation that it was the peasants' defender and indicates that some peasants did, indeed, view "their" newspaper as a potential weapon, an instrument that could defend their interests. In this sense, peasants hoped the newspaper would fill the void created by the loss of their own institutions of power.

Despite the concessions to peasant economic interests embodied in the NEP, peasants of all strata found the official attitude toward the village lacking in these years. Contrary to their expectations, the new regime had not, in the opinion of most, done all it could to alleviate the burdens that weighed so heavily on the shoulders of the small cultivator. Having only recently acquired land of their own, most peasants were not inclined to abandon individual farming to join production cooperatives or state-sanctioned collectives that would not only radically alter the life they knew but had yet to demonstrate their superiority. Separators from the commune, keen to improve productivity on independent holdings, encountered opposition from fellow villagers and got little support from local officials unfriendly to private farming. The slow pace of land reorganization prevented poorer peasants from increasing the size and improving the quality of their allotments. Restricted access to forests and pasturelands was especially felt by middle and poor peasants. Taxes, the unfavorable terms of rural-urban trade, and the corruption and arbitrariness of local government gave all peasants cause for resentment. The peasant ideal rested on small-scale farming supported by generous state assistance. Anything short of this ideal—however valid the reasons—increased dissatisfaction. In the mid-1920s, the Party tried to address some of these complaints.

CHAPTER 3

Smychka

The Bond between City and Village

Our real wager is a wager on our very selves, it is a wager on the working class and the laboring peasantry, it is a wager on the growth of socialist economic forms, on the growth of state industry in the first instance, on the growth of agricultural cooperation in the second. The bond between these fundamental forms is the necessary condition for our victory.

—N. I. BUKHARIN, 1925

For the Russian communists, the early years of the NEP were a time of reappraisal. During the civil war, the crushing of the German Spartacist revolt in January 1919 and the collapse later in the year of a Soviet-inspired regime in Hungary demonstrated that the postwar revolutionary tide in Europe was fast receding. The failure of the 1920 Soviet military offensive against Poland highlighted the Red Army's limitations as a bearer of revolution to Europe and ensured that for the time being the Soviet republic would stand alone, an isolated Red beacon in a capitalist sea. Reality and survival now dictated that Soviet leaders turn their attention inward. Surveying the domestic landscape, however, they found that economic devastation, famine, and social discontent required them to rethink many of their ideas about socialism and its application in Russia. The introduction of the tax-in-kind had been the first tentative step in the search for a Russian path to socialism.

From the tax automatically flowed the Party's need to redefine its relationship to the peasantry. Whereas Party policy had leaned on the rural poor during the civil war, inciting them against their better-off neighbors, now, in the new peaceful environment, Party leaders hoped to win over the broader peasant mass by demonstrating the Soviet regime's willingness and ability to improve the world they inhabited. In practice, this meant that along with appeasing the peasants' proprietary instincts as landowners and food suppliers, the regime had to convince them that there were genuine benefits to the new socialist order. To win back that peasant trust established in 1917 with the Decree on Land, the Party had to prove that war communism was not the real face of socialism. Policy makers summed up this new spirit of urban-rural cooperation in one word—*smychka,* translated as "link," "bond," or "alliance."

Speaking to the delegates at the Eleventh Communist Party Congress one year after the adoption of the tax-in-kind, Lenin defined *smychka* and emphasized its importance:

> This bond must be pointed out so that we may see it clearly, so that all the people may see it, and so that the whole mass of the peasantry may see that there is a connection between their grueling, incredibly ruined, incredibly impoverished and agonizing life now and the work that is under way for the sake of distant socialist ideals . . . Our aim is to restore the bond, to prove to the peasant by deeds that we are beginning with what is intelligible, familiar and immediately accessible to him, in spite of his poverty, and not with something remote and fantastic from the peasant's point of view. To prove that we can help him and that in this period, when the small peasant is in a state of appalling ruin and impoverishment and tormented by starvation, the communists are really helping him. Either we prove that, or he will send us to the devil. That is absolutely inevitable.[1]

By reestablishing the exchange between the city and the village, the tax-in-kind had jump-started economic recovery. This was but one component of smychka. Continued success in this direction depended on an even greater strengthening of all the interconnections between town and country. To maintain the smychka and avoid the fate of which Lenin spoke, the peasants had to see tangible benefits to Bolshevik rule. Economic concessions such as low taxes, loans, and credits directly aided the peasant proprietor and served as the basis of smychka. Just as important, industry and the trading network had to meet the agricultural sector's demand for manufactured goods at reasonable prices. To avoid the split (*razmychka*) between proletarian and peasant, between the socialist and the private economy, which Lenin justifiably feared, the peasant household, which in difficult times tended to seek shelter in its age-old consumption economy, had to be kept engaged in the market.

In short, for the smychka to work, the Party had to relinquish coercion and force in its relations with the peasantry. Lenin had by this time already concluded that the peasant economy could not be socialized through the immediate introduction of collective farming. Instead, he came to believe that the pressures of real life would convince peasants (first the poor, then the middle peasants) to abandon individual farming in their own self-interest and form voluntary production associations or cooperatives. For this reason, Lenin began to place great emphasis on state support for such ventures. So long as the proletariat maintained its political supremacy and its control over the "commanding heights" of industry, Lenin reasoned, then "a complete socialist society" could be constructed on the basis of the cooperative movement. This would require time, "one or two decades" was his optimistic prediction, but it held out the possibility of a socialist light at the end of the NEP tunnel. The leading Party theorist, Nikolai Bukharin, further developed this particular strain in Lenin's thinking on the NEP and became the most prominent spokesman for the idea of Soviet Russia's gradual evolution into socialism on the basis of peasant cooperation and peasant demand for manufactured goods. As Bukharin wrote in 1925, peasants who organized themselves in cooperatives for the purposes of trade would, given time and sufficient state assistance, see the advantages of large-scale cultivation and begin to amalgamate their individual holdings into production cooperatives. Gradually, through their increasing demands for industrial products, these amalgamated farms would become linked with socialist industry, which itself would be spurred by these demands. As these links multiplied and developed, both sectors would ultimately conjoin to form a unified economic whole. "And this type of economic chain," he declared, "which is organized in all its parts is, in essence, socialism." Only by means of this slow process, Bukharin argued, could the Soviet Union retain any hope of emerging, at the end of the day, a socialist country: "Thus, we are approaching socialism despite the economic and technical backwardness that continues to distinguish our country. Of course, in conditions of technical and economic backwardness, this path is a very long path. Nevertheless, it is the right path, the path by which we will reach socialism if only we carry out the correct policy in regards to the peasantry."[2]

Not everyone shared Lenin and Bukharin's faith in the efficacy of time, however. The Party's left wing, most notably the revolutionary luminary Leon Trotsky and the economist Evgenii Preobrazhensky, warned of the dangers inherent in a policy that so indulged the peasant and delayed industrial recovery. Since agriculture was bound to recover at a quicker pace than industry under the NEP, both were concerned that the state industrial sector would be unable to meet growing peasant demand for cheap goods. With manufactured goods in deficit and dear, the peasants' incentive to market grain and other products at low official prices would evaporate. In response to this situation they would find refuge in subsistence, holding on to or consuming what they produced until favorable terms of trade returned. By then, of course, the smychka would have come undone. This, in fact, was the scenario of the so-called scissors crisis in late 1923, when a shortage of industrial goods necessitated emergency measures to save the situation.[3]

Unless industry was developed at a more rapid rate through greater capital investment, the left contended, "goods famines" and all they implied would be a permanent feature of the Soviet economy. Moreover, Soviet agriculture would never be mechanized, and consequently never socialized, unless a well-developed industrial sector encouraged the formation of production cooperatives and collective farms by supplying them with tractors and other needed machinery. Lacking other sources, Preobrazhensky argued, industrialization should be financed by the peasants sooner rather than later through high taxes and low prices for produce—in other words, by a policy just the opposite of the current policy. As many Party members pointed out, however, what Preobrazhensky termed "primitive socialist accumulation" also threatened to destroy the smychka. Throughout the 1920s, policy makers wrestled with this fundamental economic dilemma, and their debates played an important role in the political battles that decided the fate of the small peasant farmer in the Soviet Union.[4]

The slow pace of industrial recovery was not the only issue of concern that the NEP raised. The confiscation and distribution of crown, church, and gentry lands following the revolution and the frequent repartitions thereafter had greatly diminished rural social stratification by increasing the number of middle peasants, a phenomenon rendered by the Russian term *oseredniachivanie*, or "middleization," of the peasantry. Between 1917 and 1920, the number of holdings with no arable land in densely populated Moscow province, for example, had been reduced from 22 percent to 9 percent of total farms, with small farms of no more than two desiatinas increasing by over 20 percent (57 to 77 percent). Nationwide, farms with no arable land fell from 11 to 6 percent. Holdings of up to four desiatinas had increased from 73 to 89 percent in Moscow province and from 58 to 86 percent in the country as a whole. Conversely, holdings of more than eight desiatinas of arable land had declined from 9 to 2 percent nationwide.[5]

Reducing the number of better-off peasants and raising the poorer into the ranks of the middle not only accorded with the peasants' propensity to equalize landholdings but also strengthened the foundation of small-scale farming. The NEP, by reintroducing the market, promised to reverse the trend toward social equalization, for the more enterprising peasants would invariably take advantage of the new conditions to increase their wealth and landholdings at the expense of their less capable neighbors. Party leaders, and not just those on the left, although they were the most vocal, worried about the detrimental effect this would have on the correlation of forces in the countryside. They were, therefore, highly attentive to indications of economic differentiation and the growing kulak political influence that it was assumed to portend.

Within the peasant class, Party doctrine identified four general categories or strata: laborer (*batrak*), poor (*bedniak*), middle (*seredniak*), and wealthy (*kulak*). Because laborers' landholdings were meager or nonexistent and survival demanded that they rely primarily on their labor, Party doctrine identified batraks as proletarian. The bedniaks, who eked out a precarious living on inadequate allotments, needed to rent tools and draft animals from better-off neighbors, often hired themselves out as laborers, and in hard times had to purchase grain, were considered semi-proletarian. The regime viewed these two strata together as its social base of support in the countryside, the natural constituency for collective agriculture. On the other hand, as practitioners of moneylending, labor employment, land leasing, and other exploitative and revenue-generating measures, kulaks were deemed rural bourgeoisie. The majority of peasants fell into the middle, or seredniak, category. This was the most diverse, least homogenous peasant stratum, and it defied easy classification. In many cases, the only distinction between the poorer (*malomoshchnyi*) seredniak and the bedniak was that the former owned a horse. At the other extreme, the wealth, practices, and landholding of the well-off (*zazhitochnyi*) seredniak may have differed little from that of a kulak.[6]

These categorizations, despite their static nature, reflected a certain reality. Economic divisions within the village were real enough, as was the hostility that weaker peasants often felt toward the better-off, but peasants did not necessarily see their more successful neighbors as "class enemies" in any meaningful sense. Nor were poor peasants seething to carry out a "rural October" against the village bourgeois in the name of socialism. Peasants tended to attribute another's success to a combination of hard work, shrewdness, and luck. Their envy went hand in hand with admiration and the desire to emulate. Economic categories that emphasized intra-class divisions underplayed the resilience of the peasantry's class cohesiveness. Likewise, envisioning the village in terms of an urban "bourgeois-proletarian" continuum distorted the reality of rural social relations. As a result, outsiders arriving in a village often found their preconceived notions of village life crumbling as ideology collided with experience. This is neatly expressed in the epiphany experienced by the Party organizer Zharkov in Fedor Panferov's 1930 novel, *Bruski:*

Before he came to the Alai district he knew village life only by
hearsay, by the reports of village delegates at conferences, by casual
conversations with different peasants. He always imagined a village as
a large, dark lump divided into three sections: the poor peasants, the
middle peasants and the rich peasants, the Kulaks. The Kulak had a
big head and wore leather boots; the middle peasant had ordinary
boots and wore a jacket; and the poor peasant ran about in bast shoes.

That, at least, was how villages were depicted on posters, and that
was how Zharkov thought of them: on the one side the enemy of the
revolution, the Kulak; on the other side, the defender of the revolution,
the poor peasant, while the middle peasant stood aside, biting his lips.
But after he had lived in [the village of] Shirokoye a few days, all his
ideas and impressions grew blurred and confused, and when he came
across such poor peasants as Petka Kudeyarov, Mitka Spirin and
Shlenka, doubts were born in him.

"If this is really village poverty," he wrote in his notebook, "then we
are building our village policy on sand."

. . . What astonished him most of all was the realization that it was
not the poor who held leadership of the village, but such strong
individuals as Kataev and Plakushchev.[7]

The economic concessions made to private farming in the areas of
leasing land and hiring labor in the early 1920s could not but benefit
these "strong individuals." Many Party leaders, however, objected to
what they considered a new and dangerous "wager on the kulak." Like
the private traders, the NEPmen, and the old-regime holdovers working
in Soviet institutions, the kulaks continued as the Party's bêtes noires
throughout this period. This was the case even though the debate on
whether the new conditions were fostering kulak growth remained in-
conclusive. For those who opposed the new concessions, though, the
size and even the productive share of this stratum were not necessarily
the crucial issues. As Charles Bettelheim noted, the kulaks presented the
regime with a number of other challenges: in the "commercial relations"
they maintained with other peasants; in the example they set—especially
to middle peasants—as successful individual farmers; and in their consequent ability to influence decisions taken at peasant councils. In other
words, the kulak was not reproached simply as self-interested opponents
of collective farming; by their very existence they served as a political and
social alternative to the Party's agrarian program, an alternative to which
many peasants naturally aspired.[8]

To meet these challenges, the Party had, relatively early, settled on a
strategy of courting the middle peasant. By breaking up the large estates
and distributing this land among the peasants, the rural revolution had
greatly expanded the ranks of the seredniaks. In 1919, the Party's new

program recognized this development and made relations with the middle peasant its linchpin in the countryside. This strategy involved detaching the middle peasants from kulak influence and winning them over to socialist construction not through coercion but "by paying special attention to their needs."[9] The NEP provided the opportunity for more concrete measures to be taken in favor of the seredniaks, a tactic the Party followed through 1925. But the middle peasants remained problematic for the regime. To the extent that they worked their allotments with their own hands, they remained laborers and allies of the proletariat. If by dint of hard work and fortunate circumstances middle peasants improved their situation, enlarging their holdings and employing the labor of others, then they suddenly became aspiring kulaks and potential opponents of the socialist transformation of the countryside. This seemingly arbitrary sociological taxonomy also explains why many peasants resented their division into different categories. As one correspondent from Kirghizia wrote, the middle peasant "would bury you in grain" if left to his own devices: "just don't say kulak if he makes good" (Document 45). Relations with the middle peasant encapsulated all the contradictions of the Party's rural policy.

If Communist Party influence in the countryside was at the mercy of a deceptively complex social structure and mentality that often confounded simplistic class analyses, the extreme weakness of rural communist organizations compounded the problem. Rural Party membership stood at only 137,000 at its nadir in January 1924. Full-time Party officials tended to congregate in provincial towns, blunting Party influence still further. Of the 14,630 rural Party cells in early 1924, more than half—8,000—were located in rural district (*volost*) administrative centers, leaving the remaining cells to contend with tens of thousands of far-flung Soviet villages and settlements. In a famous study of the Smolensk region, a predominately agricultural territory, Merle Fainsod found that in 1924 there were only 16 Party members for every 10,000 inhabitants of working age, or about one for every ten villages. Although rural Party membership grew steadily throughout the remainder of the decade, until collectivization the permanent Party presence in the countryside remained inadequate to its tasks. Moreover, the growth that was achieved did not necessarily coincide with the Party's social program. The purge of the Party rolls in 1921 hit peasants harder than any other group. Following Lenin's death in 1924, the Party sought to increase peasant recruitment. In 1925 and 1926, peasants made up 30 and 39 percent of all new Party recruits, respectively. The recruitment of batraks, however, lagged behind recruitment from the other rural strata, and by 1927 the majority of rural communists came from the wealthier sections of the peasantry.[10] Accusations made by some peasants that a new "Red" gentry was gaining influence in the provinces (Document 17) were not entirely hyperbolic.

Thus, even while enabling economic recovery, the NEP and the smychka presented Party leaders with a host of concerns. Nevertheless, the years 1922 to 1925 saw increases in both sown area and harvest size, and after Lenin's death in April 1924, this gave the dominant Central Committee faction under the leadership of Stalin and Bukharin a seemingly sound basis on which to reject the left's criticism of its pro-peasant course. In October 1924, to strengthen the smychka, further concessions were granted to the peasantry as the Central Committee lifted restrictions on renting land. The next two years were the high-water mark of NEP practices.

To win over the peasant to Soviet power, the smychka could not be limited to the economic sphere. Lenin had stressed that for "NEP Russia" to become "socialist Russia" the countryside, and the nation as whole, had to undergo a "cultural revolution." Lenin conceived of cultural revolution narrowly and within very well defined limits. He rejected the notion that under the guidance of intellectuals workers could develop an identifiable "proletarian" culture; he advocated instead what amounted to the popular acquisition of the existing store of knowledge and expertise—that is, "bourgeois" culture. This meant, in the main, extending literacy and education to the masses and introducing the latest techniques into industry and agriculture to increase efficiency and productivity. The achievement of cultural smychka, like the achievement of economic smychka, implied the long-term tutelage of the backward peasant village by the advanced, enlightened, proletarian city.

To accomplish the cultural transformation of the countryside, the Party, beginning in 1923, encouraged the establishment of *shefstvo,* or patronage, agreements between urban enterprises and villages. Ideally, in this arrangement, groups of factory or office workers would undertake to provide material and cultural assistance to a selected village or villages in the form of financial contributions, technical expertise, books, agronomic literature, and newspaper subscriptions; they would also provide the means to set up reading huts and libraries. The exercise had the advantage of giving flesh to the worker-peasant alliance on which the state was allegedly based. No less appealing from the Party's point of view, it reinforced the vanguard role of the working class and the city's dominance over the village. References to the practice in the letters below indicate that to the extent that a shefstvo relationship supplied concrete benefits to village inhabitants, the peasants appear to have welcomed the affiliation. However, many culture bearers from the city approached

the village—its institutions, practices, and religious beliefs in particular—
so condescendingly that they alienated the peasants. All too often, empty
agitation and propagandizing substituted for practical activity and aid.
In emulation of national leaders, local agitators felt compelled to address the peasants on international events and the prospects for world
revolution. This occurred often enough for one peasant, doubtless expressing the feelings of many, to complain, "We are sick of hearing
about the international situation."[11]

For the Bolsheviks as for the tsars, the greatest stumbling blocks on
the road to cultural and economic advance remained mass illiteracy and
a poorly educated population generally. For the Bolsheviks, moreover,
an illiterate population remained politically "unconscious." While mass
illiteracy inspired the effective use of visual media like posters, to successfully mobilize the population in support of Soviet programs, the
Bolsheviks required a population of readers. Literate workers and peasants were also desperately needed to fill clerical and administrative positions in the Party, state, and industrial apparatuses. To this end, the
Party initiated crash campaigns to eradicate illiteracy and provide rudi-
mentary education to the masses. By 1919, these campaigns, which were
often combined with crude political indoctrination, were proving successful. According to official figures, somewhere between five and seven
million people learned to read and write in these years.[12]

By and large, the benefits of these early undertakings accrued to urban residents and Red Army soldiers (millions of whom, upon demobi-
lization, returned to their villages) and had little direct impact on the
countryside. In rural areas, literacy rates varied greatly depending on location, age, gender, and ethnicity, making generalizations difficult. Nevertheless, it can be said that throughout the 1920s, despite the state's
determination to spread literacy by means of "liquidation points," reading huts, and organizations like the Society "Down with Illiteracy!"
(ODN), adult illiteracy persisted at high levels in many rural areas.[13]
During the NEP years, lack of funds, texts, and instructors permitted
only modest gains. According to census results published in December
1926, 55 percent of rural citizens over the age of nine could not read or
write (34 percent among men; 63 percent among women) even by the
loose standard of being able to sign one's name and recognize a few
printed words. On the other hand, more than three-quarters of the urban population were literate.[14]

Disruptions in the production and distribution of reading matter
greatly hampered efforts to advance rural literacy in these years. As Jeffrey Brooks has demonstrated, by the mid-1920s state publishing houses
and distribution networks had yet to match the productivity or effi-

ciency of prerevolutionary publishers and purveyors. Newspapers, books, and other literature remained in short supply, and the political and propagandistic nature of the available material failed to satisfy the peasant appetite for practical works on agronomy, lively reporting on everyday life, and entertaining fiction. Peasant letters to newspaper editorial boards often bemoaned the dearth of engaging reading matter and frequently contained pleas for back issues or full subscriptions.[15]

In the end, then, for the mass of peasants the reality of smychka did not match its promises. They remained keenly aware of their inferior status in the worker-peasant alliance, a humiliating situation that inspired hostility toward both the state and the working class, a hostility the peasants increasingly gave voice to as the decade progressed. For all the talk of change and progress, for many the new regime approached the village in much the same way the old had—as a place to collect taxes, recruit soldiers, and extract food. The NEP had changed the conduct but not the logic of urban-rural relations.

Documents

The Worker-Peasant Partnership

In an April 1924 letter to *Krestianskaia gazeta,* the peasant I. N. Demidov, from Popasnyi settlement in Voronezh province, expresses a variant of the official view of the smychka. His use of reigning platitudes and clichés to describe peasant-worker relations and the emerging future society shows the degree to which official rhetoric was penetrating the countryside. Embedded in this formulaic presentation, with its closing homage to V. I. Lenin, is a reference to an issue of overriding concern for the peasants: the terms of urban-rural trade.

· 32 ·

Letter to *Krestianskaia gazeta* from the peasant I. N. Demidov on the smychka, 2 April 1924. RGAE, f. 396, op. 2, d. 18, l. 293. Original manuscript.

After reading a few of your newspapers, I was deeply touched by a spirit of happiness and hope for a better future existence for the peasants and the workers. I see that we, peasants and workers, are triumphantly building a better life, and I fully expect that we will overcome collapse once and for all only if we have a close connection [between] the workers and the peasants. All the peasants of our region are now convinced that we are setting out on the sure road of a life of brotherhood. The peasants are saying, "Now we are seeing in fact what we used to just dream about." The administrative order has arranged everything for the better. Education is reaching the broad mass of the population. Now, in any tiny settlement, one finds newspapers and pamphlets. The population is thoroughly engrossed in the construction of the country's new life. Cooperation is growing stronger and is overtaking the private-trade market. And, more and more, rural products are fetching prices in line with factory and plant [goods]. And we expect that soon all goods will be regulated according to their value, especially with the issue of the new, hard currency.[16] And so we soon hope to celebrate the occasion of the restoration of our economy. Only now it's a shame that we do not have our father, Comrade Lenin. He would also have done something for our future holiday.

DEMIDOV, IVAN NIKITICH, peasant.

In a letter to the *Krestianskaia gazeta* editors, A. Kechuneev provides a vivid account of the ceremony symbolizing the union of worker and peasant by which a shefstvo relationship was formalized in the village of Arefino, Vladimir province.

· 33 ·

Letter to *Krestianskaia gazeta* from the peasant A. Kechuneev describing the ceremony formalizing a patronage (shefstvo) agreement between his village and the district Party committee, 10 November 1924. RGAE, f. 396, op. 2, d. 129, ll. 4–5. Original manuscript.

The smychka of town and country

On 9 November in Gorodishche district, Gorokhovets rural district, [Vladimir province], a great event occurred—the declaration of the smychka of the worker and the peasant. A smychka not of words but of deeds, as the dear beloved leader of the toilers, the late Vladimir Ilich Lenin, bequeathed to us.

We peasants have never seen anything like it in the confines of our, let's call it remoteness, where a vital idea rarely penetrates. It was so solemn and touching as to make one weep. Worker, party, Komsomol, and trade-union organizations came from the city with banners and music. The peasants received them very affectionately, shouting, "Long live the smychka of town and country!" In the school, a celebratory meeting was organized at which the comrades who spoke explained the great significance of the smychka of the peasantry with workers. The smychka that Vladimir Ilich bequeathed to us. Much of the time was also spent honoring the proletarian revolution. The peasants listened very attentively as the speakers explained both the goal and significance [of the revolution]. They declared that should the foreign comrades manage an October [revolution] for their bourgeoisie, then we are always prepared to help them.

The most solemn moment of the day was the district Party committee's proclamation of shefstvo over the peasants of the district and the presentation of the shefstvo banner to the peasants.

The speakers noted that the peasants of Gorodishche district are extremely conscious both culturally and in political relations, and that the goal of shefstvo will be to support the smychka of workers and peasant in developing among the backward peasantry cultural and political awareness and, most importantly, to strengthen and secure material support for the poor population.

The district committee, noting the political and cultural consciousness of the peasants, considers it its duty to entrust the Red Banner to the peasants. Accepting the banner, the non-Party peasant Comrade Ignatiev[17] thanks the patrons [shefa] for the honor of its bestowal and expresses the hope that in the future the peasantry will follow in the footsteps of Vladimir Ilich Lenin. The Red Banner is all the more dear to us because it is anointed by the blood of workers and peasants.

Long live the international union of workers and peasants! Long live the Russian Communist Party! Long live the USSR! Long live Soviet power over the entire globe! Long live the legacy of Comrade Lenin!

A non-Party peasant

Like the previous letter, this account employs official language, phrases, and imagery indicating that even among non-Bolshevik peasants, state and Party efforts to set the terms of political discourse were making headway.[18]

Without a doubt, the villages were in dire need of whatever cultural and educational support the cities could provide. The Bolshevik leadership endorsed "cultural revolution," and the smychka was able to partially satisfy the population's seemingly unquenchable thirst for knowledge and education. In this regard, the peasants accepted smychka wholeheartedly, as A. Stikharyov declared in his letter to the *Krestianskaia gazeta* editorial board in 1925. The letter lacks any indication of geographical origin.

· 34 ·

Letter to *Krestianskaia gazeta* from the peasant A. Stikharyov requesting
more cultural aid from the cities, 1925. RGAE, f. 396, op. 3, d. 43, l. 41.
Original manuscript.

The village from the town!

Step by step the countryside is falling behind the cities. If we were to glance
at the city, we would see a whole array of clubs, Red corners, libraries, and
reading rooms that have a tremendous abundance of books. But in the
countryside, we encounter very, very little [of this], and if we should happen
across [it], then [it] is in the house of some kulak who has abandoned the
countryside. The village is still in no condition to build new Red reading
huts—it is too poor. From the city, agitation [reaches] the village in connection with the building of reading huts. This agitation is well stated, and we
like it [not clear in original]. In the cities these things are progressing well;
the cities are well supplied with the houses of runaway White Guardists
[and] bourgeoisie—a Red corner can be built here. But in the village, the
peasant only has the corner where a whole array of rural icons hangs. These
are of no use to us. But we are in no condition to understand this, since
there are no books. For the lack of a reading room, we have no books. [I]t
is very, very sad because we also want to read and to study them.

Thus, the village requests that the city share with us the books that you
have read several times. Books about politics, which we do not see in the
village, and also [books] to improve our agriculture. A lot of words [have
been written] about the four-field [system], but we have no material [on it].
This is also a smychka of the city with the village. Until a Red reading room
[is built, we will find] a place for the books in the school or the village soviet, and we will entrust this affair to good administrators.

The village.

A. STIKHARYOV

In the early NEP years, the cities could not satisfy the cultural needs
of the villages. In rural areas there was a tremendous lack of schools,
teachers, and books. Campaigns and mass propaganda compensated for
the shortages. Exhortative slogans poured out from the pages of newspapers and widely distributed pamphlets. Often, contrary to official
intentions, individuals assimilated these ideas in highly questionable
and idiosyncratic ways, as seen in a disturbing letter from the peasant

I. M. Zakotnov, from Pleshakov settlement (*khutor*), Kamenka district, Shakhty region, North Caucasus territory.

· 35 ·

Letter to *Krestianskaia gazeta* from the peasant I. M. Zakotnov describing his troubled emotional state and his conflicts with local officials, 7 April 1926. RGAE, f. 396, op. 4, d. 26, ll. 715–16. Original manuscript.

My mind is seething and gives me no respite, I'm prey to illnesses, my head hurts constantly, and I cannot tolerate this chaos (drunkenness, hooliganism, all kinds of foulness, swindling, brutality, tax fraud, etc.). Dear beloved brothers, if even one honest political comrade would come to me, I would expose all the machinations and the dangerous underground [activity of my] enemies from 1917 to the present. Beloved brothers, I began to fight in 1916 at the time of the uprisings in Moscow and the city of Bogorodsk, when the workers of the Morozov factory and the Russian-French works in Pavlovsky Posad rose up.[19] I still had connections with the workers then because I served right there. And dear comrade brothers, since then I have not taken one step back. I am fighting honorably and sincerely with the injustice of the bourgeois laws under which the worker and the peasant were shackled. And, comrades, I am very impatient, I am always agitated and angry, I engage in rancorous arguments, I simply can't control myself because we still have people who are rich and proud. And right off I found myself in the clutches of the politically unconscious people. They pressure me from all sides, dear comrades, with all their strength, threatening me with murder, criticizing me, etc.; they mock me and damage my farm. As a grain grower, I, dear brothers, am a very passionate, fervent person, and when facing injustice, I am even more determined. All my enemies, from the very day of the revolution, burned my mother's [house], and every day my mother took a club and drove [me] from the house and intended to kill or poison [me]. I wanted a landholding to practice progressive [agriculture], but there was never a minute's respite from my emotional struggle, [because] it is in my soul. [I am against] injustices, this is the first thing, and a landholding (it is the truth) is dearest of all because for a hundred years the tormented people have been working for this truth, [the truth] of the dictatorship of the proletariat. And I, my dear, dearest comrades, a peasant grain grower, firmly and strongly affirm that this is the people's will. The beacon of communism is far more powerful, far brighter and more scientific, than any man or machine . . . [word not clear in original] and the machine is the callus, a working hand. And so, dearest comrades, I am dying from impatience at all the doings of the new [Party cell] leadership, at . . . [word not clear in original] of the organizations of the different associations, the organized talks, the cobbled-together artels and societies [that] aren't useful. Why is this? That's the reason I don't go here. When I begin to talk, they laugh at me; they also ridicule me in every possible way. The cell secretary is no different: "What, do you have Lenin on the brain? You've gotten stupid." There are never any serous discussions in the cell, and nothing in it makes sense; it's always trifles and mockery. And so, my dear comrades, they make my head spin, and it is impossible to work; my head always aches terribly.

IVAN MIKHAILOVICH ZAKOTNOV

To [my] dear brother-specialists of labor in the national economy, the all-union brothers of Selmash, I am sending a heartfelt greeting. In the depths of my heart I wish for the brothers of the whole world, for the workers, the creativity of production for a new machine and perpetual motion. My true brothers, the thinking of those who labor are wiser [only] in a free country [where] thoughts and all the machines are made stronger, sturdier, without underhandedness, insincerity, or capriciousness. Everything will become more honest through work. The free path forges more machines than people.[20] And I attest that the free path of labor will smash bourgeois technique and [its] theoretical tail because it became accessible, it was dull, it was large-scale, and [it involved] furious exploitation and profit by means of oppression and sucking the workers' blood. And this bourgeoisie ruled for centuries and tightly fettered, hobbled and defended itself from head to tail. And with this worker's blood [the bourgeoisie] fashioned themselves an instrument for the destruction of this very worker. Now, dear comrades, we must turn this [gun] barrel into their fat belly and wide ass and get back our fraternal, scarlet blood. This is our truth of the dear leader, V. I. Lenin.

IVAN MIKHAILOVICH ZAKOTNOV

The message that freedom and dignity can be achieved only through individual labor resonated deeply in the writer's consciousness, but the letter, in its plaintive incoherence, also expresses the psychological toll exacted by Russia's revolutionary apocalypse.[21] Unfortunately, such letters, similar in style and using the same expressions, were not uncommon.

A village librarian (*izbach*), M. R. Biriukov, from Luk village in Gomel province, describes improvements in shefstvo work in his letter to the *Krestianskaia gazeta* editors.

· 36 ·

Letter to *Krestianskaia gazeta* from village librarian M. R. Biriukov on the benefits of shefstvo, 20 May 1924. RGAE, f. 396, op. 2, d. 91, l. 47. Original manuscript.

Patrons are helping the countryside

In the fall of 1923, the Zhlobin local tractor service No. 1-a initiated shefstvo over our village. Before, shefstvo went as follows: they came once or twice a month to deliver some kind of report and brought old, dated copies of the newspaper *Gudok*[22] that were unintelligible and uninteresting to many of the peasants, and as a result, the peasantry found this type of shefstvo unsatisfying. But now the situation has changed. Patrons have begun to visit more often, they are acquainting themselves with the rural situation, they explain problems raised by the peasants, and, most important, they order newspapers: six copies of *Krestianskaia gazeta,* two of *Bednota,* two copies of the journal *Krestianka,* and so forth. Now, if the peasants come up against unintelligible problems or see some sort of unintelligible occurrence in the village soviet or the cooperative they say, "[When] the patrons come— I'll ask!" The peasants look forward to the arrival of the patrons in the village very much and are grateful for the newspapers they've ordered, which are quite useful for the countryside. They say: "If the patrons would come to the village even more often, then there would be a normal relationship between the worker and the peasant."

Izbach BIRIUKOV

In a letter from late 1924, D. T. Yesipov from Konchino district in Moscow province also describes shefstvo in positive terms and informs the *Krestianskaia gazeta* editors of the significant aid that patrons rendered to his village.

· 37 ·

Letter to *Krestianskaia gazeta* from the peasant D. T. Yesipov describing the shefstvo relationship between the People's Commissariat of Foreign Trade and his village, 27 December 1924. RGAE, f. 396, op. 2, d. 93, l. 6. Original manuscript.

How our patron works

Since April 1924, Konchino district has had a patron—the People's Commissariat of Foreign Trade. Shefstvo work has enjoyed successes for nine months. I will speak briefly about this. The main thing the patron has done is this: In the fall, the district cell of the Russian Communist Party, with the help of the patron, conducted a political course for the teachers of our district. The patron helped supply the school with textbooks and, through its work, put the cooperative on a firm footing. The patron conducted all its work in accord with the district communist cell and during all the campaigns sent out, and continues to send out, speakers and literature. Because the Commissariat of Foreign Trade cell takes the great tasks of shefstvo seriously— the union of town and country—we peasants now have, instead of two reading huts (before shefstvo), twenty-two cultural Red corners, and instead of one district Komsomol cell in December, there are already around ten Komsomol organizations. The district cell and the patron have also begun to work with women. Women peasant-delegates have been elected. With the help of the patron, the district reading hut is improving its work and is providing the countryside (village reading huts) with newspapers, magazines, and books. Teachers are playing a large role in the work with the young and in the reading huts. This year, during the reelections to the village soviets and committees of mutual aid, our peasantry nominated honest administrators, and in the elections for the district executive committee [they] entrusted the leadership of the district to reliable people, which will help with the management and administration [of the district]. The patron helped our district organizations buy two tractors so that in the spring, with the implementation of land reform, it will ease the move to an improved economy.

The district communist cell has gotten stronger and larger; the patron correctly plans and carries out its work. Through the fraternal union of town and country, we are heading toward a better life.

Worker-Peasant Friction

Behind the idyllic picture of the developing relationship between city and village painted by the regime and its rural supporters lay many problems and contradictions that could provoke extremely sharp commentary, as in the letter of F. Morozov from Vysokovo village, Tver province, written in June 1924.

· 38 ·

Letter to *Krestianskaia gazeta* from the peasant Fedor Morozov demanding the formation of a peasant union, June 1924. RGAE, f. 396, op. 2, d. 18, ll. 412–13. Original manuscript.

Deeply respected citizen editor, we would like to hear your answer! Why do all the praiseworthy officials and builders of the Soviet republic take up arms against the peasantry and try to wring the last drop of blood from them? The peasants sustain everything: the upkeep of the army, industry, the dual administrative apparatus, elected and Party; and all the organizations in the republic, without exception, hang on the neck of the peasant alone (in two years, if the harvest is good, he may buy [himself] one calico shirt). And what do you, *Krestianskaia* (perhaps not entirely) *gazeta* do? [You] set the peasants against each other. I would think it better to print the advice of agronomists than to disorganize the peasant masses. There, you dream you are only doing and saying what is just, but here they only call out "slander" and "provocation." Here is an example: Comrade Kalinin gave a report at the Communist Party Congress on the peasantry![23] What does a former corkscrew from the bureaucracy know about the peasantry? It's simply a mockery. But the last rural tax, 1924–1925, is also a very terrible mockery. The peasantry will not forget this! Tsarism beat, and robbed, and made drunkards of the peasantry, and what have you, its defenders, done? The tax is ten or more times bigger, and you are grabbing [the peasant] by the throat. Who is worse? Give the peasantry their union, that is, a peasant [union], and we will know that you are for us! Otherwise you are a mercenary party of Yids. But if the peasant mentions the union, he hears [his] defenders answer, "It smells of 1918." But don't forget that everything comes to an end, and so will the peasants' yoke.

I await [your] lame answer.

MOROZOV, FEDOR, peasant

Morozov clearly rejected the official view of state-peasant relations and refused to adopt the acceptable form of discourse. On the contrary, he makes no bones about his political hostility to the Bolsheviks, and his letter enumerates the many ways in which Soviet power caused dissatisfaction among the peasantry, especially the continuing pressure that high taxes placed on the village. In Morozov's opinion this could be explained only by the enormous cost of the huge and, in his view, parasitical Party and state bureaucracies. Given the progressive nature of the Soviet tax system and his resentment at official efforts to distinguish various economic strata within the peasant class and to set them against one another, it might be supposed that this peasant numbered among the better-off. Morozov makes a special point of showing that he sees the pretensions of the title *Krestianskaia gazeta;* for all the journal's catering to the muzhik, he emphasizes that it is not the peasants' but the government's own newspaper. In a similar vein, he also seeks to undermine the

official image of M. I. Kalinin, the state and Party figure who embodied
the union of the working class and the peasants. The Party celebrated
the spectacled, goateed Kalinin as the peasants' "fellow countryman"
(*zemliak*) and "the all-Union village elder." Kalinin's social origins were,
in fact, peasant, but he had also worked for many years in the factories
of St. Petersburg. Following Yakov Sverdlov's death in 1919, Kalinin
had been raised to the chairmanship of the All-Russian Central Executive Committee—a post that carried the formal title "head of state"—
precisely because of his rural roots. Despite Kalinin's standing within
the Bolshevik leadership—which Morozov describes in anti-Semitic
language—as a peasant expert, he was, in the letter writer's opinion, just
another privileged bureaucrat who had no grasp of peasant affairs. In
any case, the mere presence of a Kalinin at the Party's summit did not
satisfy Morozov that peasant voices were being heard. He demands the
creation of a peasants' union.

The call for a peasants' union was not new. As a political organization, the All-Russian Peasant Union (VKS) came into being during the
1905 Revolution. Strongly influenced by the Socialist-Revolutionary
Party, the agrarian program that emerged from the two VKS congresses
that year proposed a radical expropriation of noncommune lands and
their redistribution to the peasants. It also called for the abolition of
private property in land and the equalization of individual holdings via
the communes. By 1907, as a result of the heavy state repression that
followed the revolution, the VKS collapsed. Its leadership, consisting
primarily of intellectuals, resurrected it after the February Revolution in
1917. After the October Revolution, however, partly in recognition of
the soviets' popularity among the peasantry, the VKS disbanded. During
the war-communism period local unions emerged; particularly in Western Siberia, they took part in peasant uprisings against Soviet power in
the summer of 1918. This, evidently, is what the author meant by writing, "But if the peasant mentions the union, he hears [his] defenders answer, 'It smells of 1918.' In the spring of 1920, some local unions in
Tambov province joined the Antonov rebellion. During the NEP years, a
few members of the central Bolshevik leadership flirted with the idea of
a state-sponsored peasant union. In April 1924, the OGPU noted the
union's continued popularity among kulaks and middle peasants in many
regions of the country and reported that in some localities, forming a
peasant union had the support of village communists. As an organization capable of defending peasant economic interests and raising their
status within the smychka to that of the workers, the union retained an
important place in peasant thinking right up to collectivization. In
March 1929, at the height of peasant resistance to forced grain requisi-

tions, a secret Central Committee report admitted that the demand for a union had arisen in every province in the country.[24]

The well-off segments of the village were not the only ones who expressed dissatisfaction with the current state of affairs. P. I. Bazhin from Staryie Kopki village, Viatka province explains in his letter why the poor peasant had no place in the worker-peasant union.

· 39 ·

Letter to *Krestianskaia gazeta* from the peasant P. I. Bazhin on the economic burdens placed on the peasant by the Soviet state, 13 June 1925.
RGAE, f. 396, op. 3, d. 212, l. 77. Original manuscript.

I am a peasant and have already addressed these very topical questions to the office of *Krestianskaia gazeta*. I, Bazhin, Prokopii Ivanovich, have read books on the tsarist government before, but hardly ever got political books. I saw that in the tsarist government, land captains [*zemskie nachalniks*], district police chiefs [*ispravniks*], the police, and the like ran things. The memory of 1905, when I saw people in our settlement, exiles from the Caucasus and Ukrainians, honorable, conscientious people, is vivid because it was clear that these people had been deported to our Viatka province because they had a conscience and were truthful. In 1905, Nicholas II published the Manifesto of 17 December on freedom of speech and the press.[25] And what happened? Instead of this, every village got fifteen mounted police gendarmes, and every hamlet a policeman on foot and even more exiled people. As a result of the manifesto granted by Nicholas II on 17 October, the first state Duma, with peasant representatives, assembled and deliberated in the interests of the laboring people—it was disbanded. A second Duma was called with representatives of the landlords and merchants headed by the tsarist minister Stolypin. It strengthened [private] property, which proved distressful for the laboring masses. Prokopii Bazhin learned of this and was also unhappy with it and, concealing this from others, turned against tsarism.[26]

When the world capitalist war broke out, then for two solid years the people pinned their hopes on the tsar, but by the end of the war the people began to worry; they understood that there was no justice in the war and that treachery was everywhere. The war caught up to me in 1916, and I was compelled to take part in the Romanian front. Rumors of the Constituent Assembly began to circulate, and I felt compelled to vote for Bolshevik [list] no. 6. When I was demobilized, I went home and told the village that Soviet power would be good for the peasants, but the elders and young people—defenders of the bourgeoisie—paid no attention.

When 1920 came, we were all registered and forced [to do] grueling winter work: the men had to saw twelve cubic [meters] of firewood, and the women eight cubic [meters]. I happened to do twenty cubic [meters], but the sons of the bourgeoisie who had joined the Party kept the accounts and lined their own pockets. What is there to say? A person was ashamed to show his mug to those who had said: "So, Soviet power is good, eh?" When 1922 and 1923–1924 came, then once again the poor muzhik had to pay more, while the rich had pedigreed cattle, etc. All the same, they will find a way out [of this situation].

Now in our territory there are organizations like the agricultural association [selkhoztovarishchestvo] and the village cultural union [Kulselsoiuz][27] that traffic in everything the peasant produces. They buy the peasants' products on the cheap and sell manufactured goods dear. And once again the bourgeois who ridicule the authorities are in charge and are getting rich off it. And now the small smiths, who are vital for the village, for the peasants, are required [to purchase] a license, etc., and the water mills are also obliged [to pay for] a license, [as well as pay] a leveling duty and state rent, and this all falls on the backs of the peasants. How long must the poor man wait for justice?

We request an explanation.

It would also be interesting to find out if France is going to pay its debt? This also falls on the peasant.

It is with genuine faith that I write.

BAZHIN, PROKOPII IVANOVICH.

———————

The definition of the Soviet state as a "proletarian dictatorship" also fostered dissatisfaction among the peasants. Peasants often expressed bewilderment about this formulation, as in this letter from S. Gogoi, a peasant from Ternovka village in the Moldavian autonomous republic.

· 40 ·

Letter to Krestianskaia gazeta from the peasant Sergei Gogoi requesting an explanation of the peasants' role in the proletarian dictatorship, 13 June 1926. RGAE, f. 396, op. 4, d. 24, l. 408. Original manuscript.

———————

The first word.

On reading Comrade Mokhov's letter in issue 23 (130) of Krestianskaia gazeta, I remembered that at a cell assembly in March of this year a question was openly put to the chairman of the assembly, a Party member: Why is 'Proletarians of all countries unite' written here, there, and everywhere, but not 'Proletarians and peasants of all countries unite'? That is, what distinction between the proletarian and the peasant prevents a joint call to unite against the bourgeoisie?" We know that the workers and peasants don't have a pot, just their working hands—this is a proletarian. But just because the peasant has a little bit of property, does that mean he is not a proletarian but a petty property owner, even if the peasant happens to be a poor peasant? But for us they write, "Long live the union of workers and peasants," inviting even middle peasants to stand together as one not only here but in the whole world.

The answer of the Party member to the peasant:

1) The boss forcefully said that the slogan could not be changed. That the proletariat, more than the peasants, can be depended on to support and raise up Soviet power in other countries because peasants are petty property owners.

2) Comrade Zviagin, chairman of the village soviet and a Party member, said that the peasants must be proletarianized, and then the peasantry can also be placed in the aforementioned slogan. I ask for your explanation because from the foregoing it follows that the peasants must question [whether everyone is] equal and [whether everyone's] life is improving in the country, and [this] concerns us at the present time in the USSR.

The peasant, GOGOI, SERGEI.

The second word

It is not clear to us peasants why, at party gatherings, the party members say "the dictatorship of the proletariat." If we are not mistaken, then it seems to us that the proletarian (worker) dictates according to his needs, and the peasant obeys and fulfills [his command] whether or not the peasant wants to.

We peasants would like to be proud of the words and deeds "worker-peasant power," "the state," "the government," "the Red Army," and "the worker-peasant dictatorship," and not the proletariat alone.

If we peasants have misunderstood all of the above, we ask that you explain it to us.

The peasant S. GOGOI

Similar sentiments are expressed in a letter written sometime in 1925 by the peasant G. Masiura from Mikhailovka village, Amur province, in the Soviet Far East. Masiura expresses the anger that many peasants felt, but his letter is also of interest for its historical and literary references.

· 41 ·

Letter to *Krestianskaia gazeta* from the peasant G. Masiura on the
Communist Party's exploitation of the peasantry, 1925. RGAE, f. 396,
op. 3, d. 105, l. 12. Original manuscript.

More and more when reading *Krestianskaia gazeta,* especially its most
recent issues, one comes across the Bolsheviks' appeal to the peasants for an
alliance with the workers around state construction. In *Krestianskaia
gazeta,* no. 60, this is how Comrade Yakovlev[28] begins his article, "From the
peasant one has occasion (sometimes) to hear the opinion that the Communist Party does everything for the worker, that the Communist Party only
talks about the [peasant] union, etc., and even that Comrade Lenin said:
'Nothing is stupider than when people who know nothing about agriculture
take it upon themselves to teach the proprietor' and, furthermore, 'It is not
necessary to order the peasant about, but to learn from him and help him.' "

Maybe there, in the center, the Communist Party does think this way, if it
gives the peasants any thought at all. But here, among the people, this idea
has not occurred to anyone, because as long as the peasantry has existed, it
has not felt the kind of hostility toward anyone that it feels for the worker
in general and for the communists in particular. The source of this hostility
is not here, it was not born locally, but came out of the very heart of the
Communist Party, from its sacred slogan—dictatorship of the proletariat—
and that is the root of the hostility. There are no generals or landlords to
establish a dictatorship over. This leaves only the peasants. The peasant sells
his last cow for taxes, not out of goodwill, but only under the pressure of
the dictatorship. And you are seeking union and fraternity with him! The
second source [of hostility], to tell the truth, is the Party itself, because no
party thinks of anything other than itself. [It sees itself] as a small band and
the rest as a non-Party whole. In other words, the Party in the localities has
turned into an *oprichnina,* although it does not tie a dog's head to its saddle
like the *oprichniks* of Ivan the Terrible, but it always stands apart [*oprich*]
from the people.[29] The peasant provides food and drink, shoes and clothes,
for everyone except himself; he often goes about hungry and naked. But the
communist takes [for himself what others have] prepared and not only is of
no help, but everywhere he only obstructs, and his willfulness knows no
limits. Curiously, Comrade Yakovlev's article concludes by saying that in
the future *Krestianskaia gazeta* can have no more honorable mission than
to achieve the union of peasants and workers. No doubt much will be written about this. Many resolutions will be passed at assemblies and congresses, but the peasant and the worker will never cooperate, because, in the
first place, when the worker works he gets [something] for his labor, but
when the peasant works he pays for his labor, and because, in the second
place, it's as if the Communist Party is trying to turn all the peasants into
beggars. It only encourages the poor [peasants]; it doesn't offer any help,

just encouragement. And if the peasant should succeed in improving his holding and scrambles out of poverty, then they brand him a kulak and consider him an enemy of Soviet power. But there is no peasant who would not try to improve his holding and become, in the eyes of the Party, a kulak. In other words, there is no peasant who would not try to become an enemy of Soviet power.

A lot of people are writing about changing this or that in the agricultural tax, but they are saying this only because they're afraid to say straight out, "Don't limit free speech, cancel censorship, don't obstruct the independent progressive press," and then you will come to see that in gathering acorns you are undermining the root of the oak, you are chopping the branch on which you sit, for you the oak [is] the peasant, but come what may, you'll have no pity: to you they are tax-acorns, and you will get fat on them . . .

The author openly expresses his belief that the Bolsheviks need the "proletarian dictatorship" to rule over the village and to extract whatever possible from the peasants through taxes. He effectively illustrates his accusation that the Communist Party was little more than an organization dedicated to robbing the peasants and providing its functionaries with sinecures by invoking a familiar historical episode (Ivan the Terrible's oprichnina) and referring to Ivan Krylov's fable "The Pig under the Oak."

Underlying the specific charges is the commonly held resentment of the constitutional limits placed on the peasants' political rights. The constitution of 1918 had established the category of "disenfranchised person" (*lishenets*). Included in this category were those whose income was not derived by their own labor and those who employed the labor of others. In addition, the constitution also discriminated in favor of workers by granting them one representative to the Congress of Soviets for every 25,000 of their number, whereas peasant representatives were elected at a ratio of one for every 125,000. Memories of peasant self-rule following the 1917 Revolution, when local peasant assemblies established their autonomy throughout the countryside, must have made this second-class status all the more galling.[30]

The maintenance of these constitutional strictures into the 1920s, despite the rhetoric surrounding the smychka and the worker-peasant state, exacerbated peasant dissatisfaction and strengthened the impression that Soviet power had "sons" and "stepsons." This is the point of I. L. Chibutkin's March 1927 letter from Yefremovo village in Yaroslavl province:

· 42 ·

Letter to *Krestianskaia gazeta* from the peasant I. L. Chibutkin on the
Soviet state's "sons" and "stepsons," 12 March 1927. RGAE, f. 396, op. 5, d. 30,
ch. 1, l. 459. Original manuscript.

Sons and stepsons

As I look through the constitution, article 9, where a city resident is given
more advantages than the peasantry, unconsciously catches the eye. It says
there that the workers are represented at a congress [at a rate of] one representative for [every] 25,000 residents, and the peasantry at one [representa-
tive] for 125,000 residents. As a peasant, this seems very strange to me, and
it seems that here we have [a case of] sons and stepsons. I know the answer
I'll get. That the workers are the vanguard of the Revolution. That the
worker made the revolution, and that the worker carried all of the revolution's hardships on his shoulder. That is why he is given an advantage in
representation. They may even add that the peasant's political training is
not as advanced as a worker's, etc. But I think that just as the worker raised
the revolution's banner, the peasant saw it through. After all, the peasantry
also sacrificed everything on the altar of the revolution. He sacrificed his
son, his horse, grain, and everything that was needed. Didn't the partisan
brigades in Siberia and Ukraine drive out Kolchak, Denikin, Wrangel, and
the whole gang? They drove them from our revolutionary country. Didn't
the peasantry subsidize the construction of freedom? Yes! But what did he
get for these services? A one-fifth share in voting rights. Where is the
smychka of town and country in that! Where is the slogan "Face to the village!"? It is not here! On the contrary—only one-fifth of a face to the village. And my article "Sons and Stepsons" will definitely prove how just this
is. I know what they will tell me now. That the peasant has still not overcome his individual-proprietary views, and as they say, he has far from a
social view of life. The peasant, they say, has a house, a horse, a cow, and
considers them his property. But, after all, he needs these for his work just as
a worker needs a machine, and with them the peasant raises industry, raises
the welfare of the country. But if one looks deeper, then it appears that even
the worker and the city dweller in general have a taste for property. To be
precise, the workers have houses, a gramophone, a bicycle, clothes. Not one
worker would deign to wear the clothes a peasant goes around in. And you
should be aware that just for a bicycle, I would give up a horse with a
wagon and throw a calf in too. Therefore, insofar as the peasant is an individual, then; in this respect, the worker does not lag far behind the peasant.
Knowing workers as I do, they are not all that different from the peasant.
Moreover, the worker and the peasant make up a common family, one chain
forged together. Why, then, this stratification stipulated in article 9 of the
constitution? That is why I am putting my article "Sons and stepsons" to the

judgment of the masses themselves and am requesting a review of this question at the next All-Russian Congress of Soviets. We are already on the threshold of the ten[-year anniversary] of the October Revolution. The peasantry has fully matured and can fully arise, hand in hand, in socialist construction.

I am completely familiar with the worker's situation since I was a worker myself for fifteen years.

Ivan Lvovich Chibutkin

The humiliations arising from legal inequalities were only a part of the picture. Tensions between workers and peasants made themselves felt in myriad informal ways. The regime's endless contrasting of worker consciousness and peasant ignorance—not to mention peasant mendacity— reinforced time-worn notions of the city's superiority over the village. Inevitably, workers, who may not have been that far removed from the village themselves and, as a result, were eager to display their superiority, looked down on the peasantry as a mass of rubes and hicks opposed to progress. This attitude was put to use by regime later in the decade. As Lynne Viola found, a number of workers cited their past familiarity with rural life as their motivation in volunteering for the collectivization drive in late 1929.[31]

One expression of this condescending attitude crops up in a June 1925 report by the village correspondent L. Yarovoi from Bogoslovskoe village, Akmolinsk province, Kazakh republic.

· 43 ·

Letter to *Krestianskaia gazeta* from the village correspondent Leonid Yarovoi describing the condescending behavior of Komsomolists toward the peasants, 27 June 1925. RGAE, f. 396, op. 3, d. 83, l. 51. Original manuscript.

During the Atbasar fair, which is held on 2–3 June, the people came in droves. The peasants travel 209 kilometers from the remotest villages, where they never read much and rarely see things like a musical show or a band. And so, at the end of the first day of the fair all the young peasants, and even the grown-ups, all burst into the garden to see the sports and music. Admission [cost] fifteen kopeks. Not cheap, it's true, and not everyone has the money. And so those short of cash forced their way in without paying—but nothing doing. The Komsomolists from the city of Atbasar took on the role of gendarmes and shouting, whooping, and whistling, they began throwing the young peasants who had no tickets out the gates or snatching their caps and tossing them over the gates. And they [the peasants] understand and go under a wagon to sleep, never having been anywhere in their lives. That's how the city Komsomolists approach the rural youth. Such views of the urban public about the peasantry must be mercilessly eradicated.

LEONID YAROVOI

At the end of 1925, additional strain was placed on the smychka by the decision of the Fourteenth Party Congress to proceed with industrial expansion. Those leaders like Bukharin who advocated equilibrium in the growth of industry and agriculture recognized that the demands that industrialization would make on the budget threatened peaceful relations with the peasants. Indeed, consistent with this turn in state policy, were changes in the tax code for 1926/1927 that reversed the reductions of the previous fiscal year and increased total collections from agriculture by over 100 million rubles. Wealthier peasants were made to shoulder a significantly greater tax burden as marginal rates increased to 25 percent while rates on poorer peasants were reduced, signifying that toleration of the kulak was on the wane.[32]

As the decade came to a close and pressure on those proprietors who had somehow managed to extricate themselves from poverty increased, resentment and confusion over social categorizations rose. Identification as a kulak became dangerous from both an economic and a political standpoint. Independent farmsteaders and peasant-traders were now in a very precarious position. In a letter of 21 April 1926, T. V. Shevchenko, from Borisovko settlement (*sloboda*), Voronezh province, details this process for the chairman of the All-Russian Central Executive Committee, M. I. Kalinin.

· 44 ·

Letter to the VTsIK chairman, M. I. Kalinin, care of *Krestianskaia gazeta,* from the peasant T. V. Shevchenko requesting a definition of a kulak, 21 April 1926. RGAE, f. 396, op. 4, d. 27, l. 507. Original manuscript.

Mikhail Ivanovich Kalinin! Please explain this business. Currently meetings of poor peasants are held in the villages, where only poor peasants and invited middle peasants may attend them. During [these meetings] kulaks showed up. Therefore, I ask you to answer me in writing, Who may be considered a kulak, who well-off, and who a middle peasant? You understand that the situation is insulting—if I know that I am not a kulak, but they, for all purposes, lump me in with them. Thinking that [I am] a kulak, not once did they allow me into the meeting of the poor peasants. I will describe for you what I had in the old days: for twelve souls, one and one-half desiatinas of my own land and twelve desiatinas rented from the landlord. Now for nineteen souls I have twenty-six desiatinas, twenty-four sazhens [of land], and livestock: four bulls, two cows, one horse, eight sheep, and one pig. Buildings—one hut, one grain store, one threshing barn, and two barns. I work the land by my own labor, and at certain times members of my family take on wage-work. Another farmstead of three souls has two bulls, one cow, one hut, one barn, and he is counted as a poor peasant. This forces the large families to break into smaller ones just not to be considered kulaks. The result is that we are heading not to socialism but to small poor farms that will forever be seeking state aid. That is what I think, although my letter is not eloquent and not so informative, but all the same I ask you to answer me in a letter and explain who to consider a kulak, a middle peasant, and well-off, and how the peasantry is advancing to socialism. This is occurring so quickly in the villages that it resembles "war communism." All the desire to improve one's farm is passing. Where's the good in it if by improving one's farm, one will be considered a kulak?

I request, Comrade Kalinin, [that you] answer my letter.

SHEVCHENKO

The letter provides evidence of the changing mood in the village that, in Shevchenko's opinion, recalled the hostility of war communism. Preparations for soviet elections scheduled for early 1927 also revealed the center's growing turn against the kulak. The explicitly stated goal of the elections was to promote the participation of the propertyless and the poor in rural politics by increasing their representation in soviet institutions. To this end, Moscow instructed local officials to more strictly enforce the restrictions on the disenfranchised than in previous elections.[33]

Evidence for the deteriorating social atmosphere in the village may be found in letters from the latter half of the 1920s. Increasingly, we read expressions of hostility directed at other peasants as well as workers. Such is the case in the following anonymous letter, which has little good to say about either kulaks or poor peasants. The original was replete with colloquialisms and spelling errors and was probably written by a peasant of Ukrainian origin. The letter is from Karakol canton, Kirghiz autonomous province.

· 4 5 ·

Letter to *Krestianskaia gazeta* from the peasant S.M. on the detrimental effects of the state's lack of faith in the middle peasant, 11 August 1928. RGAE, f. 396, op. 6, d. 114, ll. 771–772. Original manuscript.

Citizen editor, I request that you place [my] letter [in the newspaper]. I am getting my beloved *Krestianskaia gazeta* [for] the third year and read [it] closely, never skipping an issue, following state policy and thinking with my block of a head. That higher authority also makes mistakes, especially in an economic policy that is resting on the proletarian and poor peasant in the peasant groups, but as for the middle peasant, it has no trust in him as a shaky element. But if you look at who supplied the cities with food and raw materials. The middle peasant . . .

The middle peasant, this is the provider for the mass of people. The kulak—this is the plunderer of honest labor; the poor peasant, [the] proletarian in the village—this is the lazy exploiter of honest laborers. The kulak underhandedly tries to exploit the honest laborer.[34]

But the proletarian lays around, does nothing, and reckons like a lout how to enjoy someone else's labor while saying, "Power is ours." "We'll do with you whatever we like . . ." And every year the government is getting closer to them, [gives] all the help it can; they are getting used to this, and the middle peasants look at this and bit by bit move over to their group. Like they say, it's hard to lift up one's situation, but it can all be lost in a heartbeat, and they point at the poor peasants: Look, they do nothing, but they don't live any worse than us, and they don't pay taxes. In the newspapers you see, supposedly, the poor are moving up, but it's really the opposite, like you see with us, here in the Kirghiz autonomous republic, [where] they've created twice as many and more poor peasants.

Here's an example from us in 1921: they did a distribution. They broke the people up into three groups according to their living and dead inventory—kulak, middle peasant, and poor—and they began to allot [land]: the poor peasant—three shares; the middle peasant—a share and a half; and the kulak—only a half-share. The same with the large and small livestock, and the remaining tools and grain.

And what happened in practice? Among the poor, who got three shares, rarely does anyone have a cow left after a year despite the fact that they got as many as twenty head of cattle and a plow, but after three years the kulaks have to be split up again because as he was, so he's become again. They made it by their own labor, without exploitation of somebody else's labor, because he works day and night. And the middle peasant was a middle peasant and stayed a middle peasant.

And now this is the conversation going on among the peasants—we'll sow less so not to pay taxes, and we'll fall into the 35 percent of the poor; till then we'll be working for them, for the lazy ones, but we, too, will become poor peasants from our work. And that's how it happens that, according to statistics, the sown area is supposedly increasing, but there's still not enough grain. And that's what it all comes down to. As a middle peasant, both now and under the old order, I know how every peasant breathes, not just what he thinks. Before, our middle peasant sowed ten or twelve desiatinas but only registered six or seven for the books, but now, in accordance with article 107, they have begun to register down to the furrow.[35]

I'm a person who sees little, a true bumpkin; in my opinion, the government is trying so that everyone will be proletarian and poor peasants, but how will the state exist then? If we peasants turn out only enough food for ourselves, then how will the city and the worker exist, and without them the peasant also won't be worth anything. Or in the case of a sudden disaster or crop failure, then we, without reserves, will all drop dead like flies in a frost. Then the state won't get us all back on our feet 'cause we're too many.

Or if the government rests on the *kolkhozes*—it's a beautiful thing on paper, but ain't near like what they write; I tried that too. [I] lived in a collective and got out only with difficulty. First off, you gotta remake all the people and then build kommunas. In peasant work, like in the factory, not everyone can do everything. There's a type, takes three hours to harness a horse, and off went the rig, overturned, broke up, but listen to him talk; he knows everything better 'n you and orders you around.

The opinion of the middle peasants, and mine, concerning the agricultural tax [is to] give the poor peasants up to three or four desiatinas and then assess each desiatina at five rubles or ten rubles, and [apply] this [same rate] whether [the peasant has] five desiatinas or fifty desiatinas and [let them] pay ten rubles tax for each [desiatina]. Then we middle peasants would bury you in grain, and no other workers are needed, just give ours freedom and machines and we'll show you—without your agronomists, at whom the state throws money in vain. In every locale every peasant grain-farmer knows how to work his land, what to fertilize it with, more than your agronomists. Just give the middle peasant the privilege, so you don't say kulak if he makes good. Then won't you be happy, and there won't be nowhere to hold all the grain. I'm done for now, I'll save the rest for later.

Kirghiz autonomous republic, Karakol canton, S.M.

For this writer, state policies favoring the poor do not raise productivity and only discourage industrious middle peasants from working hard. His views of the kulaks are also revealing, expressing both hostility and admiration. In one breath, he describes the kulak as a devious exploiter of other peasants' labor, in another as a hard worker destined to succeed regardless of government discrimination. In opening his appeal in a humble and self-deprecatory manner, the writer also employs a tactic frequently encountered in communications between the less and the more powerful. Whatever feelings of individual inferiority this may actually reveal, it also serves those from the lower social ranks as a form of protection from retribution for being bold or uppity. That the writer, for all his grammatical shortcomings, carries anything but a "block of a head" (*durnaia golova*) on his shoulders becomes evident on reading his letter, which accurately points to the contradictions in Soviet rural policy, at least as far as stimulating peasant productivity is concerned.

The peasants' hostility toward workers sharpened despite the persistent newspaper rhetoric on the success of the NEP and the importance of the union between city and village. This is seen, for example, in the letter from A. F. Shklinov, a peasant living in the village of Boguslavskoe, Samara province.

· 46 ·

Letter to *Krestianskaia gazeta* from the peasant A. F. Shklinov on the relative economic condition of the worker and the peasant, 4 January 1928. RGAE f. 396, op. 6, d. 97, l. 71. Original manuscript.

Comrades, first of all, I apologize for my clumsy letter. Life has left me ignorant and given me the calloused hands of a poor peasant. Comrades, I had the chance to see Comrade Khotiev's speech in the peasants' newspaper no. 904/2049/, wherein Comrade Khotiev gives an example of how, allegedly, under Soviet power the peasants have begun to eat better than under the old system and are not sending white flour, butter, and meat to the city—everyone is eating the same [food]. Comrade Khotiev, you are deeply mistaken in this. The workers are exploiting the peasant's labor. True, the muzhik has all sorts of provisions, but he doesn't end up with any of them; the comrade worker eats for the muzhik. Comrade Khotiev made it out as if the worker's labor is cheap, giving as an example nails and sugar, which sell cheaply, but Comrade Khotiev did not consider the textiles needed by the peasants, and he did not take into account at all the most important role, which is that the entire burden of the state is borne by peasant labor. For every meter of sateen the peasant must pay a pud of grain, and for every funt of sugar, twenty funts of grain. Yes, the worker has served his whole life, the worker has offered his hand to the peasant, but he sits on his back [like a weight]. The comrade worker thinks that only his labor is of high value and that of the peasant lower. [...] With compassion, Comrade Khotiev stresses poverty: how the poor lived under the old regime, how they paid high prices, down to the very last pud [of grain]. Comrade Khotiev, how do the poor buy things now? He's a member of the cooperative but does not get a full share [in the cooperative]. He also can't get goods because of the shortages when the rich buy up what he needs. Then they [the rich] sell to the poor peasant, and all because he does not have a full share. Comrade Khotiev mentioned what they paid for land before the war, but now, supposedly, it is cheaper. Comrade Khotiev, it is still not easy for the peasant [to buy land], even though the peasant shed his blood for the land, even though it was the peasant's provisions that they commandeered for the army, all to install the peasants over the land, but the peasants got nothing: it all turned out to be a fraud. Capital and the bosses [*zavkhozy*], even the state fund, rule the land, and the muzhik is again forced to climb into the yoke, into the *sovkhoz* [state farm], so that the muzhik will work as if on barshchina— one part to him and [one part] to the sovkhoz, but you don't want to reduce the state fund so that the meadow campaign would cost fifteen rubles.[36] That's the sort of life of the poor peasant and the middle peasant. The request of the peasants to the higher authorities is to lower [the price] of goods in accord with the prewar [price] for grain and for the state to divide the land among the peasants—not two desiatinas per soul, but more. The land should be taken from the sovkhoz.

I request an answer in *Krestianskaia gazeta*.

A. F. SHKLINOV

P.S. Excuse my bad writing. I am hardly literate, beaten down by the ignorance of poverty.

Like the previous letter writer, Shklinov accompanies his sharp and astute criticisms with humble apologies for his alleged "ignorance." The price differentials between industrial and agricultural products, a cooperative system deemed unfair, and the privileges that urban workers supposedly enjoyed greatly angered this poor peasant. In early 1928, a similarly contentious attitude toward workers inspired the following anonymous letter from a peasant in Kursk province to the editors of *Krestianskaia gazeta*.

· 47 ·

Letter to *Krestianskaia gazeta* from an unknown poor peasant on the favoritism the state shows toward the working class, 27 January 1928. RGAE, f. 396, op. 6, d. 114, l. 15. Original manuscript.

From a poor peasant

I'll get right to your way of thinking and would like to offer a few words on your outlook. How good and free our life is at the present time, under

Soviet power, that is, and how good and comfortable for us workers this Soviet power is, how equal and free. Free and equal is our Soviet power, but not entirely equal and not entirely liberated. The government has equality, but people live unfairly, and they [the government] don't pay attention to the people equally. The government has devoted most of its attention to the worker, the factory hand [*khvabrikantov*], and the plant worker [*zavod-chikov*], that is. They raised his wage and shortened his day, and there is certainly freedom for him. He works eight hours, sometimes even seven, and the rest of the time he is free to relax, take a stroll, and get some enlightenment. But they pay very little attention to the poor peasant. The peasant does not get a short day or a big wage. The factory hand [*fabrikant*] works eight hours a day and he is called a worker [*rabochii*]; the peasant works seventeen hours a day and is not called a worker. The factory hand receives sixty rubles a month and pays no tax; the peasant works for seven rubles a month, seventeen hours a day, and he pays the tax. But the peasant really wants to help Soviet power, and he is prepared to help the state with all his might. But Soviet power does not allow him to develop, does not give him independent powers. They don't give the peasant the forests that are so necessary for the peasant. They fix the price of peasant grain, but there are no fixed prices on what the peasant buys. If he buys boots, he must fork over fifteen rubles, but grain is collected at seventy [kopeks] per kilo: how many puds must the peasant [turn over]? For the peasant there is no other source [of money] besides grain—the government does not allow [a peasant] to earn money, doesn't allow [a peasant] to develop his farm, doesn't grant freedom or liberty to the peasant. As soon as someone improves his farm, they deprive him of his right to vote. So, the government wants everyone to live as beggars. No, this sort of thing isn't freedom; they need to help, they need to give freedom to the peasant, so that he can develop, so that he can enrich his landholding. After all the peasant feeds all of Russia, all of Russia [depends] on his support, [so] how can they not grant freedom to the peasant? But what we call freedom we don't see. The state needs grain, and where can they get it? [They can] take it from the peasant; the tax comes from the peasant; grain—from the peasant. The peasant gets nothing but what he earns from working the land. So, what's your opinion now: how bad does the peasant have it? You would have them live differently, collectively; in general, you would have them live together. It is you who are hanging a heavy stone on the peasant's heart; it is a heavy burden you put on the peasant's neck. How can one live under compulsion? Give the peasant freedom, give him liberty, let him live as he wants. Whoever has a good head on his shoulders, let him live.

I would like my letter to be printed in the newspapers *Krestianskaia gazeta* and *Kurskaia pravda*.

To the editorial board, Moscow city 7, Vozdvizhenka 9. I request that you print my article in the newspapers *Krestianskaia gazeta* and *Kurskaia pravda*. I beg and appeal to you to print my letter in the newspapers, but you won't print it; I am sure that you won't print my letters, but I beg you to please print it in the newspaper.

Despite the NEP and the smychka, this poor peasant accuses the state of neglecting the well-being of the peasant in favor of that of the worker. The unfair terms of urban-rural trade are, in his view, no more than a way to gouge the peasant. Convinced that no benefits can be expected from the state, he reiterates the pleas of previous peasant generations to be allowed to live in freedom (*svoboda*) and to enjoy individual liberty (*volia*).

Peasants harbored a degree of resentment for those who left the village to take up nonagricultural occupations and who, in the process, adopted a disdainful attitude toward the village and its inhabitants. In the following short letter the peasant A. A. Shchipakin from Krugloe district, Orel province, demands equal treatment for workers and peasants. He expresses the transformation from peasant to worker symbolically, calling attention to a change in footwear from the traditional bast sandals (*lapti*) worn by peasants to the type of rubber or leather boot worn by miners known as *chuni*.

· 48 ·

Letter to *Krestianskaia gazeta* from the peasant A. A. Shchipakin calling for equality between workers and peasants, 14 October 1927. RGAE, f. 396, op. 6, d. 210, l. 518. Original manuscript.

Dear comrades! Many peasants already have no need for property since they have already gone hungry for the ten years of the revolution. They can hardly turn chaff into grain. He who is in state service is full to bursting. Not for nothing was the slogan "Down with bast sandals," and now there are chuni. Whoever goes off somewhere soon forgets the life of the poor peasant. I will only say one word, "Long live equality and brotherhood both for the workers and for the peasants." If one of you says that some stupid blockhead has written [this], then he has no faith, and he is not a brother to the poor peasant and to the entire laboring peasantry.

Against the Bureaucrats

For the peasants the state was not an abstraction but was personified by the officials who governed the countryside and administered rural affairs. Many village dwellers viewed the salaried state officials—who produced nothing, yet earned a guaranteed wage—as obstacles to the well-being and productivity of both peasants and workers. Overt hostility toward salaried state employees (*sluzhashchie*) can be found in many letters from the 1920s. On this point at least, peasants believed that their interests intersected with the workers'. Various writers discuss the need for workers and peasants to wage a joint class struggle against bureaucrats. This is particularly evident in the following letter to M. I. Kalinin from I. A. Rusov of Sukhanov settlement, Nizhny Novgorod province.

· 49 ·

Letter to the VTsIK chairman, M. I. Kalinin, care of *Krestianskaia gazeta*, from the peasant I. A. Rusov on the conflict between bureaucrats and those who labor, 30 March 1927. RGAE, f. 396, op. 5, d. 30, ch. 2, ll. 721–722. Original manuscript.

A letter to the editorial board of *Krestianskaia gazeta* for the chairman of VTsIK, Comrade Kalinin, M. I.

Dear Comrade M. I. Kalinin! I am writing you a letter concerning socialist construction. Will there not be two hostile classes, and will there not be a struggle between these classes? The first class is [made up of] the workers and peasants, who are the two laboring classes. The second class is [made up of] the soviet employees, namely those who are well paid, [getting] fifty or more rubles, etc., who even get one hundred or two hundred rubles. I, of course, won't dwell on the big shots. The struggle among the state employees will be precisely between those who have grown a big belly on the working neck of the proletariat under Soviet power. I won't go all the away to the center for an example, but will take the example of our peasant Chistiakov, N. V., who serves in the village of Uren, Nizhny Novgorod province. When he lived in the hamlet, he ate bread and water, his favorite food, but when he began to serve in soviet institutions, his belly grew, and he turned into a bureaucrat [*zabiurokratilsia*]. This winter he came back to our hamlet and even spoke badly to the muzhiks. And this goes on in every institution; even elected officials do these things. In our opinion, the peasantry mustn't allow this boorish yoke into the administration during socialist construction because this breed hinders the work and building of socialism in one country. The struggle between these classes is already beginning; the peasantry has begun to hate those administrators who receive a big paycheck. You get little work out of them—like he-goats: no milk, no wool. [Their] salaries should be cut as much as possible because then there will not be this hatred toward state servants, then there will be fewer parasites, and then the belly of people such as Chistiakov, N. V., will get small, and the taxes on the peasantry will be much less. Almost all the taxes are a result of the high pay scale, and this will be a savings to the national wealth. All the wealth that we will get by cutting the pay scale will go to our industry. Whoever gets one hundred rubles will have to make do not with one hundred rubles but with seventy-five rubles. I think the chairman of the All-Russian Central Executive Committee will answer my letter or put [it] up for discussion by the worker-peasants themselves.

RUSOV, IVAN ARKHIPOV

Based on his experience, Rusov arrives at a class analysis of bureaucracy in the Soviet Union that identifies politics as a struggle over privileges between bureaucrats, and in this sense, he anticipates Trotsky's writings on the subject.

Organizing Agricultural Production

To many workers sympathetic to the plight of the village, the problem of agriculture presented itself, first and foremost, as a problem of organization. Progress, in their view, demanded that peasants abandon individual farming and pool their resources. But what organizational form should collective farming take, and how could peasants be convinced to join collectives? In a letter written to the editors of the railroad workers' union newspaper, *Gudok*, A. I. Sechko, a resident of Sot station on the Northern Railroad in Yaroslavl province, proposes organizing collective farms on the basis of communal principles.[37]

· 50 ·

Letter to *Gudok* from the peasant A. I. Sechko on the proper way to organize a collective farm, 14 July 1928. RGAE, f. 396, op. 6, d. 61, l. 291. Original manuscript.

Especially for the peasant's corner.

The peasant's labor now [is done] on the separated, small farm and on the collective farm. On which principles should the kolkhozes be built?

I live in Yaroslavl province, where there was not enough grain before and there still is not enough grain. You should see how the peasants stand in line for a crust of bread at the cooperative. You don't believe [me]? [You find it] funny? That's right, peasants stand in line for a crust of bread—this is a fact! (Predtechevsky cooperative, Danilov county.) The Yaroslavl peasants could not have it any worse now. They survive on milk alone (the cow is rescuing [them], thank you), and that is the trouble. They are dying of hunger because there is no bread, no potatoes, no meat, no nothing. The reason for this poverty lies in the fact that holdings are small, the land is wretched, and outside work is scarce. Now the peasant works without the benefit of labor laws; he works like a convict and for his labor receives just enough to keep from starving. His work is thankless. It would be a different matter if the peasants were organized in kolkhozes. For example, let's say two or three neighboring hamlets were organized, merged into one collective farm. They would select the administration, they would appoint the head of the farm, and business would proceed quite differently. One thing is certain, that in a large farm, as in any large concern, work is much more profitable. The peasants should be forced to think about the formation of kolkhozes. Kolkhozes are organized on cooperative principles, and they promote a type of productive cooperative. But I would propose an organization of kolkhozes on communal principles [na kommunalnykh nachalakh] even though I know there will be a mass of opponents against my proposal; nevertheless, I will try to prove the utility and advantage of the communal conduct of a farm. Let's take two to three (the more the better) neighboring hamlets; we amalgamate them and inventory all available livestock. We select and hand over to the administration the equipment and all the necessary [items]. We sell off the unnecessary [items] and buy what's needed with the money. As is well known the state gives generous credit to these organizations and helps in any way it can. Work and life in these *kommunkhozy* (this is what I will call them) should also be based on the communal principle.[38] That is, every member does a definite job assigned to him and for his labor receives everything he needs from the organization's warehouse, and all the output is likewise given to the warehouse. The organization's administration would figure its due, and all the extra it gives over to the state in exchange for what it—that is, the given farm—needs, either products of state industry or money or various and sundry things in accordance with its needs and the circumstances. Through the liberal aid of the state, which undoubtedly will be given to such farms, and through the proper management of the latter, rich results may be achieved in the course of a short period of three to four years. The farm will have its own agronomist, all fieldwork will be mechanized, cattle rearing will be developed to the required degree as well, and, most important, the labor of the members of the farm will be rewarding. It won't be necessary to work from sunrise to darkness anymore and exist in [a state of] semi-hunger. [Farm] labor will be protected by those same laws protecting [nonfarm] labor, by those [laws] that protect us in the cities involved in production.

Danilov [county]
ALEKSEI IVANOVICH SECHKO

Sechko's advocacy of the most highly socialized form of collective farm organization—the kommuna—recalls the utopianism of the war-communism period. The following letter from the worker A. N. Kuznetsov, Kresty village, Panino district, Rzhev county, Tver province, is also reminiscent of the civil war. It is even composed in the spirit and style peculiar to that time.

· 51 ·

Letter to *Krestianskaia gazeta* from the worker A. N. Kuznetsov detailing his civil war experience and his later efforts to organize collective farms, 15 March 1928. RGAE, f. 396, op. 6, d. 61, ll. 14–15. Original manuscript.

Respond

Four years ago I worked on the Moscow-Briansk-Belorussian rail line, as a metal worker [*slesar'*] on the service section for the twenty-ninth track. There was a staff reduction, and I asked myself, Why don't all the peasants join together in collectives when social plowing is so beneficial to members of the collective? I decided to quit voluntarily, to leave my job during the reduction. I have never lived in the countryside. I decided to go to the country and share my practical knowledge in regard to the building of socialism because in 1919, [during] Kolchak's first offensive, I voluntarily left the Izhev factory when the Communist Party called and went to the front on 15 February despite my family situation. Izhevites respond. On 18 April 1919, for the newspaper *Bednota*, for reading it, I was brutally beaten by White Guards during the retreat of the Reds. On 19 April 1919, I joined up with the Reds. *Krasnyi voin* of 25 April 1919, respond.[39] In 1920, I was initially commandant of the guard [not clear in original] of a factory in Tver under the commissar of special missions, Comrade Saksa. Who was first to go voluntarily to the Polish front? Comrade Saksa, respond. Who, in fact, in the political administration of the revolutionary military soviet of the western front fed the political workers in 1920, and in 1921, at the time of the evacuation of Minsk, who, on his shoulders, [carried] one hundred puds of foodstuffs so that the Polish White *pans* [landowners] would not use them?[40] Political administrators respond. On 5 August 1921, I went on an indefinite leave with my collected verse, written about my recent past. In 1924, I convinced a peasant to go to his home in the village of Kresty, Panino district, Rzhev county, Tver province, [and I] will teach his two children a trade on the condition that we divide our earnings between us. The peasant was happy for this little bit. In the village, [I practiced] my trade: (1) a smith for all types of agricultural implements and goods, (2) a mechanic for autos, clocks, typewriters, and handicraft wares, (3) a coppersmith, (4) a tinker. And I was able to be useful in social life because my skills were needed in peasant life. For three years I tried to create collectives, but not one peasant signed up. They were selfish. Everyone was called upon; I invited credit and land improvement societies, and even they did not join. But they envy those for whom I work and say [to me], "Your boss was born in a shirt."[41] I won popularity in a large circle of the population. For two years I was a member of the village soviet. My family consists of seven persons: four toilers and three young ones. I am looking for a collective farm so that they can use me, an honest, helpful administrator. Hire me; after all, I can be of use to all laborers. I give my word as a citizen and fighter that I will be faithful to the worker-peasant state unto the grave. Whoever is needed, answer the call, even in Siberia.

N. KUZNETSOV

The smychka of town and country. The legacy of Lenin.

I am sending a heartfelt greeting to the Red Army, which stands on guard for the whole world, for socialist society. Comrades, and my fellow old Red Army soldiers scattered throughout every corner of the worker-peasant Soviet republic, we protect the republic, we did not forget to hold onto our rifles, and if needed we will strike [our enemies] in the eye and brow.

NIKOLAI MIKHAILOVICH KUZNETSOV

———————————

The "face to the countryside" policy and the attempts to tie the city and the village more closely together by means of smychka did not, in the end, bridge the gulf separating proletarian from peasant Russia. The peasants, given their material deprivations and their indignation at their second-class status, could not accept at face value the Soviet state's claim that it was working to improve their lot. No amount of books, pamphlets, or agitators from the city could make up for the tax burden, the unfavorable terms of urban-rural trade, and the shortage of affordable goods. In turn, the hardships the working class experienced in the 1920s—rising unemployment, periodic food shortages, high prices, and the regime's frequent attribution of some of these problems to the peasants' undeveloped socialist consciousness shaped the workers' opinion that the peasants benefited most from the smychka. As a result, by the eve of mass collectivization many workers subscribed to the view that the road to rural socialism ran through the collective farm. In this sense, despite the genuine efforts during the NEP to realize Lenin's instruction for a mutually beneficial alliance between worker and peasant, the stage was already being set for the drama that would climax at the end of the decade in the city's assault on the village and the forcible socialization of agricultural production.

CHAPTER 4

Was Society Transformed?

What will the revolution offer that is new, how will it change the bestial Russian way of life, and will it bring much light into the darkness of the people's life?

—MAXIM GORKY, December 1917

The writer Maxim Gorky posed questions about the revolution—how it would change Russian lives and whether it would bring "light"—in his newspaper, *New Life,* shortly after the Bolshevik takeover, knowing full well that similar questions had been uppermost in the minds of idealistic Russian men and women for decades. Only if the revolution succeeded in liberating the people from ignorance and crushing material deprivation could those dedicated to revolutionary change justify to themselves and to posterity the suffering they were willing to endure and to inflict on others. Such validation could not be found in the simple act of overthrowing one ruling elite to replace it by another but instead required a higher, transcendent purpose. Although the Bolsheviks were hardheaded and ruthless practitioners of class warfare, they were also heirs to this tradition and rationalized their actions, even the most heinous, by invoking the good they believed they would ultimately bestow on the people.

The Bolsheviks were convinced that their revolution had, in fact, broken the shackles that prevented the mass of people from developing to their full human potential. For them, the events of October 1917 had cleared the way for the laboring classes to partake at last in the material and cultural advantages previously enjoyed by a tiny and privileged minority. The revolutionary expropriation of that minority in the name of the many had initiated the process. Henceforth, under the Communist Party's tutelage, the people's share in the nation's wealth and store of knowledge would steadily increase. Yet the devastation visited on the country between 1914 and 1921—coupled with its international isolation, archaic agricultural practices, and some of the Bolsheviks' own policies—placed severe restrictions on the scope for both material growth and the cultural melioration of the lower classes. Economic recovery not only demanded an inordinate share of the budget, crippling local administrators' ability to introduce or improve badly needed daily-life institutions, but also required enlisting the services of the remaining cohort of old-regime educated and technical personnel—members of the previously privileged minority—and granting them, in exchange, a favored position in the new social order in terms of salaries and other prerogatives. For these and other reasons, in the 1920s, the early optimism that the laboring classes would be lifted from the realm of darkness into that of light on the wings of the country's simultaneous material and cultural advances was put to a sobering test.

As Party spokesmen themselves admitted, wartime destruction, the postrevolutionary international blockade, and the civil war made immediate enactment of much of their program for social and cultural change impossible. With the end of hostilities, however, the ruling Party was seemingly poised to effect change in nearly all areas of public and private life. Even had the wartime devastation of Russia's infrastructure and productive capacities been much less absolute, however, the Bolsheviks still would have had to contend with the enormous obstacles presented by Russia's material poverty and entrenched traditions. In a pattern that repeated itself time and again, attempted change foundered on the shoals of scarcity, ignorance, cultural intransigence, red tape, and venality. For much of the 1920s, the Party's social engineers found themselves at war not so much with a class enemy as with Russia's destitution and Russian *byt*—that totality of social, economic, and cultural patterns that determined daily existence. To narrow the enormous gap between their revolutionary ambitions and Russian reality, the Bolsheviks would have to transform byt itself.[1]

In a series of articles written for *Pravda* over the course of 1923, Leon Trotsky mapped out the road the Bolsheviks would have to travel if they were to translate their social vision into social reality. In these, Trotsky repeatedly made the point that overcoming backward Russian habits, customs, oppressive familial relations, bureaucratic practices, and even vulgar speech depended on the socialist transformation of the economy. "It is all the more obvious," Trotsky wrote, "that a radical reconstruction of daily existence—the liberation of the women from the condition of a domestic slave, the social education of children, the freeing of marriage from elements of economic compulsion, etc.—are realizable only to the degree of social accumulation and of the increasing ascendancy of socialist economic forms over capitalist."[2]

By 1920, Lenin had come to see widespread electrification as key to this change. In that year, a two-hundred-man committee of specialists headed by the Old Bolshevik, and prerevolutionary-trained engineer, Gleb Krzhizhanovsky, began working out a plan for the construction of power stations throughout the country. A rationally planned economy was a deeply held Bolshevik goal, and the GOELRO plan represented the first serious effort at comprehensive state economic planning. At well over six hundred pages, the committee's report was an enormous document that encompassed a number of industries, including metallurgy, coal, fuel, and construction.[3] As a blueprint for the future, the plan also represented the practical, political, and symbolic ways technology would carry backward peasant Russia literally and figuratively out of darkness. Lenin expressed this line of thinking in December 1920 at the Eighth All-Russian Congress of Soviets. In addition to his famous slogan "Communism is Soviet Power plus the electrification of the entire country," he also declared,

> What we must now seek is to turn every electric power station we build into a base of enlightenment used to make the masses electricity-conscious, so to speak . . . It should, however, be realized and remembered that we cannot carry out electrification while we have illiteracy. Our commission will endeavor to stamp out illiteracy—but that is not enough . . . Besides literacy, we need cultured, enlightened, and educated working people; the majority of the peasants must be made fully aware of the tasks before us. This program of the Party must become a basic book to be used in every school. You will find, along with the general plan of electrification, special plans written for every district of Russia. Thus every comrade who goes to the provinces will have a definite scheme of electrification for his district, for the transition from darkness to a normal life . . . You must see to it that when the question "What is communism?" is asked in each school and in each study circle, the answer should not only be what is written in the Party program but should also say how we can emerge from the state of ignorance.[4]

The GOELRO plan projected a twentyfold increase in prewar generating capacity and the complete electrification of the country in ten to fifteen years. In the event, this estimate proved wildly optimistic. Not surprisingly, given industrial priorities and the large number of small settlements scattered across the huge country, the rate of rural electrification lagged far behind that of urban electrification.[5] But, as Lenin stated, the plan was more than just an economic blueprint; it would also serve as a mobilization tool whereby the population would be drawn into socialist construction. Far-flung factories and power stations were to become "bases of enlightenment" where workers and peasants received, in addition to instruction in communist fundamentals, education and training in science and modern production methods. In Lenin's vision, by implementing the GOELRO plan the Bolsheviks would carry the light of cultural revolution to the entire county.

GOELRO was not the only vehicle for enlightenment. Article 17 of the July 1918 RSFSR Constitution guaranteed free schooling to children of workers and poor peasants. Later that same year, the Commissariat of Enlightenment (Narkompros) expanded the promise of free elementary education to all children aged eight to seventeen. Narkompros officials, like A. V. Lunacharsky and Lenin's wife, N. K. Krupskaia, intended to provide more than just free tuition, however. They aspired to nothing less than the remaking of humanity in accordance with the ideals of classlessness and the collective. To accomplish this they set out to rebuild the entire educational system by replacing all existing schools with co-educational "unified labor schools" that employed progressive instructional methods intended to counteract the spread of "bourgeois egoism." According to their plan, the unified labor schools would form an integrated, compulsory, nine-year primary and secondary general education system. The so-called complex method favored by Narkompros eschewed instruction in traditional subjects in favor of systematically exposing pupils to the interrelationships that formed the material world of nature, society, and labor. To this end, practical activities, excursions, and workshops took the place of memorization, homework, and testing.[6]

In the conditions of 1920s Russia, the Narkompros program proved highly unrealistic. Not only were resources like schoolrooms, books, and paper in short supply, but provincial officials, teachers, parents, the Komsomol (Communist Youth League), and advocates of vocational training found the complex method incomprehensible, impossible to put into practice, too at odds with the traditional goals of education, or itself "bourgeois." In most places it was never employed. Rural parents, in particular, expected their children to acquire basic literacy and numeracy after one or two years of schooling, then be done with it; they certainly could not see the justification for nine years' attendance. Aggregate statistics for the 1920s are dismal. Children averaged only two and a half years schooling in 1925. Failure and dropout rates were alarmingly high. In rural areas only one-half of the children entering the first grade in 1924 reached the third grade.[7]

In addition to problems of attendance and continuance, the Narkompros prohibition against the teaching of traditional subject matter helped ensure that illiteracy and innumeracy remained serious problems throughout the decade. The census conducted at the close of 1926 found that one-half of children aged eight to eleven could not read or write. The figures were higher in rural areas. These numbers were all the more disturbing considering the Soviet government's determination "to liqui-date illiteracy." Lenin had famously said that the illiterate person "stands outside politics," and there is no doubt that the Bolsheviks saw worker and peasant illiteracy as an obstacle to their effective political indoctri-nation. But the implications of Lenin's observation are much broader. If Lenin truly believed—as he also said—that every woman who slaved in a kitchen should be freed to participate in state administration, then at minimum they had to be able to read, write, and do sums. In this sense, too, socialist construction depended on cultural advance.[8]

Until the second half of 1919, however, efforts to eradicate adult il-literacy made little progress outside the ranks of the Red Army, where they did achieve a good deal of success. By the end of 1920, the literacy movement began to make inroads in society as well. By that time, there were over 12,000 "liquidation points" (*likpunkty*) where nearly 280,000 individuals were gaining instruction. In a year, these numbers had in-creased to 37,163 and 854,746, respectively. Many of those involved in the movement were highly dedicated activists. Decades later, one peas-ant, S. L. Grachev, recalled how in winter, when peasant women tradi-tionally spent their time at home spinning wool, a young man, probably a civil war veteran, whom he identifies as "a hero of the Soviet Union," came to the village from a nearby town three times a week to teach the girls to read and write. The movement suffered a serious reversal with the introduction of the NEP, for central funding dried up and local bud-gets would not or could not take up the slack. In rural areas hit by the 1921 famine, the campaign virtually ceased. As of April 1923, only 3,649 liquidation points with a combined clientele of 104,341 remained in operation.[9]

The development of the Soviet health-care system also followed a pattern of soaring aspirations frustrated by cruel reality. To centralize health-care administration, the Council of People's Commissars created a Commissariat of Health (Narkomzdrav) in July 1918. Under Commis-sar N. A. Semashko, the Narkomzdrav leadership dedicated itself to es-tablishing a unified socialist system committed to supporting well-trained physicians, delivering free care to all, and preventing the spread of ill-ness by improving general hygiene and sanitary conditions. It also estab-lished special facilities to treat chronic "social diseases" like tuberculosis, venereal diseases, and alcoholism. During the civil war, the medical system was overwhelmed by military needs and the spread of epidemics like typhus and cholera. After the introduction of the NEP, as support for medical facilities devolved to overstrained local budgets, financial restraints prevented the implementation of the Narkomzdrav program, which opened the door to social inequalities in health-care delivery and quality. In the first half of 1922, the number of hospitals in the Russian republic actually decreased by 16 percent, and the number of beds by 29 percent. The situation improved together with the economy in the mid-1920s. The percentage of local budgets devoted to health care in these years rose steadily but still remained below that of prerevolutionary times.[10]

In addition to straitened circumstances and Russian underdevelopment, organizational weakness and labor shortages hindered the implementation of the Party's various social measures. The Bolshevik leadership recognized early on that communist hands were too few to effectively carry out the Party's program of change. This was particularly the case in the countryside. To supplement the Party vanguard, the Central Committee supported the formation of "social organizations" among specific groups. Workers, women, and the young were special targets of Bolshevik organizational efforts.

Trade unions provided ready-made channels through which Party activists could reach and mobilize workers. Other groups required new, Soviet forms of organization. The first congress of the Komsomol met in October 1918. Though formally an independent organization, it was closely related to and heavily dependent on the Communist Party. During the war-communism period, Komsomol members took a militant stance toward the peasantry, enthusiastically participating in grain-requisitioning brigades and antireligious activities. The Third Komsomol Congress in October 1920 abandoned the militant approach and decided to extend the league's reach into the village by organizing peasant youth, but for the time being the number of rural organizations remained small and their activities lax. This situation changed after the shift to the NEP. By the second half of 1922, two-thirds of Komsomol cells were located in rural areas, and peasants already made up half the organization's membership. While this growth greatly facilitated one of the Komsomol's chief tasks—spreading Party influence across the under-governed countryside—many Komsomol leaders, who tended to be urbanites, opposed what they saw as the organization's "peasantification." In the NEP-era countryside, Komsomol activists conducted political education among the newly recruited membership, worked to popularize modern agricultural techniques and collective farming, and tried to counteract the influence of traditional folk practices and religion.[11] Success in these endeavors often depended on the tactics employed. Antireligious activities in particular frequently resulted in direct and often nasty confrontations with the peasants.

To organize political work on issues relating to women, the Central Committee Secretariat established a department for work among women (*zhenotdel*) in September 1919. Utopians of the women's movement had theorized that once proletarian revolution brought about socioeconomic changes, a new morality would emerge. In combination, the changes would destroy the essentially proprietary arrangement that marked bourgeois male-female (as well as parental) relations and complete the destruction of the family already occurring under the pressures of capitalism. Zhenotdel leaders like Inessa Armand and Aleksandra Kollantai strove to create a "new woman" by means of sexual liberation and the abolition of the patriarchal family. They expected that women would then emerge as free individuals, truly equal to men and capable of participating in all areas of public life, including politics and production. In private life women would be free to pursue sexual relations in accordance with the tenets of free love. Additionally, by overturning Russia's highly patriarchal sexual order, the reformers hoped to eliminate a fundamental obstacle to class solidarity and socialism. As Kollantai put it, "In place of the individual and the egoistic family, a great universal family of workers will develop, in which all the workers, men and women, will above all be comrades."[12]

As in other spheres, the NEP forced a suspension of official sponsorship for utopian schemes to reconfigure sexual relations and the traditional family. According to the historian Barbara Evans Clements, the Party's new emphasis on stability and economic reconstruction resulted, in fact, in a practical demotion of women's issues to secondary status. By 1923, the Zhenotdel had muted the more radical visions propounded by theorists of sexual liberation, focusing its energies more on organizing women and the institutions that could alleviate the burdens daily life placed on them. To socialize the domestic tasks that traditionally fell on female shoulders, Zhenotdel activists promoted the creation of nurseries, kindergartens, laundries, and communal dining halls. Lacking sufficient state support for such projects, organizers sought to cultivate local initiatives and female self-reliance to achieve their ends.[13] In the village, certainly, efforts to reconfigure family relations and traditional practices faced an uphill slog. But as seen in several letters included below, the new morality did make inroads in the countryside, and the establishment of daily-life (*bytovye*) institutions like nurseries, when they were well and conscientiously run, could have a profound effect on the lives of peasant women.

Overall, the NEP marked a hiatus in militant and utopian activity to transform byt in a socialist direction. Given the strategy and mechanisms by which the NEP operated, this could not have been otherwise. Although official state efforts to transform society did not cease, the enormity of the task itself, popular exhaustion, limited finances, simple inertia, and a host of other reasons meant that it now took on a more modest and prosaic coloration. In the countryside, new ideas and institutions were having an impact, and according to the testimony below, cultural change was occurring, albeit unevenly and slowly. Continued progress along these lines, however, required what would increasingly become, by the second half of the 1920s, the scarcest commodity in Soviet Russia—patience.

Documents

In evaluating the changes that Soviet power was introducing, letter writers frequently employed the comparative device of discussing life as it was "then" under the tsars and "now" under the Bolsheviks. In the following letter, dated 10 November 1927, S. Ya. Kozlov of Kostroma province praises Soviet power for raising the cultural level of his village by introducing a reading hut.

· 52 ·

Letter to *Krestianskaia gazeta* from the peasant S. Ya. Kozlov describing how a reading hut transformed the life of his village, 10 November 1927. RGAE, f. 396, op. 5, d. 1, ll. 396–397. Original manuscript.

What caused ignorance to disappear in our village?

Far, far away from the heart of the Moscow republic, forgotten amid the thick forests and ravines, [stands] our poor, desolate village of Samylovo, Manturovo district. Officials from the district and the county happened upon her rarely, but the peasants took no offense at this: "So what if we don't hear from them! We live as our fathers and grandfathers did, and that's just fine." And it is true, our village of Samylovo was a most ignorant and backward little borough. In the village life went on as before; there was nothing new, good, or bright.

For three years the drunk Aleksandr Ivanov sat in the village soviet. Village affairs did not interest him. His only interest was drinking foul homebrew until he passed out. It is well known that such officials are completely good for nothing and are themselves accomplices of the moonshiners. It is rare for them not to have a still to turn out dizzying moonshine. Before, the young loafed around. In the evenings they roamed the village, engaged in hooliganism, and sang vulgar songs to the accompaniment of an accordion. On holidays, after morning church service, fights would break out in Samylovo. From the early morning, both the young and the old would fight—not out of anger but just because it was a holiday and there was nothing to do, so why not show off their strength? They beat the hell out of one another. For weeks after this "holiday amusement" the young would go about with bruises on their faces and black eyes. In the fields the Samylovites worked as in olden days. Scraggly [horses], straining, dragged the ancient wooden plow [*sokha*]. Dripping with sweat, the peasants cursed their unenviable life. And if some misfortune should befall the village, if a horse or a cow should fall ill, they would turn, not to a veterinarian, but to the priest. They would bring the good father eggs or even a hen and request that he sprinkle the animal with "holy water" on the chance that it would cure it. And the fat father would come, sprinkle the horse, suggest that the owner come to church more often and not forget God. But the father's "holy water" did not help. The cow would breathe its last, and bitterly the owners would mourn her, [their] livelihood. We had an old woman in the village—Kiriana the healer.[14] She cured all sicknesses and cast spells. She blows on water, spits all around herself, blesses the sick with the sign of the cross, and makes him drink: "Drink, cousin, and everything will be put right." Kiriana fed on the stupidity of Samylovo. She made fools of the honest folk. Before, neither newspapers nor journals made it to our village, and no one [could] read besides the shopkeeper, Fadiev. For the Samylovites, the only news and merriment was had in church. Before, when the kulak-merchant would arrive here, he would sing a medley of tales to our peasant men.

And our village of Samylovo would still live as in bygone days if not for the reading hut. And now the end has come to all that was before, and village life has become new, good. Instead of the old wooden plow, new iron plows from Briansk have taken their place. Each advance brings more. Our village has seen all types of people. The Party and Komsomol cell did not pass us by. It tried its best to tell us peasants how people in other places live, what Soviet power is doing for the peasants so that life will improve. We Samylovites listened to everything and mulled it over. Now the entire village gathers in the reading hut to listen to reports and talks. Every day we read all the newspapers. The description given above says that our village of Samylovo is no longer lagging in cultural relations. Who, apart from Soviet power, cares for us? It has organized the poor. It has tried to take [each] illiterate village "*baba*" [peasant woman] cook and turn her into a literate, conscious woman-delegate.[15] And our Samylovo youth have become aware politically, culturally, and economically and have been strengthened. And why do you think this has happened? Because we have all sorts of newspapers, and they are bringing all sorts of benefits to our village. If not for the reading [hut], the Samylovites might still be living as before.

KOZLOV, STEPAN YAKOVLEVICH

Kozlov portrays Samylovo before the introduction of the reading hut as a static backwater, a breeding ground for blind habit, boredom, alcoholism, brutality, and superstition—in short, the stereotypical Russian village languishing in darkness and ignorance. Written on the occasion of the October Revolution's ten-year anniversary (now celebrated in November), the letter's overall tone, and the writer's use of expressions like "kulak-merchant," show that some peasants understood how and when to employ the new "Soviet" language. Even allowing for exaggeration in the description of Samylovo's benighted condition and the quasi-official account of the village's transformation, the letter does indicate that relatively small innovations could have a profound effect in remote areas. In other locales, however, the impact of the reading hut was less dramatic. Many peasants found the huts cheerless and uninviting and organized their own meeting places. I. Vasiliev from Gdov county, Leningrad province, reported that "in the village of Spitsyno there is a reading hut. The peasants don't go there but prefer a different house, where they go every day. They call it 'The House of Peasant Thought.' They declare that here it's freer: we can swear at whom we like—that in the reading hut smoking is forbidden, and that the responsible officials snatch up all the newspapers. If you talk too much in the reading hut, then they quickly brand you a kulak." Reports from other nearby villages confirmed that the peasants avoided the reading huts and organized their own meeting places, which they called "peasant dumas."[16]

Historically, gathering places like taverns have served as arenas where subordinate groups could escape the strictures of the dominant culture and maintain or generate their own, potentially subversive countercultures. Avoiding the reading hut and congregating elsewhere is one conspicuous example of peasants rejecting a form of social organization imposed from above in favor of their own. However beneficial the reading huts, they remained state-controlled spheres. Free speech (swearing) and free thought—implying criticism of the state, its policies, and the official culture it fostered—could truly be practiced only in surroundings of the peasants' own making.[17]

Alcohol

Many letters touch on the problems associated with the Russian propensity for excessive drinking. As hostilities with Germany approached in 1914, Nicholas II's government banned the production and sale of spirits. Predictably, the "dry law" served to stimulate the proliferation of illegal distillation. By the end of 1919, the new Soviet government, following its predecessor's lead, also instituted what amounted to total prohibition. During the war-communism years, peasants turned to home brewing as an effective and profitable means of keeping grain out of the hands of state requisitioners. Efforts to enforce the ban following the civil war led to massive police crackdowns on the illicit distilling of *samogon* or *khanzha* (homebrew). Shortly thereafter, however, recognizing the limited effect of prohibition, and unable to resist the revenue potential of spirits, the government gradually reinstituted the tsarist-era state monopoly on vodka over the objections of its own health officials and portions of the Party Central Committee. By August 1925, just as before the war, vodka at 40 percent alcohol content was once again legally available.[18]

Even with the lifting of prohibition, samogon production for the purposes of sale and personal consumption continued to flourish in both the cities and the countryside. The fact that legal spirits rarely reached the village and, when they did, cost much more than homebrew made illegal distilling very attractive to the peasants. Profit from sale was another attraction. It took approximately two puds, or nearly thirty-three kilograms of flour to make a bucket of samogon, which could then be sold for about seventeen rubles. By contrast, in its nonliquid state, a pud of flour fetched only one ruble, thirty-four kopeks. The financial benefits that could be derived from bootlegging were glaringly obvious. As one peasant from Penza province in the Middle Volga region observed, "You can brew three or four times, as you like, and then, maybe, you can buy a horse."[19]

Contrary to Party dogma, which placed blame for the practice on commercialistic kulaks and merchants (the NEP bourgeoisie), bootlegging was largely conducted by the urban and rural poor; in cities large numbers of poor women could be found operating stills. According to a 1928 survey in the RSFSR, 35 percent of all peasant households practiced moonshining (probably an underestimate). Drinking, on the other hand, knew no social boundaries. It seemed that everyone imbibed—the old and the young, men and women, communists and noncommunists, government officials and average citizens. One study conducted in a Moscow elementary school in 1927 found that 10 percent of the students were daily drinkers, and only 11 percent never drank. By the end of the decade, sales of alcoholic beverages generated 12 percent of state revenues. At a village wedding in 1929, Maurice Hindus observed that nearly all the guests openly brandished bottles of samogon despite its contraband status. The apparent increase in spirit consumption over the course of the 1920s raised concerns about public health, worker productivity, social order, and, perhaps most important, the diversion of precious grain from food reserves to the still. During the grain procurement crisis in January 1928, the authorities initiated a campaign against moonshiners that produced over 200,000 arrests in its first four months.[20]

The state's dual (some might say hypocritical) approach of profiting from the alcohol monopoly, on the one hand, while demonizing samogon and arresting moonshiners, on the other, did not escape popular commentary. A discussion on this theme is found in F. I. Privalov's letter from March 1925. Privalov lived in Dmitrovskoe village, Akmolinsk province, in the Kazakh republic.

· 53 ·

Letter to *Krestianskaia gazeta* from the peasant F. I. Privalov on the harmful effects of alcohol consumption, 7 March 1925. RGAE, f. 396, op. 3, d. 83, ll. 5–6. Original manuscript.

The struggle against moonshining.

In recent times, especially in 1925, moonshining [*khonzhovarenie*] has greatly increased among the peasantry. In this arena, the district executive committee, the district Party committee, the militia, the village soviets, and the troikas are engaged in a vigorous struggle against the same.[21] Citizens who are caught are fined and brought to trial. Citizens on trial are sentenced to hard labor in addition to their fines. Now, if we approach [this problem] from a broad-minded point of view, [can we say] this is how we should fight it? Can we destroy this evil by these means? Of course not. Why? This is why. The Russian people are infected with hereditary alcoholism. This is the first thing. Second, no matter how they are punished or treated, they are calm [i.e., sober] only when shaken every which way. Given any leeway, their infected organism starts to plague them, and they again get down to work [i.e., to drink]. In this regard, [they are] acting unconsciously, ignoring the fact that alcohol is bad for a person, [that it] affects the brain and saps [one's] strength, and that ultimately [the individual] becomes moronic and even loses his mind or, under strong abuse, dies. Now this raises the question What should be done? How can alcohol be destroyed? This is how. Comrade Lenin laid three tasks before us: first, study; second, study; and third, study. Only with the help of science can we eradicate this evil, only if every one of us understands that the alcoholic is always sick, less intelligent, that his children are cripples, weaklings, dull-witted, and useless in our construction [work]. To do this it is necessary to attract the attention of our center and to attract medical officials for the study of the evils of alcohol. Very well, we will get down to work and begin to study. The question immediately arises: "But why does the state allow vodka [*russkaia gor'kaia*] if they are definitely aware that it is bad for all the people of our Russia? Then they shouldn't allow it." We answer: "In the first place, the state competes with moonshining. In the second place, it [liquor production] supplements the budget." They answer us: "True, for supplementing the budget it is a very good thing." Besides which, among those who value it [liquor], the authority of Soviet power is greatly enhanced. I will address all the questions and answers expressed above. Of course, it appears to us that we are receiving great benefit from it. We compete [with moonshiners] and raise the prestige of Soviet power [among the people]. No comrades! We take in thousands from it as profit for the state, but we lose the minds of millions of our people. Because of the spread of alcohol, the village has not and will not flourish. Everyone understands this. As long as the village [is trapped] in ignorance, our state construction will proceed poorly. For example, there are people, even geniuses, who would be of enormous help to our construction [work], but they flood their minds, their talent, with wine and vanish into the depths of the dark masses. He has not studied and does not wish to study because he has a special profession— alcohol. If this is so, slavery will rule on this soil. The kulaks will oppress the poor, and they [the poor] will be forced to drown all [their] grief, all their hardship, in wine. Here is the situation regarding competition: one bottle of vodka sells for one ruble, seventy-five kopeks; a pud of grain—one ruble, twenty kopeks, and from a pud you get ten bottles of hooch of about 35 percent alcohol. At this rate, competition is not feasible. There is no advantage [for the state] in gaining the respect of those who value this [moonshine]. You need the respect of independent peasants and workers, scholars, writers, poets, and artists of the whole world. Only then will we rise on a militant leg and advance to socialism. I have already said that for a few thousand rubles we are losing the minds of millions of our people. Now they ask me the question: "Would this mind be beneficial for the state?" Of course, everyone understands this. One mind is good, but two are even better. If all of us were educated, then our state would not need to stop before anything. It would develop industry, transport, technology, agriculture, etc. Then we will receive billions [of rubles] from free, healthy labor and not these miserable thousands that we take in from vodka and that obstruct a theory of humanity. Our center needs to think this problem through and to lead our citizens living in the USSR to the correct and vital spot. Hopefully, Comrade Kalinin will address my point of view.

F. I. PRIVALOV

Just because the state tried so hard to control the sale of spirits, tales of inebriated officials were, no doubt, especially satisfying for the citizenry, and even more so when they illuminated corrupt behavior. One such account describes how a village-soviet chairman, a certain Zorin, took a large sum of public money to the county center, where he proceeded to drink most of it away. Zorin's coachman also got so drunk during the binge that he somehow lost the chairman later on the road home. The fiasco was complete when a peasant from Zorin's own village found him stumbling along the road and took what remained of the funds for safekeeping.[22]

Peasants and petty officials were not the only ones given to outrageous behavior under the influence of alcohol. In such cases, otherwise suppressed hostility to Soviet power—its representatives and its symbols—sometimes broke to the surface. The following account from a 1927 police protocol describes the misadventures of two drunken Red Army soldiers on the outskirts of Moscow.

· 54 ·

Report from the Moscow military procurator's investigation into the drunken behavior of two Red Army soldiers, 1927. TsGAOD g. Moskvy, f. 3, op. 11, d. 522, l. 28. Original manuscript.

From an inquest of the Moscow Military Prosecutor into the case of drunkenness, debauchery, and hooliganism by Red Army soldiers of the fourteenth squad of the rifle platoon, Antipin, Vasily, and Ovchinnikov, Ivan.

It is established that the aforementioned Red Army soldiers were extremely drunk on "Defense of the Country" Day and engaged in a whole series of hooligan outbursts in the village of Bochmanovo, such as:

Being drunk, they emptied their bladders on the front steps of one of the houses; they took the chairman of the village soviet by the scruff of the neck with the intention of striking him; they tore (slightly) the shirt of the chairman's brother (a Party member since 1920), whom they also planned on hitting. They roughed up the guard at the railroad crossing whose shirt they tore because he had not given them milk (he did not even have a cow). One of them, by the name of Ovchinnikov, tore his own collar tab and ripped off the Red Army star, trampling it in the dirt. Further, they accosted passing women and girls, and, employing every possible indecent word, cursed both the Red Army and the tribunal. Even though the village soviet chairman, his brother, and one Klimonov, who was passing by and who had been drinking together with them, continually begged them not to behave disgracefully, they listened to no one and continued on their way, attracting a large crowd of people (no fewer than eighty individuals).

Ovchinnikov and Antipin denied everything enumerated above except the fact that they had been drunk.

The conduct of a fuller inquiry is rendered impossible because no one identifies in their depositions exactly whom they injured, which women they accosted, and what they tore from the clothing.

Chief Clerk POPUTCHIKOV

Though supported by the Politburo (Trotsky, who suggested that the cinema would serve as an effective substitute for both alcohol and religion in the daily life of the masses, being a notable exception), the vodka monopoly had its detractors among the Party rank and file. In November 1926, the OGPU recorded the following comments during a discussion devoted to the struggle with drunkenness at an open Party meeting in the Moscow district of Rogozhsko-Simonovsky, the site of the "Sickle and Hammer" metallurgical plant:

Why not drink if it benefits the state?

Soviet power, like the tsar, has a drunken budget.

Soviet power tries to get the workers drunk so that they'll look into soviet shortcomings less, and, second, it [vodka] is a very profitable item.

Hasn't Soviet power committed crimes against the revolution and the working class by allowing 80-proof vodka, since the worker, instead of gathering in the club or in school for political literacy, goes to buy vodka, collapses, and then doesn't give a thought to the revolution and the productivity of labor?

More clinical was the question posed by a man who may not have been unfamiliar with the bottle.

Is it true that the corpse of a drunk doesn't decompose for a long time after death?[23]

Education

Like middle-class reformers, Party members not only wanted to promote sobriety among the laboring classes but open the doors of schools to them. In the decades leading up to World War I, through a combination of local efforts and state support, great progress had been made in extending primary education to the Russian population. By 1914, approximately one-half of the school-age population (eight to eleven years old) living in the European parts of the empire were attending schools. In some provinces the figures were much higher. The world war and the civil war interrupted progress. Narkompros took up the task of extending primary and secondary education to the entire school-age population by means of the nine-year compulsory unified labor school. The obstacles were great, and to the detriment of the primary system the Narkompros program did not succeed. Figures from the 1927–1928 academic year show that, as in 1914, one-half of the eight to eleven age-group were still not attending school.[24]

More success was achieved at the secondary-school level. During the 1920s, interested parties like the Enlightenment Commissariat, the Komsomol, the industrial leadership, and the trade unions debated the merits of general, polytechnic, and vocational education and put into operation various forms of popular schooling, including factory vocational schools (FZU) and worker schools, or "faculties" (*rabfaks*) to encourage worker study. Although the full benefits of these efforts would only come after the five-year plan was in place, already in the 1920s the expansion and democratization of adult education offered the possibility of social advancement to those who previously would never seriously have thought of improving their station in life. Despite serious obstacles, by 1927/1928, student enrollment in all grade levels and institutions had increased—in some cases dramatically—over that of the tsarist years. At this time, nearly fifty thousand workers were enrolled in the ten-year-old rabfaks while almost a quarter-million attended some form of trade school.[25]

The 1918 RSFSR Constitution had mandated free schooling for worker and poor-peasant children. The transfer of education costs to local budgets during the early NEP years, however, forced local officials to introduce fees for school attendance. Until payments for primary education were abolished in 1926 and 1927, the fees placed a great burden on poor families and undermined the egalitarian promise of Soviet education.[26] Understandably, supporters of the revolution denounced the inequalities of the payment system, as in this letter from Ural province.

· 55 ·

Letter to *Krestianskaia gazeta* from the Red Army veteran and day laborer Izersky
on the high cost of education, 10 February 1924. RGAE, f. 396, op. 2, d. 16, l. 112.
Original manuscript.

I am begging the *Krestianskaia gazeta* editorial board to explain to me
what to do, and where to find the truth, and who to turn to.

My brother resided in school commune No. 1 in Yelovo district, Sarapul region, [Ural province], but for some reason they dismissed him, saying that he has a mother and me, his brother. But our mother is in no
condition to work. She is fifty-five and still has a ten-year old sister. We
have no permanent residence, and our situation is extremely poor. I work
as a day laborer, one day after another as a peasant, but my health is bad. I
damaged it during Red Army service. I am unable to teach and support my
brother, but my brother is twelve and a half years old and is in the fourth
group. And now when he needs to continue his schooling, it is forbidden.
But to keep him in school we have to pay four gold rubles a month. Where
can we get it when we are hardly getting by? It turns out that some, it
seems, have capital and will be educated, but our brother, as before, is one
of the poor living in untutored ignorance. I was in the Red Army. I served
and defended the soviets. And when I returned from the Red Army, I
thought to get my brother some learning so that he would not turn out a
fool but be literate and understand everything. But they are asking for
money. Where can it be gotten? There are those who have their own
households, and they live in the school-commune and do not pay—why
don't they dismiss them? Whose interests is the worker-peasant authority
defending when education is once again forbidden for poor orphans? Whoever has capital may study, but my brother, who was thrown overboard,
has to go and get [money] since my mother and I cannot support him.
Summer and fall he lived in the school-commune and worked, but in the
winter [he was] on holiday. Where can one get justice? Is it possible there
isn't any?

I ask the editorial board not to refuse [me] and to print my letter in the
newspaper since it defends us day-laborers.

Without an address, the day laborer IZERSKY.

In the mid-1920s, children from worker and peasant backgrounds
rarely advanced beyond the fourth grade, so the upper levels were dominated by pupils from professional, intelligentsia, and white-collar families. In response to this situation, militant proponents of cultural revolution
in the education system began purging middle-class children from schools

despite the opposition of Narkompros leaders like Lunacharsky and Krupskaia.[27]

The conclusion of the civil war and reduction in the size of the standing army resulted in large-scale demobilizations between 1921 and 1925. Having fought for the Red cause, millions of veterans now sought roles in the peaceful reconstruction of the country. Many returned to their villages to take up the plow or to serve in their local soviets. Others abandoned their previous pursuits and domiciles and stepped into vacancies in the growing Party and state bureaucracies.[28] A good number, however, who found returning home unattractive were, to their dismay, unqualified for bureaucratic work. Here, for example, is a typical document of the period. The short letter contains numerous spelling and grammatical errors.

· 56 ·

A declaration from the Red Army soldier and Komsomol member N. I. Orlovsky to the Communist Party cell bureau of the Twenty-Seventh Omsk Rifle Division requesting assistance in gaining admittance to the Higher Party School, 6 March 1925. RGAE, f. 2097, op. 5, d. 550, l. 177. Original manuscript.

A declaration to the cell bureau of the independent cavalry squadron of the Twenty-Seventh Omsk Rifle Division, named for the Italian proletariat, from Red Army soldier [and] member of the Komsomol, N. I. Orlovsky.

I request your intercession if it is possible to assign me to the school H[igher] P[arty] S[chool][29] as it is my desire to study political instruction; if this is not possible, then I request [to be] sent to the factory of our patronage in the city of Moscow since I cannot live on my small and poor farm and need to seek aid for my daily subsistence, or, in the extreme case, [if] it is not possible [to] send me anywhere, then I ask [to] stay [and] serve in the ranks of the Red Army together with the squadron commander. Since I cannot petition for nothing more, I request that the cell bureau give attention to my stated request since I think that it may help me settle somewhere.

Komsomol member, NIKOLAI IVANOVICH ORLOVSKY
6 March 1925

The state created the rabfaky to prepare workers and Communist Party members whose schooling had been deficient to enter institutions of higher education (VUZ). The rabfaks helped satisfy working-class demands for access to education and prepared politically committed citizens to participate more fully in the making of the new society.[30] A large number of correspondents who regularly wrote to newspapers and journals were, in fact, rabfak students. In 1925, P. Kandakov, for example, spent his summer holiday in Korliaka, a village in Viatka province. The following document presents a few excerpts from the lengthy account he wrote for *Krestianskaia gazeta* under the title "What I Saw and Heard in the Village."

· 57 ·

Excerpts from the observations on village life written for *Krestianskaia gazeta* by the rabfak student P. Kandakov, summer 1925. RGAE, f. 396, op. 3, d. 212, ll. 145–146. Original manuscript.

On the Village Soviet

I attended a session of the village soviet. All the members are beardless youths. It is difficult [for them] to come to grips with the situations and relationships [within the village] and even more difficult to work in such an unfamiliar manner, but stubbornly, overcoming the difficulties, the work, slowly but surely moves forward. When a session is interrupted by a peasant who comes to get some information, they don't get flustered—they give it [to him] and continue on.

On Samogon and Hooliganism

The situation regarding samogon is not good. "For us, this is our sole diversion and pleasure," the drunken peasants say. "Everything is broken, broken, and we need somewhere to relax and have fun. On a holiday we'll drink, forget our weariness, and cheer the soul. But state hooch is weak, and there is profit in samogon." There are dozens of stills in the village. To put the whole contraption together, the village smith gets six puds [of grain]. This really jacks up the output [of stills]. On a holiday the samogon flows like a river. As a result, you won't see an outing or gathering of the young without brawling and fistfighting. The two Vedernikovsky brothers were shot, Tarasovsky was knifed, Zverevsky was murdered, etc. Survivals of the old days have not yet been eliminated and are resistant to change. Stricter administrative-legal measures are needed to deal with hooliganism and brawling. Suspended sentences, deprivation of voting rights, and the like don't deter crime.

On Cultural Work

To combat ignorance and its allies—drunkenness and hooliganism and other shortcomings—the [politically] conscious ones from among the young together with the [Party] cell set up reading huts and organize cultural-enlightenment circles in them. Kulaks and moldy-minded old folks obstruct the work of the circles by any and all means. They influence parents, and then the latter, the children. "The children are being corrupted with this communism. Good people go to God's cathedral, but your little bastards serve Antichrist. On Christ's day they put on shows! Nice things they're learning! He-he-he . . . !" This often breaks up our cultural-enlightenment circle, but enlightenment is needed. There are illiterates in every house of our village. There are thirty illiterates between the ages of twelve and thirty-five. The entire population is about two hundred. The situation is no better in the neighboring villages. The rural cell of the Society "Down with Illiteracy!" is planning a number of "liquidation points," and they soon will begin to work.

P. KANDAKOV

The Youth

From the earliest days of the NEP, young peasants had joined the Komsomol in large numbers. As the Party adopted more openly pro-peasant policies in the mid-1920s, these numbers increased; peasants made up nearly 60 percent of the total Komsomol membership by early 1926. For peasant youth, membership offered a means of escaping the patriarchal confines of the traditional family and gaining some degree of independence. Participation provided an entrée into official Soviet society as well; a successful Komsomol stint could help erase the stigma of peasant birth and open the way to Party membership. Nevertheless, even though the great majority of peasant youth remained outside the Komsomol ranks, the peasant influx was not without its critics among the central Komsomol leaders, who fought to preserve the league as a proletarian organization and questioned whether peasants—especially middle peasants—could truly overcome their petty-bourgeois mindset and loyalties.[31]

Among the letters preserved in the archival files are numerous declarations from youths seeking admittance to the Party or Komsomol and complaints and appeals in regard to exclusion or removal from one organization or another. In a large number of cases, decisions in these matters turned on an individual's "social origins." The importance that social background and personal history now assumed in Soviet life frequently induced individuals to engage in the creation of a politically acceptable identity for themselves, an exercise that often necessitated concealing or rationalizing facts about family or self that might have raised official suspicion or stricture.[32] The creation of organizations

for children and teenagers helped to inculcate this biographical awareness in the young as well as adults. A good example is this letter to Stalin, dated 17 July 1926, from the Young Pioneer, Ilya Tarlinsky, who
lived near Lake Baikal in the far-off Siberian city Verkhne-Udinsk. True
to his purpose, the clever fifteen-year-old addresses the general secretary
familiarly, as a close friend or relation, employing throughout the letter
the second-person singular pronoun *ty* rather than the more formal *Vy*.

· 58 ·

Letter to I. V. Stalin from the Young Pioneer Ilya Tarlinsky requesting assistance
in gaining admittance to the Komsomol, 17 July 1926. RGASPI, f. 17, op. 85,
d. 486, l. 103. From a typewritten copy.

Beloved Leader
   I am a pioneer from the Twenty-Fifth, the Sun Yat-sen brigade in the city
of Verkhne-Udinsk. I am turning to you for help. I know that you have
practically no free time, and every minute is dear to you since it is used to
build a new life that will liberate the working people of the Soviet republic,
but perhaps you could find a few minutes to read my letter and answer it. I
trust in you not as someone unreachable, high and great, but as my teacher
and older brother, even a father. At the present time I want to enter the
ranks of the Komsomol, but the problem is that my recommendations are
inadequate. At one time, my father had the misfortune of being a merchant,
and the high barriers and obstacles that must be either destroyed or jumped
over block my path into the Komsomol. I am speaking directly and request
your recommendation and support. You do not know me or my family, so I
will write my biography or, more correctly, [write] of my parents and, understandably, [write] a few words about myself in the manner of a Pioneer—
directly, openly, concealing nothing.
   My origins are as follows: My granddad, with whom I am currently living, was a peasant from Petrovskoe village, Verkhne-Udinsk county, Irkutsk
province, but one year, after the crops failed, he moved his family, including
my father, to Khilok *stanitsa* [large Cossack village] and opened a shop.
Everything went along quietly until 1905. When this year, one of the most
famous in the history of the Russian Revolution, arrived, my granddad and
father took part in the struggle enthusiastically. With the onset of the reaction, General Renenkampf[33] arrived in Khilok at the head of a punitive ex-
pedition. Grandfather was arrested and after a while was sentenced to be
shot, but thanks to General Renenkampf's mistake, he was dispatched to
serve a sentence in the Aleksandrovsky ravelin until the reactionary out-
burst died down and all the punitive brigades left for Siberia. My granddad

did not belong to any party and did not understand the difference between Socialist-Revolutionaries, Social Democrats, etc.[34] As they say, he was against the tsars, the gendarmes, and the bourgeoisie. When my granddad was arrested, my father ran off and hid in the Buriat settlements and was finally exiled to Verkhne-Udinsk as an undesirable element under the secret supervision of the police. At the time of the 1917 revolution, my father and his family, including me, lived in Troitskoslavsk, where he worked for the trading firm, The Russian-Asian Association. But when the Red forces occupied Verkhne-Udinsk, he went there and, unable to find work, began to trade—more accurately, he began to help grandfather. After one and a half years, he found work in Amur, where he works now, sending me money here in Verkhne-Udinsk. Now, none of my close relations engage in trade, but almost all are employed, and all the misfortune is a result of the fact that father traded from 1902 to 1905, from 1911 to 1915, and from 1921 to 1922 and that it is impossible to prove his participation in the 1905 Revolution. I joined a pioneer organization a year and a half ago, and you will see my work from the copy of my application for entrance into the Komsomol, which the brigade gave me and which I am including in this letter. Now my dream is to join the Komsomol, and all my thoughts are focused on this. I have one recommendation from the chairman of the district children's bureau of Verkhne-Udinsk and the testimonial of the brigade. Two recommendations are still needed, and I am wearing myself out and all for naught since I have no acquaintances in the Party and the Komsomolists are all young, about twenty-four or twenty-five. This is unsatisfactory to me, and so I decided to turn to you as general secretary of the Central Committee of the VKP(b) [All-Union Communist Party] and as the beloved leader selflessly devoted to the cause of the revolution, responsive to human misfortune, and as an Old Bolshevik, comrade-in-arms of the great leader and teacher V. I. Lenin. If you put your trust in me, I vow not to shake it, and will not present the recommendation of the oldest of Bolsheviks, Iosif Vissarionovich Stalin, to the bureau of cell 18 of Narkompros in vain.

Now about myself. I am fifteen and a half years old. This year I completed the third class, second level. I have quite good political preparation and am reading the fiction of [Upton] Sinclair, [Jack] London, and our contemporary writers Seifullin, Neverov, Bibek, Serafimovich, and others, as well as the classics of Russian and foreign literature. I have read the political works of Lenin, Zinoviev, Trotsky, Yaroslavsky, and your own [works], etc. Now I am done and impatiently await your answer.

Always prepared, with a Pioneer greeting[35]

ILYA TARLINSKY

Address: Verkhne-Udinsk, Buriatskaia ul., Building 19, Ilya Tarlinsky

Whether Stalin read and acted on this letter (or was impressed by Tarlinsky's familiarity with Zinoviev's and Trotsky's writings) is unknown. Here, young Tarlinsky hopes to erase the stain of his father's evidently long-term commercial activities with vague and unsubstantiated claims that his father and grandfather were revolutionaries in 1905. Notably, he makes no such claims for 1917. Recognizing the weakness of his revolutionary and working-class heritage, he turns to flattering the general secretary. His ingratiating comments regarding Stalin's Party stature—Lenin's "comrade-in-arms" (*soratnik*) and "oldest of Bolsheviks" (*stareishii iz bolshevikov*)—indicate that the encomiums to Stalin that would become ubiquitous after 1929 were known even among the provincial population well before the Stalin cult fully flowered.

During the 1920s, the Komsomol set out to inculcate its members—new recruits especially—with identifiably "communist" or "proletarian" values. Just how well the Komsomol served as a bearer of the new morality and the new political consciousness may be called into question, however. According to a 1924 report on Komsomol activities in one Smolensk province, investigators charged that "cases of drunkenness, gambling, deviations, careerism, the persistence of religious prejudices, and other negative features cannot be counted as isolated incidents." Among the unspecified "negative features" mentioned in the report, licentiousness and sexual assault must have been foremost.[36] The incident described in the following letter from P. T. Zaitsev of the village of Zaitsevo in Bashkir republic was probably not an isolated one.

· *59* ·

An anonymous letter to *Krestianskaia gazeta* from a peasant describing a rape committed by Komsomol members, 27 April 1925. RGAE, f. 396, op. 3, d. 139, l. 201. Original manuscript.

An incident in Nikolsk

The Nikolsko-Igrovsky Komsomol did not carry out enlightenment work apart from some pitiful theatrical performances that it staged in the Nikolsk settlement [*pochinok*] assembly hall, which is twenty square meters. [These performances] did not attract the village youth. Besides this, during Easter week a young girl was raped and even beat up a little in Nikolsk settlement. The assault happened in the assembly hall, and what is more, the Komsomolist Mitrokhin, Petr, had a hand in it. Some [who took part in the assault] were a little [word not legible] in the head. Altogether there were seven men. On the morning of the second day, at a general assembly of the Nikolsk settlement citizenry, the citizens just teased her [when] the victim told of the incident. They found it amusing because she cried, and to them it was funny. The Komsomol seems to consider this affair quite legal because it paid no attention to it. The poor thing was forced to leave the assembly with tears in her eyes. Nikolsk needs to be thoroughly purged so that the rapists and homebrew drinkers will be removed from the Komsomol without a trace.

A stranger.

Frustrated by its inability to affect youth behavior in the desired direction, in 1926 the Komsomol shifted its energies away from the fight against bourgeois habits and concentrated instead on improving labor discipline and raising productivity.[37]

Just as before the revolution, evening parties (*posidelki* and *vechorki*) provided peasant teens with the opportunity for unsupervised socializing. In addition to eating, dancing, and singing (often suggestively), these were occasions for heavy drinking, playing kissing games, and, ultimately, having sex. Not infrequently, under the influence of alcohol and jealousy, the parties ended violently.[38] A. Lobanov, from Gomel province, Belorussian republic, describes a party in a letter written in February 1925.

· 60 ·

Letter to *Krestianskaia gazeta* from the peasant A. Lobanov describing the evening activities of teenagers in his village, 18 February 1925. RGAE, f. 396, op. 3, d. 234, l. 33. Original manuscript.

Village Youth

In the village of Mikhnovichi, Mozyr region, lives a poor peasant of proletarian origin—Veremeev, Andrei. In the evenings, girls gather at his place as a diversion from their work. They amuse themselves, sing proletarian songs—apparently they are [politically] conscious girls. Non-Party bachelors also frequent the place. They consider themselves big, handsome, and very intelligent cavaliers: they are Lobanov, Hariton, and Veremeev, Sergei. In fact, they are ignorant guys, and the less said about them the better. In trying to show off to the girls they, on the contrary, repel them. The first is even too young for the draft and is still going to school. He always puts out their lamp, and in the darkened hut he pushes the girls around while the other one goes to the stove, takes the soot from the coal, and smears it on the girls so that they don't recognize each other, and if one [of the girls] should protest, then he pushes her so that she cannot tell what side it is coming from. That is why I am requesting that *Krestianskaia gazeta* expose such citizens and in its conclusion give them a strict reprimand, since it is impossible to do anything with them. They do not listen to reason.

Village correspondent, A. LOBANOV

Evening parties helped young people pass the time in the village, especially during the long and cold winter months. Often they were only attended by young girls who played games, sang songs, told fortunes, and discussed their fiancés. The village correspondent Mikhail Grigorievich Kuptsov from Zadvorka village, Nizhny Novgorod province, reported that in Zadvorka during Christmas tide, boys and girls dressed up in different costumes. Costumes of "baronesses" and the dead were especially popular. They often went to neighboring villages and attended parties there. According to Kuptsov, they were uninterested in studying politics or acquiring knowledge, saying that it is better "to spin yarn." "What kind of yarn is it?" the indignant correspondent asks, "when first one, then another of the girls ends up pregnant?"[39]

In the 1920s, the Komsomol attempted to supplant peasant religious and folk beliefs by disseminating scientific findings and rational explanations for natural phenomena. However, some peasants found the new information as difficult to accept as the old legends. One peasant fell down laughing at an item in *Krestianskaia gazeta* that insisted that human beings had descended from apes. Before, he said, in the same way, they used to write that the earth rested on three whales.[40] Ye. A. Koldoshova from Yermolovo village in Riazan province disapprovingly discusses one common folk practice in her March 1926 letter.

· 61 ·

Letter to *Krestianskaia gazeta* from the peasant Ye. A. Koldoshova describing fortune-telling as practiced by young girls, 31 March 1926. RGAE, f. 396, op. 4, d. 43, l. 45. Original manuscript.

The Old Way

In our village of Yermolovo, Pobedinskaia district, Skopin postal district and county, Riazan province, on New Year's Eve, 1926, the peasant girls engage in fortune-telling. Each girl pours out some water, takes a wedding ring, and drips candlewax into the ring. When she has put in three drops, she looks to see her future—a crown [symbolizing a wedding] or a coffin. Here is what two girls foretold. They took some holy water, poured it into a tea glass, broke an egg into the water, and covered the glass with a towel that had never been used before. After fifteen minutes, they took a look. One girl saw something like a church, the other a battle and several columns. This is how our peasants occupy themselves.

YE. A. KOLDOSHOVA

Women

In the hopes of breaking their attachment to the "old ways" of life, Party organizations encouraged rural women to take a more active role in political life. Peasant women like P. Ya. Novikova, who lived in the Kuznetsk coal basin, responded to the new opportunities. She sent the following declaration to *Krestianskaia gazeta* to celebrate her liberation from her husband (whom she denounces as a kulak) and to petition for Communist Party membership.

· 62 ·

A declaration to *Krestianskaia gazeta* from the peasant P. Ya. Novikova on her intention to join the Communist Party, 28 October 1924. RGAE, f. 396, op. 2, d. 26, l. 53. From a typewritten copy.

---

28 October 1924

To the Prudkovsky communist cell of the Russian Communist Party
From citizen Pelagea Yakovlevna Novikova
Shcheglovsky county, Verkhtomsky district
Age, thirty-two years. Divorced.

A Declaration

I am expressing my desire to the Prudkovsky communist cell to join the Russian Communist Party. I consider myself fully prepared for admission, prepared in every sense. I have an average education, although my political [education] is weak. I have no household and therefore am a fully proletarian woman. I have no criminal record. My decision to join the Party is an expression of gratitude to our great leader Vladimir Ilich Lenin, from whom I received complete freedom.

Having lived under the heavy weight of marriage for ten years, I could find no way out, having feared sin all my life, and thinking it [divorce] was inadmissible. But thanks to the decree published by dear Ilich, a woman is able to escape from kulak husbands and to live freely, and I firmly decided to offer my whole life in defense of women and the people and, with open arms, proudly move toward the proletariat to fulfill the request of my dear Vladimir Ilich because his name will not die in my heart, and in his name I offer my life to defend women to the end. That is why I ask the Prudkovsky Party cell to send me a questionnaire to fill out, and that is why I am signing with my own hand: P. YA. NOVIKOVA

I am enclosing my address: Taiga district, Borisov village soviet, Lanskoy ward [*uchastok*].

I ask you not to delay.

I enter the party under the slogan "Long live Vladimir Ilich's decree liberating women."

––––––––––––––

The Bolsheviks proclaimed that their revolution would end the subservience imposed on women by capitalist social and family relations and raise them to truly equal status with men. Utopian Bolshevik theorists anticipated that during the advance to communism the family would disappear, domestic chores would be socialized, and the state would assume the responsibility for child rearing. The highly progressive Code on Marriage, the Family, and Guardianship, adopted in October 1918, addressed the needs of women during the "transitional" stage to communism by extending individual rights to women. It established gender equality, sanctioned civil marriage, and introduced on-demand divorce at the request of either spouse. The code provided Soviet Russia with some of the most advanced legislation on the "woman question" of any country in the world. Yet, in its spirit and particulars, the code clashed with the religious and traditional underpinnings of Russian society. In the countryside, customary practices and economic underdevelopment prevented women from fully realizing the rights granted by the code.[41]

Specially created women's organizations and their activists—female organizers, or *delegatki*—struggled to increase the participation of women in public affairs. The difficulties they encountered among the peasantry are discussed in a letter from Voronezh province received by *Krestianskaia gazeta* on 3 April 1926.

· 63 ·

Letter to *Krestianskaia gazeta* from Dviako describing the difficulty in organizing peasant women, 3 April 1926. RGAE, f. 396, op. 4, d. 43, l. 185.
Original manuscript.

Who would liberate women?

At a citizen's rally, in the settlement of Novoosinovka, Ostrogozhsky district, during the Red Army's anniversary, various organizations offered their greetings to the Red Army: from the village soviet, from the school, from the cell of the Komosmol, from the women's organization. In short, everything was good but for one bad thing. After the rally, the teacher and women's organizer let fly at the husbands who accompanied their wives to the rally: "These [wives] of yours never come to a meeting."

Apart from this, at a general assembly in March, there were elections to the non-Party conference.[42] There was an instruction from the higher organs to the effect that half the elected delegates to the conference should be women, of which there were none at the assembly, except for teachers, and Komsomol girls, and the children's labor colony.[43] But all the same, they elected women even in their absence.

DVIAKO

Despite resistance, changes were occurring in gender relations. As a result of the new divorce law, marriage no longer bound couples together for life. In the first half of the 1920s, the number of divorces increased dramatically nationwide. Divorce was most common in urban areas, but even in the countryside divorce rates exceeded those of any European country. A popular ditty, or *chastushka,* of the time expressed some women's, especially urban women's, new attitude toward marriage: "With Soviet power, I don't fear my hubby. If life goes wrong— then it's a divorce for me."[44]

Not everyone found comfort in such a cavalier attitude toward divorce, however. Even men who professed to support the Soviet effort to emancipate women found it difficult to adapt to the new situation, as we can see in I. Ye. Polishchuk's letter, sent from Oshukhino village, Barabinsky region, Siberian territory.

· 64 ·

Letter to *Krestianskaia gazeta* from the peasant I. Ye. Polishchuk on marriage and women's freedom, 3 April 1926. RGAE, f. 396, op. 4, d. 43, l. 156.
Original manuscript.

The situation of the young in relation to marriage

Soviet power is very good in all aspects except one: the marriage of young guys. Women have been given more latitude. I will offer some personal examples. On 16 January 1925, owing to the shortage of agricultural workers, I married. I was only eighteen and married a grown-up gal of twenty-four. She lived about two months in the customary manner, but by the third month she began her hanky-panky. In the evenings, after her household chores, she would leave and stay out until midnight or later. And if you would say anything, she would shut you right up: "Now we have Soviet power. I'll go wherever I want." I suffered with her, but all the same, by the fourth month of the marriage we had to divorce.

Back now to 1926. On 18 February I got married again to another girl who had already divorced her husband, and it is exactly the same story as with the last one. Nothing helps—neither persuasion, nor kindness, nor cursing—because women are completely undisciplined. A new law should be issued regarding this problem, though somewhat more narrow so that women do not have such a vigorous understanding of Soviet power. But don't think that I am an oppressor of women. I am a prime defender of women's interests in the village; nevertheless, my advice is to rein in an unbridled woman.

Polishchuk

The following letter, dated 22 November 1926, from the peasant I. T. Potapov, from the village of Kuvai, Alatyr county, Chuvash republic, expresses deep anxiety concerning the effects that women's liberation will have on the traditional division of labor within the peasant family.

· 65 ·

Letter to *Krestianskaia gazeta* from the peasant I. T. Potapov on women's liberation and the division of labor in the family, 22 November 1926. RGAE, f. 396, op. 4, d. 43, l. 134. Original manuscript.

To the *Krestianskaia gazeta* editorial board.

Esteemed citizen editor. Here and there I have read in the newspapers, the almanac, and other books about the alleged oppression of women, that it is impossible to draw the masses into politics without also drawing women into politics because the female half of the human race was doubly oppressed under capitalism, that they did not enjoy full rights, that mainly they remained, allegedly, domestic slaves and carried out the pettiest, most laborious and onerous kitchen work and household chores in general, etc., and the like. Is this correct? Once in a while, it helps to talk with the peasants to reach the correct conclusion that this is not entirely correct.

First, if it was true that they did not enjoy equal rights with men and did not have the vote, then it is necessary to equalize them in law. Second, but it is radically wrong, and we peasants do not agree, that a woman was oppressed, allegedly, and that they remain domestic slaves and are given the pettiest, most laborious and onerous work in the kitchen and household chores in general. The question is, If a wife will not cook and care for the children, is the man supposed to do all this himself? The man is supposed to plow, to feed the livestock, to go to the bazaar, to work, and, in general, to engage in agriculture or, in winter, engage in some sort of handicraft trade, etc. [Should he] bake bread, make soup, fix the tea for the family, look after the kids, and the woman or wife should do nothing in the kitchen but sit? Is this how it should be? And is this right? What if [we are talking about], not a peasant, but a commissar in Moscow? Let his wife go from the kitchen to the office, and he himself can make stuffed breads, soup, and tea in the kitchen, and be a nanny. Would this be right? No, it is not right; it is just an incitement to keep women from working in the kitchen. Not everyone can work in an office; someone has to work in the kitchen. After all, not every man sits [in an office]; some work constantly, now in the field, now around the homestead. Therefore kitchen work isn't so bad—baking stuffed breads. In a family the burden of labor is carried more by the man and not by the woman. Of equal rights, I say that a woman should be equal, but in the distribution of labor the woman should be in the kitchen and not in the office. Editorial board, I request that [you] publish my letter or answer this letter in the next issue of your paper. I will be watching [to see if] it is correct or incorrect.

With respect, IVAN TIMOFEEVICH POTAPOV, peasant.

Sometimes recognition of the need to recast gender relations came from surprising quarters. Because of the Cossacks' adherence to traditionally defined roles for husband and wife, this letter from a Don Cossack in answer to a complaint that women had received too many rights under Soviet power is of particular interest.

· 66 ·

Letter to *Krestianskaia gazeta* from the Cossack S. T. Naumov regarding the law on divorce, 8 December 1925. RGAE, f. 396, op. 4, d. 43, l. 160. Original manuscript.

To *Krestianskaia gazeta,* regarding the article "The Wailing of a Man's Soul," [published] in *Krestianskaia gazeta,* no. 104, 8 December of the year 1925.

I am a Cossack from the Zhukovskaia stanitsa, Tsimliansky district, Salsk region. I wish to say the following to the young peasant in relation to his letter: Brother [*bra*], I read your little article, and as I thought about it I became frightened that our Soviet government might heed your plea and use legal measures to place women [back] within the bounds in which they were before 1917, that is, before we had a revolution in Russia. I will give you an example and maybe it will have an effect on you, brother, and maybe you'll be frightened and won't think so hard about women's rights. Here is the example, brother: with us Cossacks there is a practice, and it has a place at events everywhere. When a Cossack's wife is at a general get-together, at a small banquet, say, and you, young peasant, or I dance with her and say, as is the custom, "Whoever dances together should kiss," well, it is well known they would kiss. For this trifle, even though it is common, a Cossack would beat his wife half to death out of jealousy, even though she could not avoid the kiss of the other Cossack. But she must be dumb as a fish and cannot complain to anyone, and if she should complain, they would say, "Listen to your husband." Or even simpler, "You get what you deserve." There it is, brother; it makes your skin crawl. And so you are frightened that they said your wife will leave and take a share [of your property]. I propose that in such circumstances the court should find grounds for divorce and award everyone what they deserve. As for the child, of course, you, as the father, should raise him; otherwise, you know, we're men of the world . . . and then good-bye.[45]

NAUMOV, SERGEI TIMOVICH, Red Cossack

To the chagrin of the reformers, the liberal divorce law made it possible for men to abandon their wives and families at a stroke to marry younger or otherwise more desirable women, a maneuver some men employed on a serial basis. V. S. Goncharenko from Akmolinsk province in the Kazakh republic denounces this practice.

· 67 ·

Letter to *Krestianskaia gazeta* from V. S. Goncharenko, a member of the "Red Rose" kommuna, describing Communist Party members' abuse of divorce laws, 25 March 1925. RGAE, f. 396, op. 3, d. 83, ll. 25–26. Original manuscript.

Very recently, there have appeared in the pages of the periodical literature items about the new life in general and marriage in particular. But one very rarely encounters articles on divorce and on the struggle with divorces when it is assuming mass proportions. An especially strong epidemic of divorces has spread in the modern village in Siberia and the Kirghiz republic.[46] By and large, village divorces are initiated by the man and mainly by Party members. This, most certainly, places a blot on the Party and elicits from non-Party members a highly unambiguous ridicule of the new marriage. The main thing is that such divorces serve those husbands who have lived with their wives for ten to twelve years and who, one way or another, find themselves in the city, in the [soviet] executive committee, or in general some type of situation, and discover that their wives are inadequate for them because of a lack of culture and, rather than educate her, find an easy way to separate from such a wife with the help of divorce, which in current conditions is very simple, and in turn marry some city "prima donna." Often the old wife is even left with infant children and without any means of support. There are also other sorts of divorces. The divorce of Comrade Leliuk, the cell secretary of the "Red Rose" kommuna, Kokchetav county, Akmolinsk province, will serve as an example of these divorces. Here divorce took on a chronic character. He and his wife arrived at the kommuna in 1921, and he quickly separated from her because he cooled toward her and needed a change of "diet." After this, his wife and his baby had to move to another kommuna. According to a previous arrangement, Leliuk married another kommuna member. Some time passed, and his wife became pregnant, and like any woman during her pregnancy, she lost her good looks. On seeing the change in his wife, Comrade Leliuk remembered that his first love was living in Ukraine, and he planned to divorce once more. He mailed a letter to Ukraine with a proposal and got back an acceptance. At this time, the wife who lived with him gave birth, and immediately after the birth she received the divorce proposal.

Such conduct is unworthy of a member of the Russian Communist Party. The surrounding population sees the kommuna as a model and disapproves of these divorces. It would be desirable to expose in the pages of the newspaper similar deeds [committed by] individual officials of the Party or in the sciences, as well as [by newspaper] readers.

A communard.

Member of the "Red Rose" kommuna, Akonburlug district, Kokchetav county, Akmolinsk province, GONCHARENKO, VASILY SAVVICH.

I request that my last name not be placed in the newspaper; otherwise, it will be impossible for me to live in the kommuna.

V. GONCHARENKO

To free women from the burdens of domestic chores and to enable them to participate actively in public life, the Communist Party committed itself to creating various types of child-care facilities. The need for such institutions was made all the more urgent by the enormous number of orphans created by years of war, civil war, and famine. These orphans (*besprizorniks*) populated the cities and roamed the countryside. Following the revolution the number of child-care institutions was minuscule. During the civil war, the number increased dramatically although it remained insufficient to deal with the scale of the problem. During the NEP, financial constraints forced many of these facilities to close. Rural areas in particular suffered from the lack of child-care facilities. In the following undated account, A. I. Pukhov from Bakharevo, Matveevka district, Kologriv county, Kostroma province, paints a disturbing picture of rural child rearing.

· 68 ·

Letter to *Krestianskaia gazeta* from the peasant A. I. Pukhov on the need for children's nurseries in the countryside, no date. RGAE, f. 396, op. 4, d. 43, l. 209. Original manuscript.

To *Krestianskaia gazeta.*

Reading in *Krestianskaia gazeta,* I have not come across special articles on the rearing of children in the village, but it's important, this matter: the village needs correct child rearing more than anything. Rural children, especially in the summertime, are left to their own devices. The lack of nurseries in the countryside forces mothers, at busy times of the day, to place nursing children with young children of five to twelve years of age. Clearly, these nannies are stupid and need supervision themselves—they make poor nannies. Left to their own devices, they do not give a thought to watching over their little brothers and sisters. They want to play themselves, all the more since no one is supervising them. But in their games there is no intelligent leader who would present amusing and useful games for the development of children. Because of this, children in the village play games with hooligan tendencies, but mainly [there are] stone-throwing war [games], rousing songs, foul language, and how to ravage flower and kitchen gardens. In our area it is absolutely impossible to plant peas in a detached farm or any other vegetables that may be eaten raw, such as turnips, carrots, rutabagas, cucumbers, etc. Gardens especially suffer from these juvenile invasions in which they [the children] completely destroy still-green, no-where-near-ripe fruit and destroy young tree shoots. But worst of all is the tobacco smoking and moonshine drinking, which is often accompanied by the red rooster [i.e., arson]. These bad tendencies are very widespread among the children, and because of this, one has to fear for these future citizens of the USSR. It is not so very difficult to establish correct child rearing. Where there is a small village near another village, the two may be merged in order to build playgrounds, and where there are two large villages or hamlets, playgrounds can be built in each. The teacher leading the children in play on the playground may give the initiative to the children if they do not exceed the bounds of propriety. In the countryside, one has the option of going into the forest and through the fields explaining to the children natural phenomena and how plants grow and are nourished and the significance of the forest for the peasant and the state. And instead of bad tendencies and hooligan habits, there will appear among the children an interest in proper physical labor, development of the means of proper observation of nature, and useful games, gymnastic exercises, and useful knowledge through an intimate familiarity with nature. In this way, instead of a long, uncorrected, unsupervised rural upbringing for the young generation, we will see our successors healthy in body and mind, worthy of the name of free citizen of the USSR.

I, a peasant, devote part of my time to my peasant holding and the rest to seasonal work in the city as a construction worker. Aware of the difference in the upbringing of children between city and country, I have written this article based on my personal observations. Maybe the editorial board will find it needs to be printed on the pages of your newspaper.

Annual subscriber to *Krestianskaia gazeta,* A. I. PUKHOV.

It is worth noting that the program for "correct child rearing" that Pukhov recommends sounds very much like the complex educational method promoted by Narkompros.

Where nurseries, kindergartens, and playgrounds did exist, they made a positive impression on local inhabitants. Women especially appreciated their value, as is evident in the letter of K. Pliusnina, a peasant from Perm region, Ural province.

· 69 ·

Letter to *Krestianskaia gazeta* from the peasant Klavdia Pliusnina describing the
workings and benefits of a children's nursery, fall 1926.
RGAE, f. 396, op. 4, d. 45, l. 6. Original manuscript.

Notes on the Senkino children's nurseries.

On 28 August 1926, I, Pliusnina, Klavdia, a peasant woman from the village
of Ust-Tul, Senkino village soviet, visited the Senkino children's summer
nurseries. On arriving at the nurseries, I immediately noticed that everything in the nurseries was in order—the rooms were clean, all the children
were clean and washed and had haircuts, and everything was clean. All the
white linen on the beds was clean, and the children lay on dry sheets. The
nannies follow the children into every corner and look after them quite well
without any yells or noise and in no way frighten the children. And in the
linen closet everything had been ironed and laundered well, and cleanliness
reigned here and everywhere. Then I waited for the preparation of dinner.
The table at which the children dine is covered with a white tablecloth and
oilcloth. For dinner they served a very rich soup with meat, a millet and
milk porridge, and potato cutlets. Before dinner the children washed their
hands and put on very clean little bibs, and I asked them if they were home-
sick and did they get enough to eat, but to all the questions I received the
very same reply; in a very satisfied manner the children say, "We are better
off here than at home." And I am finding that such nurseries are serving us
as mothers. In the summertime [it is] a large aid in the sense of protecting
a child's health. A mother, too, can be more useful freed from her child. That
is why it would be desirable to have more such nurseries. The director of the
nurseries obviously gives her all to the affairs of the nursery. And it would
be desirable if there were more such directors of our nurseries.

I attest to these notes, KLAVDIA PLIUSNINA.

The nurseries operated from 9 June 1926 to 24 September. The number
of children [who attended] daily was eighteen to nineteen.

Thus, innovations in traditional village practices, when carried out
attentively, could make a positive impression on local inhabitants. For
some communities, however, tradition and financial concerns trumped
the potential for social benefit, as testified to in V. A. Khokhlov's report
from Aleksino village, Northern Dvinsk region.

· 70 ·

Letter to *Krestianskaia gazeta* from the peasant and Privodino village soviet commission member V. A. Khokhlov describing peasant opposition to nurseries and fire brigades, 18 October 1926. RGAE, f. 396, op. 4, d. 45, l. 60. Original manuscript.

Children's nurseries are not needed.

The Privodino village soviet commission was assigned the task of organizing children's nurseries and fire brigades in the villages by the Kotlas district executive committee. The commission members went out to the villages to agitate, but the peasants were stubborn and would have none of it: "This isn't for us. In the first place, the nurseries will need space. In the second place, we'll have to support the nannies and the employees from our own pockets. And, in the third place, we'll also have to provide the children's food. It's better to hire one's own nanny. She'll deal with the child, check on the oven, clean up after the livestock, and keep house. And all this can be enjoyed for no more than twenty rubles (a nanny's term is from June to 1 October, old style).[47]

"As for the fire engines and organization of the brigade, we understand that this is a good thing, but we just don't have the resources for an engine, and a brigade is entirely wrong for us. For example, let's say we put together a brigade and there was a fire across the river: how could we get there with an engine and horses? Our village lies along the river, encircled by water; we sit like hares on an island. Where could you go even in an emergency? And if you didn't go, you're done for. They fine you and drag you to court. How can they drag us off to court if it's a family matter! No, sir, spare us from all this."

This is how the peasants put it to the commission members, and they hold to their position despite any exhortation and explanation. And the members' attempts to explain to them the full correctness [of the policy] remain like voices crying in the wilderness.

Commission member V. A. KHOKHLOV, peasant.

Please print my note. If the peasants see it in the newspaper, they may catch their mistake.

With a communist greeting,
Secretary of the Prokursky cell of the Komsomol, V. KHOKHLOV.

The success or failure of the new establishments depended on a variety of factors, including the degree of official support, the availability of adequate facilities and financing, and, as Pliusinina noted, the commitment of the staff. In 1923, the Children's Commission appointed an American, Anna Louise Strong, to found and serve as the patron or sponsor (*shef*) of an agricultural commune for adolescents in a former convent located on the Volga River. Despite the children's enthusiasm, hard work, and initial successes, after two years the John Reed Children's Colony collapsed owing to official neglect and the corruption of its various directors.[48] Finding honest and devoted personnel to run the new institutions was no easy task, as the following letter from Chudovo district, Novgorod province, explains.

· 71 ·

Letter to *Krestianskaia gazeta* from the peasant Kolosovsky describing the mismanagement of the Lezno village children's nursery, Novgorod province, 30 July 1926. RGAE, f. 396, op. 4, d. 45, ll. 282, 283, 285. Original manuscript.

To *Krestianskaia gazeta*

In the Lezno children's nursery we have one nanny for twenty-five children. The director of the nursery busies herself with dressmaking as if caring for older children is not her concern. Because of this neglect, there have been cases of children falling down the stairs from the second floor. Lastly, the reduction of staff and rations due to the "regime of economy" has turned our nursery into a scarecrow [that frightens] children.[49] The poor-peasant parents who have nothing to feed [their children] and have no one at home with whom they can leave their children in the summer bring their children to the nursery unwillingly, and the rest [of the peasants] are fearful that in the event of some freak occurrence "we will [have] to take [the children] to the nursery." But the children, even though they are only two years old, know how good it is in the nursery—they are thrown in a bed or a basket and wail until evening. The nanny says, "I don't have twenty-five hands to take care of everyone." And the food is not very good either. In the morning they are given tea with white or black [bread]. At twelve o'clock they have fish soup, and from twelve to six o'clock, nothing. At six o'clock, if anything is left over from dinner, they're fed, and if not, then that's it. The director says, "They are fed at home"—as if she is unaware that several working widows have only bread themselves, while others do not [even] have enough black bread. It has to be said that at the prescribed twenty kopeks per child, mothers also provide a bottle of milk and an egg every day and half a kilo of butter each month. At this rate it is possible to feed the children very well. But our director is expending no more than ten kopeks a day per child. Eggs are not served, although she makes a couple for herself, and the butter has sat so long that it's gone rancid and has to be tossed out.

If the nanny doesn't suggest taking the children out for a walk, then the director does not send [them], and in general, the director, citizen Yeliseeva, is indifferent toward the nursery's affairs. As a result of these disorders and this inefficiency, mothers remove their children from the nursery, and the employees quit. In the nursery's twenty working days the cook quit and the director Yeliseeva discharged the laundress—a poor widow (peasant-delegate)—because she had the temerity to engage the director in conversation, to tell her how well run the nursery was last year, when the children were well fed and there were no such cries of hunger. [Then] they were under the caring eye of the director, and everyone understood that nurseries are useful for us, and the children weren't taken home from the nursery, but now they don't go [to the nursery]. What's more, in the conversation she [the laundress] indiscreetly called the director *ty* [the informal "you"].

Yeliseeva is a former teacher (removed in a verification of school officials), the pupil of some general or other. She says that Novgorod appointed her, and she is not obliged to report to village babas, and "lastly, I am not their friend that they may address me as *ty*. I did not study so that any dirty baba could start to give orders." When it was suggested that she get a little closer to the peasant women, she answered that in this area she is well aware of "definite limits."

The district executive committee was reminded of the abnormalities in the nursery's work. They came, looked around when there were no children about, found it clean, and the beds were in place, and away they went.

Sensing that the district executive committee is far away, twenty kilometers, and that they show up only when summoned, and then [only check] superficially, director Yeliseeva sees [this] and is definitely not using the regime of economy in its favor. What's more, she told a clerk, "I'll just sign over a little [extra money to myself], then hit the road, and that's it; then I'll go to Novgorod." But the peasant women say that "Novgorod has sent us a little baroness." And it seems that any simple peasant baba knows how to care for infants and would cope with running the nursery much better than citizen Yeliseeva, who is afraid to touch our children and calls them "grubby."

KOLOSOVSKY

In the mid-1920s, Soviet health care began to recover from the crisis years of the civil war. Between 1924 and 1927 per capita expenditures increased, but despite the promise of a socialist medical system with free care for all, access to quality health care remained glaringly unequal. Party members and the government elite had access to closed facilities, those with enough money could pay for private services, and approximately twelve million people—mostly workers and their families—had

insurance. Expenditures on health care in rural areas remained far below expenditures in the cities. Still, efforts were made to make sanitariums
and rest homes available to peasants. The tsar's former palace in Liva-
dia, on the Crimean peninsula, for example, served as a three-hundred-
bed treatment center for peasants afflicted with tuberculosis, and other
mansions and palaces were being similarly adapted.[50]

In the following account from December 1926, the peasant V. K.
Kulikov, from the remote Siberian village of Nikolaevskaia, Ustiansky
district, Kansk region, Yenisei province, describes his stay at one such
resort.

· 72 ·

Letter to *Krestianskaia gazeta* from the peasant V. K. Kulikov describing his
stay at the Usole sanatorium, 6 December 1926. RGAE, f. 396, op. 4, d. 44,
ll. 2–3. Original manuscript.

Paradise

And heavy was the heart of the poor peasant,
Not understanding what Soviet power had already done for him.

I don't know by what good fortune I, a sinful poor peasant, ended up in the
paradise called the Usole resort, but I'll write just a little bit of what I know
of it. While I was in a rapturous [state], I was brought to the gates of this
paradise, a dirty, tattered, sinful peasant of the plow. The following picture
presented itself to me: Out of nowhere an Angel in clean white clothes appeared to me and led me through the ordeals through which I inevitably, as
one of the poor sinners, was required to pass. For the first ordeal, they
shaved off my hair and beard, and I submitted to everything. Afterwards,
another Angel led me into a warm, bright room where I was given soap and
a sort of soft substance and told to wash thoroughly under a fountain of
clean, warm water, and I submitted. After this, I looked good and neat. Having passed through this ordeal, a clean, white, beautiful little Angel appeared before me and bade me to follow her, and I submitted to her. She led
me into a clean, bright room where a place for me to rest had been prepared. Here I saw several people, sinful poor peasants just like me. Here the
little Angel handed me clean clothes, offered me a bed just as clean and soft,
and on leaving me said to keep it clean and not to throw cigarette butts on
the floor, and I submitted to this. But since it was already evening, I fell fast
asleep. O the wonder of it! How strongly and sweetly I slept in that soft
bed! The next day a new Angel was sent to the clean cafeteria, where I was

given my morning tea, pure white bread [*kalachi*], eggs, sausage, ham, and butter. I finished the tea and said thank you to someone, I don't even know who it was. Then this same Angel presented me to another, stouter Angel, who instructed me then and there to undress, and I submitted to her. After that, she began to examine me for sins. She hoped to free me from my burdensome, corporal sins. After examining me, she placed a small chart in my hands and ordered another Angel to take me into the curative water, and I submitted. After bathing in the curative water, they led me into the garden. Here, in the garden, in a clean spot were arranged tables in rows set with white tablecloths, and I was amazed by this wonderful setting. Suddenly, there appeared an Angel with an uncovered head of curly hair holding a small chart in his hand. He began to present various dishes to me that were listed on his chart, but I kept silent, not knowing how to answer him, and only after several seconds did I summon up the courage to tell him to give me my beloved Russian borscht, and he submitted to me. After this, I saw that the tree of life and the tree of knowledge—good and evil—stood in the middle of this paradise. On the tree of life grew different fruits and on the tree of the knowledge of good and evil stood various bottles, but I decided not to touch them because it seemed to me, a sinning poor peasant, impossible.

I was destined to see quite a lot of things here about which I will not write, because it seems of little interest to me. But I was interested to learn only one thing: Who had had the opportunity to enjoy this place, this paradise, nine years ago? And I found out from one of these little Angels, who told me that at that time there were blissfully pot-bellied, pious people here with purses full of gold and that poor-peasant sinners were not allowed even to pass near the fence of this miraculous paradise. Learning of this, I nearly swooned, but when I collected myself I called to mind the words of one Angel, who told me that I could only remain a month and a half in this paradise and then, then . . . once again, a sinning poor peasant, I would have to return to my local rural hell and again face my dilapidated hut and the infant children inside in their torn and dirty shirts. Then this paradise will seem as if it were only a dream. But I know that soon the time will come when the poor peasants who are waiting will see in their lives the light of an earthly heaven. As for me, nothing is as pleasurable as writing to *Krestianskaia gazeta.*

VARLAAM K. KULIKOV

So absolute is the contrast between the efficiency and sparkling cleanliness of this sanitarium and the dirt and disorder daily encountered by the peasants in their usual surroundings that to Kulikov it appears as nothing less than the antithesis between heaven and hell. It is the paradise on earth promised by socialism made real, and none but religious imagery and language could express his joy at having experienced it, however briefly.

Religion

Despite the Bolsheviks' wholesale atheism and contempt for the Russian Orthodox Church, during the 1920s there was no single Party line on religion. The 1918 RSFSR Constitution established the separation of church and state and removed primary education from church hands, but until collectivization Party thinkers continued to advocate a variety of approaches to deal with popular religious belief, ranging from organized antireligious militancy to the construction of surrogate religions rooted in science, a common humanity, or, after Lenin's death, Lenin worship, to name a few. To be sure, persecution of the church and its priests—which had been so merciless during the civil war—continued and received new impetus with the founding of the League of the Militant Godless in 1925.[51]

Religious belief came under assault from different quarters and by different means. On 6 January 1923—Orthodox Christmas—officially sanctioned blasphemy may have reached a new high (or low) with the Komsomol's staging in Moscow of a sacrilegious Christmas carnival that mocked the symbolism and solemnity of the holiday. According to Richard Stites, the negative public reaction to antireligious festivals like "Komsomol Christmas" induced the authorities to discourage such irreverent practices and to pursue a positive strategy of developing Soviet rituals as replacements for religious celebrations and observances. No less a personage than Trotsky wrote favorably about the new ceremonies, warning only that they not be imposed on the population bureaucratically.[52] Soon "Red priests," as Party and Komsomol secretaries were often known among the people, were promoting "Red" baptisms, weddings, and funerals.

The following account of a "Red" christening, or *oktiabrina* (from the Russian word for the month in which the Bolshevik revolution occurred), comes from a January 1924 protocol of a general meeting of members of the Kremenchug district committee of the Union of Woodworkers in Ukraine. As is evident from the speeches, Party members still harbored hopes for revolution in Europe despite the failure of a communist uprising in Germany the previous summer.

· 73 ·

Excerpt from an account of an oktiabrina, *Moskovskaia pravda,* 28 June 1924.

[. . .] Opening the assembly, [Comrade Radchenko] notes the significance of the celebration, pointing out that the aforementioned fact itself attests to the results of union-educational work [and] of the membership's realization of the mass absurdity of religious rites, which stupefied and oppressed the working class over the course of many centuries. Only the Great October [Revolution], which liberated the working class from the yoke of capital, has provided the opportunity to overcome our blindness and to build our life as our conscience and reason suggest to us.

Comrade Radchenko, announcing the resolution of the [union] administration on the naming of the newborn with the name of the glorious leader, Comrade Lenin (NINEL), states that we promise to instruct the child in the communist spirit, and we hope that the new member of society will proudly wear the name of our great teacher.

The resolution of the administration on the naming of the child with the name of NINEL and on enrolling her as a member of the All-Russian Union of Woodworkers is approved unanimously.

Comrade Radchenko entrusts to the parents of the newborn her [union] membership card and gifts presented to the child by chairman of the Young Sparticists.[53]

Comrade Oreshtein, in the name of the city committee [*gorkom*] of the Young Spartacists, announces the enrollment of the newborn as a candidate in the Young Spartacists for ten years. By the expiration of this time period, the Young Spartacists is obliged to train the child in the spirit of the Spartacist program for its entry into the Komsomol.

The representative of the city Komsomol organization, Comrade Mikhelson, announces the enrollment of the newborn as a Komsomol candidate until she is fourteen years of age, after which it is obliged to transfer the child to the glorious ranks of the Russian Communist Party.

Comrade Oreshtein, in the name of the city committee of the Young Spartacists, pins on the newborn a pin with the inscription "Study, Grow Strong, Fight, Unite." In the name of the city Komsomol organization, the comrade pins on the newborn the pin of the Communist Youth International.

Comrade Koval, the Russian Communist Party representative, accepting the newborn from the Komsomol representative, says that after passing from the foregoing schools of communism (Young Spartacists and Komsomol), the finale should be to enter into the ranks of the Russian Communist Party, which, tempering her in the revolutionary struggle for our cherished ideals, transforms her into a true defender of the interests of the working class. In conclusion, he pins on the newborn the ILICH pin . . .

The newborn's father, Comrade Krasnik, accepting the child from the Russian Communist Party representative, promises to raise not only the newborn but his other children in a proletarian and communist fashion.

Comrade Verbitsky announces the resolution of the administration on the naming of the newborn with the name of the great and glorious leader, Comrade Lenin, and states:

"We, the generation of the October Revolution, have carried this banner high over our heads amid blood, amid famine, amid destitution, and [amid] desperate struggle with the wolves and hounds of capitalism. Carry it further! Fight and work with us . . .

"For centuries the working class lived in slavery to the capitalists. Six years ago, among us in Russia, the working class tossed off the chains of capitalism and took power in its hands. The true leader of the working class in its sacred struggle against the capitalists of all countries was and is the Communist Party.

"You were born at the moment of the sharpening of the class struggle throughout the world and [at a time] when the workers of Germany, encircled by enemies and traitors, want to give the decisive blow to the bourgeoisie according to the example of the Russian workers. They are prepared to take power in their hands.

"Together with the workers of Germany, the entire world proletariat will fight. And until then, while capitalism is not [yet] expelled from each corner of the globe, deprivations, labors, and sacrifice await us. We will not retreat before them. We will break through all the obstacles to victory.

"The dawn of the new life has already broken over the weary land. Let the brilliant sun of communism burn. In your person we greet the bright future for whose sake we are prepared to make any sacrifice.

"We give you a name . . .

"Read these lines when your mind matures and your will strengthens and the fetters of oppression are finally smashed . . ."

Despite the disdainful reference to religious rites at the outset of this *oktiabrina*, we cannot fail to note that it was a religious ceremony in all but name. Moreover, however reverent the participants or inspiring the final speech, in this particular instance the attempt at replicating a liturgical and sacramental atmosphere comes across as pompous and bombastic. With three red medals on her infant chest little Ninel must have resembled a tiny war veteran. Be that as it may, the new rituals did fill the need to mark important occasions with ceremony and were making inroads in the village. Maurice Hindus, visiting his ancestral region in 1929, lamented the near disappearance of the customary seasonal celebrations he had known as a child, and noted that only one of four upcoming weddings would be "white" (i.e., traditional and religious); the others would be "red."[54]

In the following letter to *Krestianskaia gazeta* from distant Siberia, S. A. Ganin of Telmenko village, Cherepanovo county, Novonikolaevsk province, describes a village oktiabrina. The response to the ceremony may be taken as evidence of what Stites calls the peasants' "fantastic ability to absorb new 'faiths' and rituals when necessary and combine them with their own ways."[55]

· 74 ·

Letter to *Krestianskaia gazeta* from the peasant S. A. Ganin describing an oktiabrina in his village, 25 November 1924. RGAE, f. 396, op. 3, d. 100, l. 1. Original manuscript.

The first oktiabrina in our village.[56]

Our militia man, Comrade Karpenko, to whom a son was recently born, did not want a traditional baptism rite with a priest but held an oktiabrina, which was conducted in the People's House on 23 November. More than two hundred people attended, the majority of whom were adults. There were even some old men and women who were very interested to see "just how this new sort of communist-type baptism would come off." And when the oktiabrina began, the audience, as never before, listened attentively and was very quiet, and at the end of each speaker's address stormy applause rang out. They unanimously named the newborn Kim, and when the oktiabrina ended, voices were heard [to say] among the public, "There you are, boys. It's true what these here orators say: you can really get contaminated by these here baptismal fonts, but now all the folks can come and get rid of all the infection." That is how our first oktiabrina, which had a great effect on the people, went.

STEPAN ANDREEVICH GANIN

I ask that [this] be published.

The name Kim—derived from the acronym of the Communist International of Youth (Kommunisticheskii internatsional molodezhi)—was but one of a host of "revolutionary" names parents could give to their newborns. Red name lists drew inspiration from revolutionary heroes and ideas, scientific and technical achievements, myths, classical history, and other sources. Besides the prosaic Vladimir, Lenin's name, as seen in Document 73, provided numerous variations, such as Ninel, Vilen, Vilena, Vladilen, which attained some degree of popularity.[57]

More often than not, the new rituals, founded as they were on nonbelief, challenged the peasants' traditional faith and practices so directly that assimilation and even accommodation became impossible. In such cases, a person had to choose between the old and the new. This was especially true when the choice involved matters concerning the grave and the hereafter. In the following account from April 1925, written by the Komsomol cell secretary I. M. Gutsev, a peasant woman (who might be expected to defend tradition) is shown adapting to the new Soviet practices under the influence of her nonbelieving husband.

· 75 ·

Letter to *Krestianskaia gazeta* from the Komsomol cell secretary I. M. Gutsev on one family's adoption of the new rituals, 15 April 1925. RGAE, f. 396, op. 3, d. 234, l. 61. Original manuscript.

The ice has begun to break up.

The ice has begun to break up. The brilliant rays of the spring sun are stealing through the dirty walls [surrounding] our peasant women. Here is a telling example from our village of Zhgun, Dobrush district, Gomel county and province. A young peasant woman, on the advice of her husband, an advanced villager, cleared the "Red corner" in their hut of the boards decorated by those icon daubers from Vladimir.[58] The religious of Zhgun and all the old women and men say it doesn't matter [if you] throw out the icons and hang [pictures of] communists in their place because when the child is born you'll surely bring God's creatures to the little father [i.e., to the priest for baptism]. They say you don't want your kid to grow up a little devil. And you won't get mercy from "the holy father or from God the creator of heaven and earth." A son was born to the godless Sushanova. But the "prophetic" words of the old ladies were not vindicated. Sushanova did not go bowing before the long-hair [i.e., the priest] bringing gifts but had a new-style baptism. The unbaptized don't live long, those same crones continue to preach. But, no! like a sin, KIM grows up fast and healthy and is not dying. Then the old women and the church council of Zhgun decided to wait for the death of someone in the Sushanov family. And it did happen that one of Sushanova's son's died, only not the unbaptized one, but the one who the three-chinned "shepherd of the rural flock" dipped in his "holy water." Everyone thought, well, baptism, this isn't new by any means, but the funeral definitely can't go on without a priest. But here, too, it turned out contrary to the thinking of the religious. Iosif Sushanov, Shushanova's husband, attended a session of the bureau of the Zhgun Komsomol and requested in the name of his wife and himself the arrangement of a Red funeral. "Done!" my guys exclaimed. They resolved: To support the sprouting of the new in the countryside; to use the entire cell; to make arrangements quickly through our representatives and with the school. No sooner said than done. And the Red funeral took place the next day. It was quite active and beneficial, like a blow against the thoroughly rotten foundations of the priesthood. That is how the new comes into being—by replacing the old. Sushanova is a simple poor-peasant woman.

Komsomol cell secretary, I. M. GUTSEV.

In this account, the husband is described as an "advanced peasant," a *peredovik,* that is, an individual in sympathy with the political principles and agricultural innovations advocated by the Party. Given the dominant role the husband obviously plays in this household—he, after all, orders the icons removed—we may reasonably question whether his wife had any say at all, for or against, in her husband's rejection of traditional religious practices. Nevertheless, the lesson imparted is clear. Soviet authorities considered women the most backward and "irrational" segment of the village population.[59] If they could be induced to abandon religion and superstition in favor of rational "Red" rituals, then the countryside was indeed destined to come eventually within the communist cultural orbit.

In contrast to the success story recounted above are items printed in *Krestianskaia gazeta* that addressed the awkward fact that many rural Komsomolists opted for church weddings performed by a priest.[60] One article, entitled "A Village Drama," that discussed the case of a young man who left the Komsomol because of his marriage, elicited many responses. This one is from the village of Matveevka, Zhigaev district, Lgov county, Kursk province.

· 76 ·

Excerpt from a letter to *Krestianskaia gazeta* from N. A. Bobkov on Komsomolists who choose religious marriage ceremonies, 18 August 1925. RGAE, f. 396, op. 3, d. 352, l. 91. Original manuscript.

[. . .] Definitely, the situation of that Komsomolist was hopeless. Even under the most favorable conditions the backward rural Komsomolists abandon the Komsomol when it comes to marriage. I will give you [as] an example the Komsomol in Belitskaia stanitsa, Lgov county, Kursk province. The cell secretary, Semyon Shchelkunov, courted a baroness [i.e., a young girl]. The family's status required that she marry. The bride doesn't go in for Soviet customs. Semyon did not marry for a long time. At last, the bride was amenable. Semyon prepared everything for the wedding. At home, they came down on the bride for doing this, and she went back [on her promise to marry]. Semyon's father jumped all over Semyon. The carcass is rotting, he says.[61] Under this pressure, Semyon had a church wedding with another girl. Given this, the young are afraid to join the Komsomol. There are no girl Komsomolists. Why is this still the case? In the village, especially in summer, the Komsomolist finds himself in a tough spot. A lot of Komsomol work has nothing to do with agricultural work; therefore the parents object. [Political] consciousness among the young is still low, and they don't see any benefits [in joining the Komsomol]. They say, "Why be a Komsomolist? Soon the non-Party [members] will get to study?"[62] Intending to stay at home, girls never join the Komsomol. The male Komsomolist who plans to stay at home says, "I'll get married sometime." As a result, the young do not join the Komsomol. Levin is right; it is the same everywhere.

NIKOLAI ALEKSEEVICH BOBKOV

New nonreligious holidays marking important events in the Soviet calendar were commemorated in the village. The majority of peasants, however, kept aloof from the official celebrations. Sometimes churchmen organized religious counterdemonstrations. This could lead to violence. In 1926, the May Day demonstration held in a village located in Briansk province, for example, was broken up by a religious procession. The OGPU reported that several demonstrators were beaten in the melee.[63] Like more traditional holidays, these festivities did afford the opportunity for heavy drinking and carousing, particularly for the younger generation. In the following excerpt, I. I. Melnikov of Rudnia-Bartolomeevskaia village, Chec.hersk district, Gomel province, Belorussian republic, describes the activities of youth gangs in his village at the time of the ten-year anniversary of the October Revolution.

· 77 ·

Excerpt from the peasant I. I. Melnikov's letter to *Krestianskaia gazeta* describing youth-gang violence in his village, 27 November 1927. RGAE, f. 396, op. 5, d. 4, l. 572. Original manuscript.

Krestianskaia gazeta.

Here is how the youth of our village greeted the October Revolution holiday: they organized their banditry into two camps, that is to say, into two gangs of marauders. The gangs are big, made up of simple youths as well as demobilized Red Army men, and they are ravaging the villages [unclear in original] because there are thieves in these gangs who are capable of anything. What they need is an elder who'll give it to them in the neck so hard that their legs will buckle. At night they destroy the barn roofs, they thieve, and steal, and drink hooch. On 30 October, they gathered at one poor peasant's place. They beat each other upside the head with his iron pots and buckets. Now they each wear their badges of bruises that they carry on their physiognomies, swollen up like pumpkins, and have knife wounds all over their hands. Weights [stuffed] in rubber were also popular [weapons]. No one can bring them to heel, because they themselves are the wickedest brothers of officials of the Soviet government—one is the brother of the chairman of the village soviet, and another of the chairman of the district executive committee. If you find it possible to print this letter, I sincerely request that you do not reveal my name or patronymic since the young have guns, there is shooting and other stuff going on, and it's easy to fall prey. [. . .]

MELNIKOV, IVAN IVANOVICH

An OGPU report from mid-1926 noted that, as a rule, participation in May Day celebrations in the villages was limited to Party and Komsomol members, teachers and pupils, and soviet officials and employees. The mass of peasants did not take part. In some locales peasants had been induced to turn out, but only through the well-timed appearance of a tractor or the installation of a radio. The report lauded these tactics as "good organization."[64] Defiantly, some people who refused to commemorate the new Soviet holidays treated them simply as ordinary working days. I. S. Chernoivanov's letter shows how one family's refusal to celebrate May Day exposed the class cleavages in a Kharkov province village.

· 78 ·

Letter to *Krestianskaia gazeta* from the peasant and Committee of Indigent Peasants (KNS) member I. S. Chernoivanov describing the conflict between a kulak and the chairman of the KNS on the occasion of the May Day holiday, May 1924. RGAE, f. 396, op. 2, d. 16, l. 183. Original manuscript.

Kulak-Robbers

In Sorokovko village, on the 1 May holiday, the chairman of the KNS was brutally beaten by the local kulak A. Samofalov. The incident occurred like this: The kulak Samofalov, in spite of the announcement that on 1 May no one should go to work in the fields, went all the same. The chairman of the KNS, seeing his brazen nature, rode up to him in the field as if to a friendly local. The chairman came up to him again and says, "Comrade, today is our proletarian holiday. Therefore I ask you not to work so as not to give the others an excuse." But it was not to be. The kulak forgot that he is in a proletarian republic and jumped down off his cart shouting, "I don't recognize your holidays, so get lost, indigent bastard." Telling the committee of the indigent where to go, he took a harness from the wagon and dumped it on the KNS chairman with all his might. Realizing his predicament, the KNS chairman took off with the kulak giving chase. But the chairman turned out to be faster, and only this saved him. The kulak shouted after him, "It's too bad you ran away, or else I'd show you today's holiday. I wouldn't let you out of the ravine alive; [then] you'd know the proletarian holiday!" Comrade kulaks, do you know where you are living? Not in some capitalist country where you can go around and beat the poor with a stick, but in a proletarian country, a country of peace and freedom. You are content to be bears living in dens, but you have to be reasonable, cultured people. Throw off darkness and ignorance; be honest toilers in a free country. We want to live peacefully with you and go about without any danger, but you, once again, raise your hands to us, the committee of the indigent. Watch out, do not draw blood, or else it will be the worse for you. We're not to be trifled with. We are well organized . . . [word not clear in original] in the ranks of the proletariat and walk arm in arm for Soviet worker-peasant power. And if you continue to raise your hand, we will boycott you.

                                                              A KNS member

----

Disorder

The NEP years struck many contemporaries as a period of social disorder. By the mid-1920s, citizens and authorities alike held the view that the rise in disorderly conduct and violent crimes against individuals posed a significant threat to social order in both the city and the village. In response, judges increasingly sentenced individuals convicted of "hooliganism" to short prison terms rather than applying the more lenient noncustodial sanctions (e.g., compulsory work) recommended by the Criminal Code.[65] Despite the stiffer penalties and an antihooligan campaign launched in 1926, some citizens believed the situation called for still-harsher measures. The following letter, dated 24 September 1927, comes from Donskaia sloboda, Prigorod district, Kozlov county, Tambov province.

· 79 ·

Letter to *Krestianskaia gazeta* from the peasant V. M. Turovtsev on criminality and how to deal with it, 24 September 1927. RGAE, f. 396, op. 5, d. 30, ch. 1, l. 395. Original manuscript.

To the *Krestianskaia gazeta* editorial board.

Just a few lines (by way of discussion) on the punitive policy [against] the criminal element. Ours is a worker-peasant state, which is why the workers and peasants face many problems because of the criminal element. It [the criminal element], beyond any doubt, is 99 percent made up of workers and peasants. Our state racks its brain over how to correct comrades who have strayed from the correct path and nevertheless has wasted a lot of money on this. How many victims [have there been already]? And it continues, and, unfortunately crime is growing more sophisticated. If one goes, or, more accurately, went, from the outlying district when it is almost dark, one has to run home so as not to be stripped. And in the outlying villages they don't move when evening comes, and if someone happens to be [out] late, then [for him,] getting back to his apartment really frays his nerves. Thieves fill the bazaars and act brazenly, and [people] are afraid to say anything or point them out because they will be beaten. And in general we must put an end to this. When will a worker be able to go freely after work to some sociable place without defending his home from an invasion of thieves? And we have many recidivists who hide [this fact] from the courts and do not say that they are convicts. They say they have large families to support, and the courts investigate just enough to be done with it and unload the usual cases to the archive. Up to one thousand cases accumulate in one district; they pile up so that they'll cause a fire. There's no way to examine [all of them]. I am inclined to raise up social opinion. How will the localities answer? I propose that terror is needed against recidivism; only the grave will correct them, and the population should report those who live by stealing.

VASILY MIKHAILOVICH TUROVTSEV

Law-enforcement and judicial experts agreed that rural crime had an especially violent character. A 1926 OGPU report on hooliganism in the countryside claimed that "not a single wedding or holiday passes without fights due to drunkenness often ending in serious injuries and even murder."[66] That violence remained a regular and commonplace feature

of rural life may be seen in the letter of S. Kugorev from Teliatovka village, located near the Sergo-Ivanovsk station on the Moscow-Briansk-
Belorussia railroad.

· 80 ·

Letter to *Krestianskaia gazeta* from the village correspondent S. Kugorev
describing instances of hooliganism, 22 February 1927. RGAE, f. 396,
op. 5, d. 1, l. 673. Original manuscript.

The Sergo-Ivanovsk station, Moscow-Briansk-Belorussia railroad, has
become quite a respectable trading stop. There are two cooperatives—a
consumer society and an agricultural credit association—as well as a grain-
collecting point for the joint stock company "Khleboprodukt." Every day
peasants come here to sell their products, such as flax, flax seed, oats, etc. All
the transactions go smoothly, and those people who transformed a station
that was all but deserted during the revolutionary years into a vital trading
post should be saluted. However, there are things and occurrences that sub-
tract from the overall good. For example, hooligans have built themselves a
nest, and no one takes any measures against their transgressions; on the
contrary, it is as if all their hooligan outrages are covered up. There are two
militia men. To the appeals of citizens for help they respond: "Calm down
and go see the medic over there in the new building by the brick factory.
He'll look you over and bandage you up. Then take your certificate of the
infliction of injury to the people's court."[67]

The agricultural association gives work to these hooligans and thereby
gives them strong support in spite of the fact that there were many instances
of assault against peasants who were transporting their grain to sell to the
association.

They can always find a good reason to give someone a beating. For example, out of the blue they'll demand a bottle of booze. And if the citizen
should find this objectionable and refuse, then he's asking for trouble.

Those close to the cooperative say that the board members themselves
fear them and therefore do not refuse them work. There are instances of the
hooligan ringleader, Sakharov, going into the cooperative store and demanding a bottle. His demand suffices because [once] he grabbed a weight
from the counter and threw it at the shelves of goods.

Village correspondent S. KUGOREV. Please omit my last name.

This final letter, highlighting the side-by-side existence of economic
progress and the toleration of violently antisocial behavior, helps illus-

trate one particular feature of the Soviet 1920s. Under the NEP, the economy was recovering and trade was flourishing. In light of the civil war destruction, these were important and necessary achievements. Reaching this point had required the Party and the state to relax the reins of the proletarian dictatorship while devoting almost single-minded attention to economic affairs. In other, equally critical areas of life ambitions had to be trimmed and efforts at improvement abandoned. This resulted in what Moshe Lewin has called the "cultural lag."[68] Economic advance simply outran the Bolshevik's ability to transform byt and raise the cultural level of the country as a whole. This imperiled future progress. Without an accompanying rise in mass literacy and education, the growing complexity of the economy and the institutions that oversaw it would only deepen the estrangement between society and state. If left uncorrected, this falling-out could lead to either of two dangers: backwardness, poverty and ignorance, small-scale agriculture, and bureaucratized and corrupt officialdom could swamp the revolution in a petty-bourgeois tsunami; or the leadership, isolated from the masses and threatened by their unresponsiveness, could attempt to force the pace of change. At the heart of the problem was the relationship between the masses and the holders of power.

CHAPTER 5

People and Power

"Yes . . . I talked a great deal with the peasants at the stations. I know how to talk to them in such a way that they never guess I'm from the Party, and talk quite straight. It's amazing how hard, alien and incomprehensible life is for this working human being, the peasant, in our workers' and peasants' republic . . . They live side by side with us, see the Revolution with their own eyes, hear it with their own ears . . . and understand nothing at all. If one could only arrange a sort of meeting of all Russia and in simple words . . . tell them about everything."

Klimin smiled at her.

"Tell them . . . They won't understand. Haven't these same working peasants killed enough of our agitators and political workers simply because they preached Communism too openly and directly? They don't read our books and they make cigarettes of our newspapers. No, Aniuta, it's all much more complicated than that. We have to rebuild their lives. They are savages living side by side with us but still in the Middle Ages . . . They believe in sorcerers and for them we are only a special sort of sorcerers, benevolent at best. We must destroy those drab villages, those groups of dirty nests in which they roost, and put in a museum their primitive plows and harrows."

—YURY LIBEDINSKY, *A Week*, 1923

In *A Week*, the communist novelist Yury Libedinsky offers an honest account of the activities of a provincial Party organization during the civil war and the motley personalities who make up its membership. From the idealist to the careerist, from the bureaucrat to the man of action, Libedinsky portrays them all through characters that appear deeply human in their thoughts and motivations, in turn hopeful, frightened, inspired, and uncertain of the future. The novel also conveys communists' isolation from their surroundings in rural Russia. Only the Red Army garrison stationed in the town ensures the organization's safety by holding the local anti-Bolsheviks at bay. The exchange quoted in the epigraph between the romantically involved Anna Simkova, a former schoolteacher now engaged in cultural-educational work for the Party, and the Cheka boss, Klimin, expresses the two extremes that marked the Party's rural tactics: the first, empathetic, with a Populist-like inclination toward persuasion and a conviction of the irresistible power of ideas; the second, predisposed to the use of force. In the end, Klimin's brand of realpolitik proves closer to the truth. While the Red soldiers are off on an economic mission, the peasants massacre the town's communists—Simkova and Klimin included—in an anti-Bolshevik uprising that takes no notice of the comrades' diverse backgrounds, reasons for joining the Party, or desire to improve rural life. The communist presence in the town is only reestablished at the point of Red Army bayonets.[1]

In 1918, the Bolsheviks' lack of influence forced them to abandon attempts to extend the socialist revolution into the countryside. During the civil war, rural administration primarily arose in response to the need to destroy the regime's enemies and extract food from the peasantry, objectives that were achieved through the application of coercive, often violent emergency measures. In contrast, the NEP and the "face to the countryside" slogan spawned conditions that demanded a more stable and regular system of rural administration. Yet until collectivization, rural Party organizations remained weak. In early 1924, there was only one rural communist for every 250 peasants. In some areas the situation was much worse. In the agricultural province of Smolensk, there were only 1,712 rural Communist Party members in April 1924—a ratio of one to 10,000 working-age rural inhabitants. The Party apparatus was infinitely smaller, numbering only twenty-eight full-time responsible officials. Central authorities could reach into the village when necessary, but the observation that during the NEP years "beneath the surface of sovietization, the life of the peasant flowed on its accustomed way" accurately captures the prevailing atmosphere.[2]

Extending organized state authority into the village under these conditions proved extremely difficult. Following the civil war, native peasant institutions filled the vacuum created by the collapse of the old state and the weakness of the new. During the NEP, communes continued to outnumber village soviets (*selsovety*)—the lowest level in the state hierarchy—by between three and five to one. Where communes and soviets coexisted, the village assembly (*skhod*) remained the more authoritative body in peasant eyes, creating a "dual power" situation in the village. Attempts to balance the formal relationship between the two proved difficult. In October 1924, a VTsIK decree declared that the soviet was obliged to carry out all legal decisions taken by the village assembly. At the same time, however, the soviet was bound to implement instructions emanating from above via the district (*volost*) executive committee, and soviet decisions were legally binding on the commune. Thus, despite the clarification, authority between the two remained muddled, and thanks to the soviets' "dual subordination," to both the local population and to higher-level executive committees, peasants had ample grounds to question whose interests the soviets actually served.[3]

The question of formal authority might have been somewhat miti-
gated if soviets had worked as they were supposed to. They rarely did. A
1925 central investigation of the lower soviet apparatus revealed that
village soviets operated bureaucratically, without popular participation,
and as a result, most village soviets played only a superfluous role in the
life of the peasantry. One report from Vladimir province succinctly sum-
marized the problems: "The picture is the same everywhere. Not a single
village soviet actually works. Everywhere, instead of the soviet, the
chairman conducts all the work . . . No questions are dealt with that
excite the vital interests of the whole peasant stratum, such as the handicrafts industry, the use of meadows and state forests, or the underuti-
lization of land."[4] Given this situation, peasants naturally turned to
their own institutions to minister to their basic needs.

As a rule, every male household head had a say in the deliberations of
the village assembly. Democratic practices in the soviets, on the other
hand, were frequently observed only in the breach. It was common for
Party organizations or electoral commissions to prepare lists of candidates in advance, a practice that turned elections into rubber stamps and
discouraged voter participation. "There is nothing for us to elect. The
communists have already chosen without us," the OGPU recorded one
discouraged voter as saying in 1924. "Let the authorities simply name
the soviets," remarked another. Despotic soviet chairmen who abused
their exorbitant power also fostered resentment among the population.
Electoral commissions applied the laws on disfranchisement haphaz-
ardly and arbitrarily in a manner that varied from locale to locale. Successful peasant farmers—that is to say, those who employed labor or
had accumulated some degree of wealth—were frequently denied the
franchise on illegal grounds. All told, undemocratic and bureaucratic
practices combined with the soviets' dual subordination to instill mistrust and ensure that peasants adopted a passive, if not outright hostile,
attitude toward the soviets. The civil war demand for "soviets without
communists" could still be heard in villages during the 1920s.[5]

Peasant apathy came embarrassingly to light in the 1924 village soviet
elections when only 29 percent of qualified voters participated. Confronted by this election boycott, Moscow sought ways to increase peasant
involvement in soviet affairs. In October 1924, the Party leaders initiated a campaign "to revitalize the soviets" that attempted to eliminate
objectionable practices and to raise the prestige of rural soviets in the eyes
of the peasantry. Reelections were called for early 1925 in those places
where voter participation had been less than 35 percent. Instructions
from the center emphasized that these elections be freely contested with
minimal interference from Party organizations and that particular efforts

be made to get women to vote. (By tradition, female peasants did not participate in village assembly deliberations.) To attract better-off peasants capable of executing administrative duties, the center insisted on scrupulous adherence to the statutes on disfranchisement: in accord with the Agrarian Code, employing hired labor was not necessarily grounds for depriving an individual of civil rights. As a result, the first attempts at soviet revitalization deemphasized class as a criterion for electors and elected.[6]

According to E. H. Carr, the revitalization campaign was the political counterpart to "face to the countryside" and "wager on the kulak" that underlay the NEP. Through the revitalization campaign, Party leaders hoped to supplement weak rural Party structures with soviet deputies drawn from literate, albeit noncommunist, peasants. Mobilizing broad peasant participation in soviet elections and activities was not without its negative side for the regime, however. Party leaders had envisioned village soviets as the means to organize poor peasants, "schools of communism" for bedniaks and batraks. Revitalization, however, meant opening the door of the soviets to peasants from the upper economic strata who were best able to perform administrative tasks. This would help alleviate the rural administrative shortfall and, hopefully, strengthen support for the regime. But Party leaders also recognized that they were handing the more powerful, better-off segments of peasant society the organizational means by which they could dominate village politics. Indeed, as a result of the 1925 reelections, the number of communists (including Komsomol members) and poor peasants elected to village soviets declined.[7]

Throughout the 1920s, the number of rural Communist Party and Komsomol cells steadily grew. As a result of administrative-territorial reforms begun in 1922, however, lower-level administrative units (particularly counties [*uezds*] and districts) were amalgamated and their overall numbers reduced. This resulted in a decrease in the number of village soviets and greatly increased the distance between muzhik and state. Between 1923 and 1924 the number of rural soviets in the RSFSR declined drastically, from 80,000 to 50,000, a figure that remained fairly steady until the end of the decade. As of 1 January 1929, the rural population per soviet stood at 1,500 to one in the RSFSR and 2,300 to one in the Ukrainian republic. This development increased the leadership's concern that elements unsympathetic to the regime would fill the void and dominate village-level institutions. In November 1925, V. V. Kuibyshev, the head of the state's main inspectorate, warned the Council of People's Commissars that "the village soviets and the regional executive committees were letting the initiative on economic and cultural questions slip through their hands." These questions, he continued, were being "spontaneously" taken up in the village assemblies, "in which the poor remain disorganized and provide an obvious opening for anti-Soviet elements."[8]

There can be little doubt that the peasants considered the rural communists an alien presence, even an occupying force, against which resistance was completely justified. This was starkly obvious during the war-communism and collectivization years, at the militant terminals that marked the beginning and end of the 1920s, when organized violence and retribution became a matter of course. During each of these ordeals, in response to violence from above, the violence frequently visited on fellow villagers and family members was turned against the invaders and their indigenous allies, often in grisly ways. In the slaughter depicted by Libedinsky in *A Week,* for example, in what surely was a vignette drawn from life, the local economic council chairman—an organizer of grain requisitions—is disemboweled, and grain is poured into his slit-open belly.

Violence and resistance continued to punctuate relations between regime and peasantry even after the tax-in-kind ushered in the truce. Working far from the centers of communist strength, rural representatives of the regime made inviting targets for terrorist attacks. Vulnerable and proximate, Party and Komsomol members, soviet deputies, and those deemed Bolshevik sympathizers were exposed to the brunt of peasant enmity. At the end of 1924, rural terror against Soviet officials and their supporters was on the rise. In December (an especially bad month), the OGPU recorded fifty-nine instances of "individual terror" including thirteen murders, twenty-one beatings, and eight cases of arson. For all of 1924, the OGPU recorded 339 acts of terror against rural representatives of the regime. The following year the figure rose to 902; it dipped to 711 in 1926 but in 1927 jumped back up to 901.[9]

According to these figures, soviet officials were the chief targets of terrorist acts, followed closely by Party and Komsomol members. Village correspondents who zealously exposed the transgressions of both the authorities and influential peasants also found themselves in a perilous position. According to a Soviet-era history, between spring 1924 and August 1925 the authorities recorded 140 attacks on village correspondents, including 25 murders. Another source claims that 51 worker and village correspondents were murdered between 1924 and 1926, and that 75 other murders were attempted; this was in addition to scores of beatings, threats, and other confrontations. Party and press publicized these attacks to the fullest, usually ascribing them to kulak or anti-Soviet conspiracies regardless of whether their origins lay in local, class, economic, nationalist, or other grievances. The murder of village correspondent Grigory Malinovsky in March 1924, after he had blown the whistle on high-handed Party officials in the Ukrainian village of Dymovka, became a cause célèbre that called attention to the criminality and arbitrariness of local authorities.[10]

The Party initiated the village correspondent (*selskii korrespondent,* or *selkor*) movement in 1923 as a rural counterpart to the existing network of worker correspondents (*rabochie korrespondenty,* or *rabkors*). By early 1924, thanks in part to the Dymovka affair, the movement began attracting high-level support. Much like the village soviets, village correspondents found themselves performing a dual function. First and foremost, the Party hoped that they would form a peasant phalanx helping to extend communist influence into the countryside by promoting the interests of the village poor. Increasingly, however, correspondents came to serve the center as irregular channels of information detailing local authorities' abuses of power and peasant dissatisfaction with the rural Soviet regime. According to Steven Coe, it was this mission that most fired the interest of the correspondents themselves. Between October 1924 and December 1927, the number of village correspondents grew from 60,000 to 193,000. The literacy required of correspondents helped limit the number of true peasants "at the plow" who served as correspondents. Young men thirty years of age or less seem to have made up the majority, with Red Army veterans composing a significantly large portion of this group. Veterans could be expected to be loyal Soviet citizens who supported the Party's efforts to transform rural life and improve rural administration. As Coe has demonstrated, contrary to the wishes of the center that village correspondents remain semiautonomous, unofficial citizen-journalists, the correspondents themselves strove to formalize the movement and to parlay their position into one of genuine and recognized local authority.[11]

The "face to the countryside" policy and the weakness of rural Party and soviet organizations elevated the importance of the press as a means of affecting peasant life. *Krestianskaia gazeta* itself began publication at this time, in late 1923. In May 1924, the Thirteenth Party Congress issued a lengthy resolution concerning the press that detailed, among other things, ways the press might further address issues of concern to the peasantry. It called for an increase in weekly publications directed at peasants and for Party and soviet organizations to "comprehensively" assist village correspondents in their work. Accordingly, peasant newspapers were now to devote more attention to letters and complaints, to provide peasants with legal assistance and agricultural information, and

to explain "general political and economic issues, in particular the problem of cooperation" in an "accessible way . . . without any false over-
simplification or unnecessary vulgarization."[12]

In addition to isolated acts of violence perpetrated against individual
representatives of Soviet authority, the possibility of larger rebellions
had not entirely disappeared. At the end of August 1924, underground
Menshevik and other anti-Bolshevik socialist and nationalist organizations combined to lead an uprising in Georgia that drew strength from
the region's economic, social, nationalist, and political discontents.
In the heavy reprisals that followed the rebellion's suppression, nearly
four thousand people, along with the uprisings leaders, were executed.
Party leaders, especially those critical of the current policy of concessions to the peasantry, portrayed the Dymovka affair and the Georgian
uprising as evidence of kulak ascendancy in the village. Stalin, still
firmly supporting the "face to the countryside," took the occasion to
argue for a further strengthening of the smychka, the urban-rural bond.
In January 1925, he imparted a lesson derived from the Dymovka affair
to local officials: "Our local responsible workers look only towards
Moscow and refuse to turn towards the peasantry . . . Many responsible
workers say that it has become the fashion at the center to make new
statements about the countryside, that this is diplomacy for the outside
world, that we are not moved by an earnest and determined desire to
improve our policy in the countryside. That is what I regard as the most
dangerous thing."[13]

He went on to reprove local officials who suppressed criticism and
complaints while mechanically feeding Moscow rosy reports on the situation in their areas. This had long been a bone of contention between
central and provincial organizations; Moscow had to constantly struggle to obtain accurate, unadulterated reports of local conditions. It was
precisely for correspondents' role as an independent source of information that Stalin valued them; he described correspondents as "impres-
sionable [people] who are fired by the love of truth, who desire to expose
and correct our shortcomings at all costs, people who are not afraid of
bullets—it is these people who, in my opinion, should become one of the
principal instruments for exposing our defects and correcting our Party
and soviet constructive work in the localities."[14]

In general, Stalin held to the official characterization of the Georgian
uprising as an "artificial, not a popular, revolt" that had been "stage-
managed" by Mensheviks and other anti-Bolsheviks. He also ascribed
economic causes to the uprising, accusing "kulaks and profiteers" of exploiting peasant dissatisfaction with low state prices for grain and high
prices for manufactured goods. But neither of these factors, Stalin argued,

would have gotten so out of control if Party and soviet organizations had been more effective and more attuned to the mood of the local population. As long as these organizations continued to work as during the civil war—in isolation from the local population, employing authoritarian methods—then effective governance of the countryside would be impossible. Stalin made the revealing point that Menshevik support was strongest precisely in those areas where communists were most numerous. He therefore called for a complete alteration in the Party's approach to the countryside and the establishment of one more in keeping with the NEP. Authoritarian methods had to be abandoned. Communists, he warned, "must not domineer, but carefully heed the voice of the non-Party people"; the peasant, he insisted, has to be "treated as an equal."[15]

Even if Stalin truly believed that rural administration should or even could be made to take peasant sensibilities more into account, given everything that stood between Party and peasantry, it is hard to envision how they could have come to anything more than the most uneasy of truces. Memories of war communism were still fresh, and peasants continued to view Party members, first and foremost, as tax and grain collectors—in other words, as adversaries who, moreover, were intent on supplanting their institutions and traditions. In turn, with few exceptions, Party officials probably thought of peasants much as Klimin had, as no better than benighted savages mired in their primitive and superstitious ways. Trying to persuade them that change would be for the better was hopeless, and abandoning authoritarian methods, as Stalin suggested, would risk losing the toehold the Party had managed to establish in the countryside.

Documents

The Great Sorrow

As Nina Tumarkin has written, Lenin's funeral in January 1924 "was the first nationwide ritualized ceremony of mass mourning" since the revolution and served to sacralize not only the memory of the dead leader but, by extension, the sufferings the entire people had experienced since 1917. The funeral and its attendant rituals and propaganda campaign had tremendous political value as well, allowing Lenin's Party colleagues to present a united front to the nation and the world at a time of crisis and uncertainty. But the most enduring legacy of Lenin's death was its quickening of the emergence of a personality cult that had been growing steadily, contrary to Lenin's wishes, since the founding of the Soviet state. The death also inspired a flood of pious letters expressing love for the late leader and devotion to his ideas that, in general, reiterated the platitudes of the nascent cult. A tremendous number of these missives were published at the time, and more still remain preserved in the archives.[16]

Throughout the 1920s, the Lenin cult enjoyed further aggrandizement. Reports from across the country detailed the monument building and extensive renaming of sites in his honor. The embalming of Lenin's corpse and its public display in a specially constructed mausoleum on Red Square dwarfed all these efforts and provided the most visible sign that relic worship had been approved at the highest state level. This aspect of the cult not only appealed to religious and superstitious strains in Russian popular culture, it facilitated the mystification of power and transferred the aura that surrounded Lenin's personality to his political heirs.

To be sure, not everyone joined the rush to idolatry. Some found the practice distasteful and protested. A peasant, A. P. Poliakov, from the village of Yurievskoe in Kaluga province, discusses the detrimental effects that the cult was having on his propaganda efforts.

· 81 ·

Letter to *Krestianskaia gazeta* from the peasant and antireligious activist
A. P. Poliakov explaining the harmful effects of the Lenin cult, 10 May 1927.
RGAE, f. 396, op. 5, d. 30, ch. 1, l. 410. Original manuscript.

Back to the old ways.

Working in the countryside, discharging antireligious propaganda, I clash with adult peasants who say straight out, "So, there's no Soviet religion, eh? Sure there is. And there are relics, and there's certainly idol worship." By relics, they mean Lenin's mausoleum, and by idol worship, the large number of monuments. "Really," they say, "don't people go there to worship the 'relics'? Otherwise, what use are these monuments if not for veneration?" Personally, I do not deny the importance of Lenin's mausoleum. In fact, it is very interesting to view the leader of the proletariat who did so much for those who labor and who wrote and said so much. But I still don't understand the desire for monuments, and I can't see the need for them. I have come across the very same opinion in the newspapers, where they say that monuments are not necessary and that they are built to no purpose. On the pages of *Komsomolskaia pravda* from the 5 May issue, [no.] 585, my eye was caught by the unveiling of a new monument to V. I. Lenin the base of which is constructed of two locomotive axles. Why and what for? This money could have been better spent in liquidating the illiteracy of dozens of people. This would have been a better monument to V. I. Lenin and a step along the path of socialist construction besides. But monuments are not needed and are already an obsolete thing. If this is not a retreat backwards, it is a detour from the straight path of socialist construction.

POLIAKOV, ALEKSANDR PETROVICH

I say we should discuss this problem and resolve whether we need to erect useless and unnecessary monuments.

Lenin's death led to the circulation of many wild rumors, the more fantastic of which bear evidence that the mass of the population had only a tenuous grasp on the dynamics of Kremlin politics. One village correspondent's report, dating from January or February 1924, attributes rumors of a royal restoration to "kulaks" who circulated them in an attempt to undermine the new regime's authority in the eyes of the poor peasants. The author wrote under a pseudonym.

· 82 ·

Letter to *Krestianskaia gazeta* from the village correspondent "A. Diletant" on rumors of a tsarist restoration following Lenin's death, early 1924. RGAE, f. 396, op. 2, d. 118, l. 71. Original manuscript.

In connection with the death of Comrade Lenin, many people who [find] Soviet power hateful—those who prefer the old order, when the worker labored for the rich man—are circulating all sorts of lies and unsubstantiated rumors among the gullible mass of peasants. In the Ereshkovo stanitsa, in Poddubsk district, Verkhne-Volitsky county,[17] a certain kulak, Nemov, along with his like-minded companions, have agitated throughout the countryside, saying that with the death of Comrade Lenin all certifications[18] have disappeared and that Trotsky took them on his escape. And [Nicholas II's] brother, Mikhail, is on the way from Germany to Petrograd [to mount] the tsars' throne, while [Kaiser] Wilhelm is heading to Germany, and soon the Bolsheviks and the soviets will be done for.

But the bird sings well wherever it sits, and so it goes with our kulak. There were people who provided outright facts to undo the lying rumors spread by the kulaks. But the Rabkrin[19] cell is finding all the guilty ones and is investigating.

Ah, how these parasites really want a tsar.

A. DILETANT

Trotsky assumes a special place in the rumors of impending economic catastrophe and political machinations that sprang to life after Lenin's death. Numerous letters indicate that peasants did not fully understand the issues that rent the Party leadership into a left opposition headed by Trotsky and an anti-Trotsky majority led by the Stalin-Zinoviev-Kamenev troika. Many questioned the motives of those now attacking the man who had so recently been celebrated as a national hero for his civil war exploits and was frequently portrayed as Lenin's second-in-command. Clearly, though, not everyone accepted the official reasons for Trotsky's anathematization, expulsion from the Party, and, later, exile. A group of Siberian peasants from Altai province, for example, requested a through explanation of Trotsky's errors. "Then," they wrote, "let the workers and the peasants judge." Another letter stated, "[Trotsky's] expulsion was pointless. Lenin dies and a schism begins."[20]

On the other hand, mention of Trotsky, a Jew, often provoked anti-Semitic outbursts and denunciations of the ruling Bolshevik elite as a "government of Yids." According to the OGPU, in the city of Dmitrov rumors abounded that the All-Russian Congress of Soviets had expelled Trotsky "for being a Jew." Elsewhere there was talk that there was "a Jewish pogrom in Moscow and that Lenin is alive and has left the country together with Trotsky" (this was rumored in Irkutsk province); that "following Lenin's death mass arrests are occurring in Moscow; that in place of Lenin it will be Trotsky, and then the Jews will take power in their own hands; that war will be the cause of pogroms and dissension; that war is unavoidable since even while Lenin was alive, Trotsky demanded war; that they killed Lenin, and Trotsky was arrested and escaped; that Trotsky sent murderers in order to take Comrade Lenin's place" (Smolensk province); that "they poisoned Lenin, they tried to eliminate Kalinin, and now there will be a government of Yids [*vlast budet zhidovskaia*]" (Gomel province); that "in connection with Lenin's death rumors are circulating among the population that Lenin didn't die, that the Yids poisoned him while trying to take power in their own hands since Lenin allegedly said that it was necessary to replace the unified tax for peasants and the taxes on traders, but that Trotsky and all the Yids didn't want this" (Tver province).[21]

Life among the Comrades

Distance from the political summit, lack of trustworthy information, ignorance, anti-Semitism, fear, and other factors contributed to people's willingness to lend credence to the most outlandish rumors. Still, in large measure, personal experience shaped peasant views of the new regime, and the attitude of individual communists toward the peasantry helped determine that experience. Like the fictional Simkova and Klimov, Party members' views were far from uniform, and this led to variations in practice. Veterans of the civil war, such as Klimov, may have been more willing to employ "administrative" and "command" methods in their dealings with the peasants. The following anonymous letter from a Kazakh village describes the effects of the attitude that Lenin frequently denounced as "communist arrogance."

· 83 ·

An anonymous letter to *Krestianskaia gazeta* describing how local communists dominate village institutions, 19 May 1925. RGAE, f. 396, op. 3, d. 83, l. 21. Original manuscript.

Why the peasants find the communists so terrifying.

It happened that I had to spend two years in our Red district executive committee in the village of Kazanskoe, Kokchetav county, Akmolinsk province. When I arrived to spend the night, some peasants were gathering around and began to inquire how things were going. "Not badly," I answer. And before long we get to the point where they are asking, "Do you have many communists [where you live]?" "Not a single one," I answer.

"[Then] the people are fortunate. We have thirty of them. They do whatever they want. We have already completely abandoned the village assembly; we just don't go."

I ask [them], "Why don't you go to the assembly?"

They answer me, "What's the point of going? It make's no difference whether we're there or not. If someone has to speak up for his interests, he is deprived of his right to vote then and there or is arrested. Eighty-eight of our people cannot vote, and more than half the village want to leave just to get away from the communist officials."

Later on in the evening, I myself, personally, had to attend a general assembly to verify all this. I saw that no more than twenty people showed up for the general [village] assembly, and those that did were like frightened crows. And the Party members, like prewar gendarmes, are on the lookout for anyone who expresses an opposing [view]!

It's resolved—the end! And who there could make a decision if they need two hundred people and only twenty show up and they can't decide [anything?] Only three do—the village soviet chairman and the two others sitting beside him. And whoever dared to oppose them trembles in his skin—they take away his right to vote. But this frightens the peasants less than arrest. It is quite apparent that in the end our communists do not try to gain the trust of the peasants. Quite the opposite. They frighten [them] with various repressive [measures] and set themselves up as commanding officials. [They] do not serve as models for the working peasantry, but, on the contrary, each tries to remain a boss in order to enjoy his privileges and to live in comfort for as long as the possibility presents itself. At present, no reelections to the lower [soviet] apparatus will help us here in the sticks because our Party officials will not allow the peasants themselves [to hold responsible positions], but appoint whomever they want. And every [communist] will say that there is not a single bad person in our Party, at least until some fraud comes to light. And even then they cover it up—not just once or twice, either. They only admit the embarrassment later and throw out [the guilty person]. Therefore, I propose that *Krestianskaia gazeta* push for the appropriate oversight in the backwater of the Red district executive committee of Kokchetav county to observe all the shenanigans of our officials. And please write about it in the next issue of *Krestianskaia gazeta*. Maybe our state workers will be ashamed and will agree to correct themselves.

———————————

Besides describing power relations at the local level, this letter goes a long way toward explaining peasants' political apathy. That the village is, in this case, located in Kazakhstan—a Muslim republic—adds another dimension to the problem and sheds light on sovietization in non-Slavic regions. Although the writer does not specify, it is probable that the soviet officials in question were not drawn from the local population but were appointed from above, and most likely they were not ethnic Kazakhs.[22]

By 1924, Moscow was very much alive to the harm that local communists' abuses of power could cause. Speaking to a gathering of rural Party officials in October, Stalin blamed the Georgian revolt in part on the arrogance and "work style" of communist cadres and demanded that the peasant be treated "tactfully" and "as an equal." Stalin also took the occasion to criticize the promulgation of antireligious propaganda among the peasantry, which many in the center were rejecting at this time. As an antidote to communist isolation from the peasant mass, Stalin called for a "revitalization of the soviets."[23] Formally, elections to village soviets were direct and, therefore, among the most democratic in the USSR, but, as the letter writer above testifies, it was not difficult for local officials to cow the population to ensure the results they desired.

The letter also raises issues related to the competency of local communist cadres. Inevitably, the transition from an underground revolutionary party to a ruling institution demanded an expansion of membership rolls, and this necessarily had an effect on the character of the Party itself.

As the Bolsheviks accepted new recruits after 1917, Communist Party membership could no longer be taken to signify fervent devotion to the cause of proletarian revolution, nor could it be assumed that a prospective Party member had even a basic familiarity with Marxist theory or Communist Party policies. A Party card now ensured for its bearer employment, privileges, access to scarce goods, and the means to professional advancement. The expansion of privileges opened the door to the uncommitted, the careerist, and others who previously would never have considered joining the Party. Once victory in the civil war had all but eliminated the risk of physical annihilation, this development received a further impetus. During the 1920s, to ensure a minimum of political acumen as well as moral rectitude among the comrades, Communist Party growth was punctuated by irregular cycles of membership reviews and the purging of undesirables. An extensive literature exists on the qualitative effects that resulted from this change in quantity. These works emphasize that as the pre-1917 Party members—the Old Bolsheviks—declined as a percentage of the whole, the Party underwent a debasement. New recruits were less worldly, coarser, and, in many cases, all but illiterate—and not just in a political sense. Some historians have argued that Stalinist blandishments and methods appealed precisely to this new cohort, and in turn, it supplied the Stalin faction with a base of support among the Party rank and file.[24]

In the following letter to Stalin as head of the Communist Party—VKP(b)—from mid-1926, T. G. Burtsev, a purge victim, appeals for reinstatement to the Party. His autobiographical sketch highlights the importance that past political and other activities now assumed.

· 84 ·

Letter to Stalin care of *Krestianskaia gazeta* from T. G. Burtsev asking for reinstatement to the Communist Party, 4 July 1926. RGASPI, f. 17, op. 85, d. 486, ll. 1–4. Typed copy.

To the Central Committee of the VKP(b)—Comrade Stalin.

From a citizen of Korovino village, Solntsevo district, Kursk county and province.

Where is there any truth? In 1917, I actively took part in the revolutionary movement in the Black Sea Fleet in Odessa, in the suppression of the counterrevolution. In 1918, I voluntarily joined the struggle against the Germans and the Gaidamaks in the October Revolution.[25] I was a member of the 1st Black Sea Brigade of Sailors under the leadership of comrade Makrovusov, [who] fought outside Odessa, and from whom I transferred again to comrade Trunov. I also fought in a partisan brigade that defeated the Germans and the Gaidamaks. In 1919, during the invasion of Denikin's White army, as an active Party member, I came up from the rear . . . [word not clear in original], despite the fact that I had been wounded three times, I headed for the battle around Kharkov and retreated to Tula, but from there, we again began to attack the White partisans.[26] Returning from there, [I found] that everything in my house had been stolen—that's what they're really after. Despite this, I set out again to fight against Markov's regiment. I [fought] against the Devil's Hundreds and drove them to the Black Sea.[27] In 1920, I was assigned to the First Division of the Black Sea Fleet as a naval artillery specialist on board the gunboat *Rosa Luxemburg*. I fought on the Wrangel front near Perekop . . . we also carried out a raid on a Cossack camp where prisoners of the White army were held . . . [and captured] booty as well.[28] With the liquidation of the front, the Army Political Bureau assigned me to work in the Special Department of the Navy in the city of Kherson, where I engaged in the struggle with counterrevolution, etc. Later, [I was sent to] our port of Kherson [to fight] speculation and counterrevolution, and from here I was demobilized and returned to my place of residence . . . [29] In 1922, I was elected chairman of the village soviet of . . . village, Nikolsk district. Later, I became a member of the district executive committee. In 1924, I also held the elected position of chairman of the Korovino village soviet, Solntsevo district, until 11 June 1925. Then . . . Comrade Mozgovova removed me, and they put me on trial for reasons unknown to me. They took me to the militia, where I stayed for five days. My case was transferred to the people's court, seventeenth ward [*uchastok*].[30] The court acquitted me because I had not embezzled, but my money, fifteen rubles, is gone, and I am still without work, but for some reason several of our Party members look [on me] badly, as if on an animal. I have also been a revolutionary since 1917, and a Party member since 1919. But I was kicked out [of the Party]; because of an illness I lost my Party card—([I was sick] for six months). In 1924, I requested that they [re]establish my [Party] tenure. They accepted me as a candidate (protocol 44), but I am again [a] non-Party [member]. The obstacle is Comrade Mozgovova. Now I have again requested [to be reinstated], but again Comrade Mozgovova blocks it [my request]. I am a union member [and] a poor peasant. I have nothing [but] one dilapidated hut. [I was] wounded three times [and] have lost 50 percent of [my] health. I am asking them to give me a job in my specialty. They won't give it to me, but they take someone who, during Denikin's advance, sold out us members of the Russian Communist Party. He put up a senior policeman and his former landlord [in his house] . . . [He] got the chance to save [them]; even though they took his livestock he did this. Comrade Stalin, what of my revolutionary service? Where are my October victories? The law offers no justice here on earth. I am asking [you] to issue an order or to give [me] an answer. Are those comrades who ignore my revolutionary service correct? I fulfilled my duty with the best comrades, fighting for the liberation of the toilers from the yoke of capital and for the unity of the toilers of the whole world [according to] the words of the late Comrade Lenin.

Long live the worldwide revolution. I ask that an answer appear in *Krestianskaia gazeta*.

Signed

––––––––––––––––

Burtsev's letter seamlessly combines several recurring genres of Soviet public letter-writing that have been identified by Sheila Fitzpatrick. First, it is an appeal for justice, a plea to right the wrongs endured by the writer. Second, its lengthy recounting of past service is a form of "confessional" in the sense of a statement of convictions that highlights an individual's loyalty to the Soviet cause. Finally, it contains an "abuse of power" denunciation in which Burtsev endeavors to turn the tables on his Party superiors by exposing their misdeeds.[31] Although its primary purpose was Party reinstatement, we can easily imagine that in penning such a multifarious letter the aggrieved individual was also satisfying his desire for both revenge and catharsis. It is also worth noting that while the letter was intended for Stalin's eyes, it was addressed to *Krestianskaia gazeta* with the full confidence that it would be forwarded to the general secretary.

Moscow Is Far Away

Rural Party organizations faced an acute shortage of qualified personnel. Fewer than 4 percent of rural cadres had any schooling beyond the primary level, and more than 35 percent were either illiterate or self-educated. Proper political training was also wanting. Instances revealing political naïveté could be sadly hilarious. In a report memorializing the 9 January 1905 massacre of demonstrating workers in St. Petersburg—the infamous "Bloody Sunday"—one Party cell secretary exhilaratingly concluded, "The cause of Gapon and Zubatov [i.e., police unions] will be fulfilled to the end!" In the exchange between a Party instructor and his charge it is hard to decide who has committed the more egregious error: "Question: 'Who takes the place of the Tsar now?' Answer: 'Before, it was Lenin, now it's Rykov.'" Purges intended to rid the Party of such artlessness were undertaken in all seriousness, but at the conclusion of a campaign institutions tended to return to business as usual. As one Party official said, "The threat has passed; now we can [put] the rule book aside."[32]

Given this low level of political awareness and the overwhelming practical problems the rural communist officials faced on a daily basis, it is not surprising to find that the momentous intra-Party struggles under way in the Moscow Kremlin throughout much of the 1920s did not concern them much at all. Navigating the esoteric, often theoretical debates was a challenge for the initiated; they were certainly beyond the understanding of individuals who were hard pressed to name the members of the Soviet government. As Fainsod observed in his study of Smolensk province, local Party organizations, in the end, simply followed the lead of their superiors, denouncing the opposition and supporting the status quo.[33]

If Party members found themselves in the dark in relation to what was happening in Moscow, non-Party members were all the more bewildered. This is brought out in I. P. Vostryshov's letter from Bolshoe Boldino village, Nizhny Novgorod province.

· 85 ·

Letter to Trotsky, care of *Krestianskaia gazeta*, requesting an explanation of his disagreements with the Central Committee from the peasant I. P. Vostryshov, 14 December 1927. RGAE, f. 396, op. 5, d. 30, ch. 1, l. 90. Original manuscript.

To the *Krestianskaia gazeta* editorial board.

We, the indigent.

I sincerely request that the editorial board forward this letter to Lev Davydovich Trotsky personally, at his address. I ask the editorial board not to refuse this request.

Respected comrade and our supreme leader, Lev Davydovich Trotsky.[34] Allow me, a poor peasant, to say a few words to you. I am a beggar—a cripple living in poverty—but I won't dwell on this. I am deprived of the ability to acquire the funds [needed] for the acquisition of articles of basic necessity and am also deprived of the opportunity of enjoying an annual subscription to a newspaper or journal. But all the same, even by reading fragments of newspapers, I have discovered in these grimy shreds and remnants of newspaper articles that disputes are going on in our Communist Party. Since the newspaper editorial boards take one side, not yours, they place the entire blame for the dissension in the Party on You and your supporters. And, if one were to trust these newspaper articles, then it seems you and your supporters are no longer the friend of the poor but its enemy. When I found out from these newspaper articles, these fragments and scraps of newspapers, I became quite sorry for you and your supporters—not, of course, [out of] an animal attachment to you as an individual, but to you as a distinguished state and public official who [enjoyed] a close bond and strong alliance with our great leader—Lenin. Together with him, you looked squarely and boldly into the eyes of death for the cause of the oppressed and unfortunate workers and peasants. You did not lose heart, but stubbornly fought against all adversities and privations. Now, it seems, that no sooner had the life and death struggle to strengthen Soviet power against all its enemies concluded—just as the entire country began to recover from the wounds of the general collapse—then deterioration began to steal into our Party. You, or as they label you, the oppositionists, are not enemies of us, the poor. After all, there were attempts [on your life], and [you have been] slandered by capitalist states and organizations. So, respected Lev Davydovich, be so good as to take the trouble to tell me in a letter exactly what you and your supporters find objectionable in our Central Committee of the VKP(b)? With precisely what actions and prescriptions do you take issue? What are you accusing the present administrative apparatus of, and what do you consider its shortcomings and mistakes? And what is Your opinion of all this? How do you and your supporters view the poor peasantry in general and the dispossessed class of peasants and workers, and [what is your general] opinion of the current situation of our Party? Comrade Trotsky, you know that there remain poor and impoverished people among our peasantry. Would you believe that I don't even have a miserable two rubles for an annual subscription to *Krestianskaia gazeta* or any other such [newspaper]? Reading newspapers over a full year is most interesting and beneficial if you aren't interrupted from day to day. But I am deprived of this opportunity, and if I get [the chance] to read, then it's like this: grab a book, open it to the middle, and read a page. What's the point of this, what can be understood from this [type] of reading, and what use do you get from this type of reading? Both the book's content and its usefulness are difficult to grasp reading like this, without the beginning and the end. This goes for my newspaper reading, too. By the way, reading like this creates the strong desire [to know] what comes next. What lies ahead? I believe that reading newspapers is interesting to everyone and especially to those who defend the rights of the poor. After all, newspapers like these are only published here in the USSR. I know that the editorial boards of our newspapers now get heaps of letters—but who writes the most? Both the kulak and the well-off peasant, and the poor, and our brothers—the indigent—write, and everyone is crying about their [situation] and themselves and those like them, and everyone is defending their rights. We, the indigent, write very little since we read so few newspapers. Because of this, Soviet power should not think that, say, if the poor are silent, it means that they are living well under Soviet power, but the kulaks [are living] badly because they sob and cry. No. The poor peasant sits cold and hungry, but is silent. He is sullen and not talkative. He is accustomed to keeping quiet and submitting to everything and everyone. Experience has taught him that tears of grief won't help. That is why he is quiet and seems satisfied with everything. He is used to debasing himself and to holding in regard all sorts of trash. So, dear comrade, be so kind and answer my letter. This is not for my personal interests, but for the interests of everyone like me.

I. VOSTRYSHOV

Vostryshov's assertion—his letter to the contrary—that the poor were less likely to write to newspapers is difficult to quantify. The letter does poignantly attest to the popular hunger for reliable facts and information about the political situation in the country. But to the extent that the peasantry took sides in the Kremlin political drama, it was not necessarily out of ideological sympathy for one or another faction. Their concerns are explained by the peasant A. F. Sdobniak, of Petrikha village, Nizhny Novgorod province.

· 86 ·

Letter to *Krestianskaia gazeta* from the peasant A. F. Sdobniak explaining the peasants' view of communist opposition groups, 5 January 1928. RGAE, f. 396, op. 6, d. 114, ll. 433–436. Original manuscript.

The peasantry and the opposition in the VKP(b)

On the whole, the peasantry has little interest in the internal Party struggle that occurred at the Fifteenth Party Congress. If at some point it expressed its dislike for the opposition, then this is not because it feels the opposition's line is incorrect, but because it does not want a struggle but wants "a calm, quiet, and peaceful life." The peasant is certain that any political struggle, a struggle at the top between the leaders, will unquestionably affect him and, in the end, his economic [well-being]. He has only one wish—to be left in peace. Naturally, this leads the peasant to protest against that which threatens to upset this "tranquility." That is why the peasantry learned of the exclusion of the opposition from the Party with a feeling of satisfaction, deciding that this eliminates a threat to its peace. Besides this, for the peasantry, the names of the opposition leaders, the names Trotsky and Zinoviev, which are better known to the peasant population than others among the leadership, are associated, like a nightmare, with the difficult memories of the "war communism" period. That is the main reason for the peasantry's hostile relation to the opposition whose leaders are Trotsky and Zinoviev, names that it cannot stomach. That is why the peasantry approves of the exclusion of the opposition from the Party. In this action, the most advanced elements of the peasantry perceive a shift in the Party's policy toward the peasantry. They are sure that Soviet power, which is under Party leadership, must lean more on the peasantry and not on the workers. They are sure that the formation of the opposition bloc and its exclusion from the Party's ranks serves as proof of this shift.

I ask the editors to tell [me] the fate of my letter, if it will somehow be printed.

SBODNIAK, ALEKSEI FEDORIVICH

Notwithstanding this letter writer's characterization of the peasantry as essentially apolitical, his letter reveals that peasants were astute enough to see in the opposition program a threat to the gains they had made since 1921. Still, his point that the village remained wary of, but detached from, the power struggle occurring at the Party summit carries weight.

Experience as a Guide

A letter from N. F. Yelichev, a peasant living in Makarovo village, Yaroslavl province, contains a sharply critical evaluation of Bolshevik rural policy. The letter appeared on page one of the 10 October 1927 issue of *Krestianskaia gazeta* in the "Miting" (Mass Meeting; Rally) column dedicated to the ten-year anniversary of Soviet power.[35] Yelichev's critique elicited a flood of responses to the newspaper editorial office. Some of these letters, mainly those censuring Yelichev, were printed in the paper; more remained unpublished. Yelichev's letter, while quite long, touches on a broad spectrum of problems that were being widely discussed and is worth reproducing in full.

· 87 ·

Letter to *Krestianskaia gazeta* from the peasant N. F. Yelichev, who is responding to M. I. Kalinin and comparing Soviet agricultural policy unfavorably to that of the tsars, 10 October 1927. From Ya. Selikh and I. Grinevsky, eds., *Krestiane o Sovetskoi vlasti* (Moscow-Leningrad, 1929), pp. 198–201.

I salute the all-union "Miting" column in which a citizen may say what he thinks is right and wrong in the administration of our proletarian state. I think that there should be no talk of censorship, that is to say, of keeping one's mouth shut. We must speak the truth.

First, I direct attention to the opening address of the "Miting" [column]'s chairman, Mikhail Ivanovich Kalinin, in which he poses the question: "Who can say that it was better under the tsar than [it is] now? If you can find such a complainer or whiner, then let him address our rally."

In this speech, Comrade Kalinin sounds somewhat imperious, not welcoming the rally but challenging [it]: "So go ahead and squawk." These challenges are unnecessary. In a word, "authority" often puts the people in a bind. If you ask the former Whites, then they will no doubt tell you that even Denikin asked them: "So tell us, then, which government is better, ours or the Bolsheviks'?"

And people, feeling that authority over them or [understanding the meaning of] "So go ahead, squawk," render divine honor to authority while damning any authority in their hearts. After all, any authority is force, and who is happy with force? So, I would suggest that Comrade Kalinin retract his words "find the complainer or whiner who says that under the tsar it was better and toss [him] out." These words may discourage the desire to express the truth. They are a victory for careerists and toadies but force the oppressed to remain silent.

I have to say that under the tsar the trading and marketing of agricultural products was much better organized. There was competition, [and] the peasant ruble equaled the trade ruble and not merely half a ruble, as now.

And now what? In prewar days, cooperation [attained] the very highest height, especially [in] agriculture. It [the cooperative movement] had factories, passable roads, steamships. It had offices in foreign countries. It supplied seeds to the peasants for 10 percent a year, but now the agricultural bank gets 100 percent and more. It buys sixteen kilograms of barley seed from the peasant for ninety to ninety-five kopeks and sixteen kilograms of oats for [between] one ruble, fifteen kopeks and two rubles, thirty kopeks. What do you say to that?

And how much do the trusts and syndicates get? Our national peasants' agricultural cooperative has been wiped out. And, in general, the entire cooperative [movement] is in ruins. It has been replaced by two state overseers of the most bourgeois type: the procurer—who takes [produce] at will, and the distributor—who directs [its allocation]. Every day our own press sings hymns to the growth of cooperation—its advantages over individual [labor]—but in fact this is not so, and cannot be, since there isn't any cooperation.

In the trusts and syndicates, members of the administration get no less than five hundred rubles a month in salary, plus travel and daily [allowances]; this isn't even counting their share of the capital. The Ukrainian trusts had a revolutionary commission with 1,255 members, and they received an allowance of 117 rubles a month for each member of the revolutionary commission. This is for just two or three days' work a month. That's why anyone can suck the peasant dry with impunity under these conditions, and why the hare shouldn't taste the bear's ear, as Krylov said. And even the miller has started to get fifteen kopeks for sixteen kilograms instead of four to five. And the smith, instead of the usual ten kopeks, has begun to charge thirty-five kopeks for shoeing [horses]. And the bootmaker gets fifteen to twenty rubles for a pair of boots instead of the usual five to six rubles. Raw materials and agricultural produce have now reached prewar [price] levels.

Who will say that my words are incorrect? Why dress up a crow to look like a peacock? Nekrasov was not mistaken when he said already in the [18]70s, "Show me a situation [*obitel*] where a Russian peasant would not moan." And [now] he has begun to moan even more.

Now I will touch on the agricultural tax that all the papers claim is so small and so scrupulously assessed that nowhere is there such an equitable tax as ours.

We all remember the slogan "All land to those who work it." And it was done. They seized [the land] from the landowners; they also seized the beggarly redemption allotments from the peasants for the state. And everything was assessed at a light rate: model settlements that use experimental seeds that increase the productivity of their fields with little extra work they assess at seventy-five rubles a hectare, but non-model farms, as well as those [peasants] who prefer outside agricultural work and peasants who happen to live near cities or factory settlements. pay [only] thirty to forty rubles a hectare.

Is this right or not? Let anyone speak up. I'll move on to agriculture. To the best of my recollection, before the war there were about thirty-two million horses in Russia; now there are about fifteen million.

There is a great shortage of horned livestock. We have failed by 20 percent to reach the prewar level. In both 1925 and 1926 we [the number] declined.

Readers of *Krestianskaia gazeta* no doubt will remember the report of one Moscow province agronomist in which he explained that in the province the number of horned livestock declined by 20 percent for the year. Further, our Yaroslavl province representative, Comrade Sentsov, reported to the All-Russian Central Executive Committee on 27 December 1926 that in Yaroslavl province 10 percent of the horned livestock were lost that year.

The decline in horned livestock deprives the earth just a little bit of the most precious fertilizer—manure. In general, we use no more than three tons of manure per hectare, while abroad, in Germany, Denmark, Belgium, and Holland, they use thirty-six tons per hectare. Besides this, they use more mineral fertilizer.

All this proves our poverty. Therefore, the authorities need to look on the peasantry not as a deep pocket [*koshelyok-samotrias*] but as something dear that is now in a weakened state.

What's more, speeches are pouring forth, almost [like] commands, [saying] that we need to increase output five times and more per hectare. Do these people think about what they are saying?

What will we give [to celebrate this] October [anniversary]? Will it be only a piece of paper adorned with all the blessings of a horn of plenty that even shames the one who draws it?[36]

N. F. YELICHEV

Yelichev's well-informed critique of agricultural policy combined information gleaned from the press with personal experience and knowledge of local conditions. His description of a cooperative movement in decline and the parasitic role of bureaucrats and traders undoubtedly found sympathy among many peasants. Noteworthy, too, is his combination of folk sayings with references to Krylov's fables and Nikolai Nekrasov's poetry, which indicates a literary bent and skill at communicating ideas to peasants. Still, others had different experiences and challenged his interpretation of events. A good example is M. V. Kiselkina, a peasant woman from Voloshko village in Leningrad province, who wrote to *Krestianskaia gazeta* in October 1927.

· 88 ·

Letter to *Krestianskaia gazeta* from the peasant M. V. Kiselkina in response to N. F. Yelichev's letter (no. 8), 11 October 1927. RGAE, f. 396, op. 5, d. 210, ll. 95–96. Original manuscript.

I salute the all-Union peasant "Miting" [column] honoring ten years of the October Revolution, and in the process I wish to respond to the faulty understanding of soviet construction and kulak views of citizen Yelichev in his report printed in *Krestianskaia gazeta,* no. 40.

Yelichev says: "It was better under the tsar, that the market for agricultural produce [was better] and the peasant ruble equaled the trade [ruble]." I recall that in tsarist days we peasants, undoubtedly Yelichev too, handed over our produce cheaply, for example: A calf sold for two rubles, fifty kopeks. A pud of cow's butter for eight to ten rubles. Eggs at next to nothing and even cheaper for one. Maybe, having the means, Yelichev could hold onto his goods [until prices improved] and his ruble equaled the trade [ruble]. Then Yelichev points to the high wages of syndicates and trusts but says nothing about the enormous sums that were needlessly spent under tsarism on the upper clergy, gentry, and others, but the Soviet government pays high wages based on one's work. Then Yelichev says that they suck the peasant dry from all sides. This is also untrue because we have only a single agricultural tax and insurance, and at that, every year they decrease and the propertyless are completely free from them. Under tsarism, there were all those debts and all those exactions that were never tallied up, and on top of this the poorest were gouged the worst. Further on, Yelichev says that at the millers' they charge fifteen kopeks to grind a pud of grain, and that in tsarist days they charged four to five kopeks. But at the present time they only charge us five to six kopeks to grind a pud of grain. The smith, he claims, charges thirty-five kopeks to shoe one hoof: he is wrong here too. I myself took a horse for shoes and was only charged ten kopeks a shoe, just like under his wonderful tsar.

Now, in conclusion, I want to say how, ten years after the revolution, our village is unrecognizable. Under tsarism, it was half filled with black huts, but now there are forty-five houses with white stoves and new outbuildings. Also, it's been three years since it changed to the seven-field [system]. Three fields already are in clover, about thirty desiatinas, thanks to the help of an agronomist, and they got clover seed from the credit association, of which they became members. About eight desiatinas of meadow have been drained. They improved the road by digging a [drainage] ditch. Before, half the village had no horses, but now they have acquired young horses so that one-third [in the village] have a pair, and only two homesteads are without horses. We are [increasing the number] of horned livestock and pigs. A school opened in our village only after the October Revolution. There has also been a liquidation point for two years, and now there are almost no illiterates in our village—only those old folks who do not want to learn. Also, a Red corner has been open for five years and a children's nursery for two. On this note I will conclude.

M. V. KISELKINA

Based on the fortunes of her village, Kiselkina counters Yelichev's criticisms in a way that underscores the diverse experiences of different regions under Soviet power. She also quickly, and threateningly, raises the class issue asserting that if Yelichev fared so well under the tsar, he must have been a kulak. This may be taken as a small sign that, ten years after the revolution, toleration for any positive assessment of the pre-Soviet past was on the wane, and class tensions in the village were rising.

Yelichev was not alone in equating political power with force. This theme had appeared in letters criticizing state policy since the revolution. By the end of the 1920s, the number of such letters had significantly increased. Here is a particularly hostile anonymous letter from August 1928.

· 89 ·

A hostile anonymous letter to *Krestianskaia gazeta* condemning the Bolsheviks
and their rural policies, 13 August 1928. RGAE, f. 396, op. 6, d. 114, l. 631.
Original manuscript.

———————————

Little mother, *Krestianskaia gazeta,* damn the tyrants and torturers and
the entire Soviet government—Rykov, Kalinin and his whole gang—to hell.
Damn you on behalf of all the laboring folk, damn you murderers on behalf
of the laboring peasantry. You seized the land from the tiller of the soil and
gave it to the loafers who neglected it, and you drove us, the tillers, into the
swamp. But even in the swamps, we have sweated away our lives and crip-
pled our children. Anticipating famine, everyone took our grain, and you
took our livestock from us. When we, weary from our daily labor, take an
evening ride on our horses, your bands slaughter our families. You and your
bands fatten up off our labor. We are hungry and have no strength to continue our heavy labor. O murderers, you've filled your bellies with the blood
of the laboring folk, but the time is fast approaching when it will ooze back
up again. O insatiable predators, you are cutting the branch you sit on. The
time is not far off when you will fall head first. In all creation there have
never been villains like you. We, the toilers, do not have the strength to suffer any more. We see that you are leading us to our doom. But better we
perish with honor than continue to live in shame. Every famished province
is fighting for itself. If only you would understand, it is the toiling peasant
alone that feeds every living thing in the world. Life under Soviet power is
for the bandit, the robber, and the loafers. Sleepy and fat, they come to our
cooperative. They make off with our produce while our wasted and exhausted wives stand [in line] all day and leave with empty arms. Japan rescued the Chinese laboring folk, and noble England and the most worthy
pan Pilsudski, together with Mister Chamberlain, are rescuing us.[37] Again,
damn you tyrants, thieves, bribe takers, and robbers of the entire laboring
peasantry. You villains pretend to build nurseries, but the ten tots there are
cared for by five prostitutes, who, along with the directors and the teachers,
are getting fat and almost every night take home a sack of grain while we
work our fields from morning to late at night. You haven't built a nursery
with our heavy labor, but a brothel. Do you not live off our labor? We hand
over our first [portion] to you, the second piece we give to your bands and
loafers, and the remaining third you take, you seize it from us. Why do you
force us, the hungry, to work for you, why do you pounce like beasts on
those who feed you, you damned torturers? You openly send your bands
and shoot us down. It is easier for us to die right now than suffer any longer.

A letter from the entire laboring peasantry.

———————————

This letter is the language of apocalypse, and the writer makes no effort to speak in an idiom that the Bolsheviks may find acceptable. In anger and desperation, the author damns not just individual leaders but the entire Soviet rural structure, which he sees as no more than the systematization of a brutal exploitation and expropriation that can be escaped only by dying.

Far from every letter to state and Party leaders spewed such venom or sounded such dark tones. Other individuals took a more nuanced, less overtly hostile approach in the hope, perhaps, of getting a fair hearing for their complaints. Before launching into a critique of rural policy and the behavior of local communists, I. G. Shokin, of Rumiantseva village, Ulianovsk province, affirms that higher Party officials, in contrast to their local counterparts, thought and acted in the best interests of the people. The following document consists of excerpts from his letter.

· 90 ·

Excerpts from a letter to *Krestianskaia gazeta* from the peasant I. G. Shokin criticizing local officials and the unfair application of the "kulak" designation to productive peasants, 10 October 1927. RGAE, f. 396, op. 5, d. 210, ll. 1212–1216. Original manuscript.

[You ask:] How is it you are still alive? And you get back the answer: Because now we have to work twice as hard, we dress worse, eat no meat, and there's no white bread on the holidays. But ask them: Which is better, the tsar or Soviet power? And you get the unanimous answer: Only Soviet. How can this be? you ask in disbelief. Because, they'll tell you, Soviet power is ours, and we can see that it is trying with all its might to build a better life for the laboring folk. It just hasn't come upon the right path, but it is looking [for it]; we see that it is looking day and night, and we have faith that it will find [it]—life itself will show the way. I think that the advanced peasant-cultivators should heed this advice. The trouble is, it's impossible to say anything to you [the leadership] that you don't like. Among yourselves at the top it's possible [to say these things], but here, at the district or county assembly, just try it, especially if you are a good farmer, mature and non-Party. For this [they'll grab] you by the collar and [drag you] to the GPU, and if you are a Party [member], they'll kick your tail out of the Party as a worthless element. And an honest person, who loves his government and brother peasants, cannot go and be heard at the congress. Maybe then these meetings would be a bit more "straightforward." Maybe this would be of more use than just singing hymns at the congresses.

There is nothing good to be said for the agricultural tax that all the newspapers childishly extol. Here there is a contradiction in the government itself. On the one hand, it cries with all its might, "Increase agricultural [production] and the country will thrive." But on the other hand, there is the tax: and they say that if you increase it [agricultural production] so much that you earn more than one hundred rubles per person per year, then we will take one-fourth of your wages away from you. But the question arises, what if you work the land so much that the loss of one-quarter of your earnings was worth it? Then you will be a kulak besides, and they will hunt you down like a wolf. They say that this is done to prevent the appearance of the dangerously rich from among the peasants. But is it really possible, on the land in particular, especially now when the grain ruble is twice as low as the trade ruble, [that] "from righteous labors will come palaces of stone," and will someone, somewhere, get them by their own labor?[38] And if, in fact, some peasants, through extreme diligence, should begin to earn revenues in excess of what they must spend on necessities, must this really be prevented, is this really dangerous, must these peasants' surpluses really be put away in the trunk? No, he too is drawn to the light, to a newspaper, to a book. He wants to teach his children. He wants a breathing space from the punishment of laboring twenty hours [a day] so that during that respite he can read about how best to sow grain. He wants to organize his land-holding in a much more proficient way [*kulturno*]. In accordance with what I've written, what this means for others will become clear. This will be of advantage to everyone: himself, the poor, and the state. In a word, when we decipher this law, it amounts to the following: "We are trying to help you peasants recover; it's just that we won't allow it."

[. . .] Village Party members, with rare exceptions, perhaps, are *oprich-niks* through and through. Hiding behind the pretext of zealousness and defense of the people's interests and authority, they recklessly bring tyranny and violence into being. In fact, they are identical to counterrevolutionaries, despoilers of that which the central authority is wearing itself out building. Even in the county, Party members are completely different, and in the province they are as different from the village [Party members] as the sky is from the earth. I have worked in the village, the county, and the province—in all [I have held] elective offices for ten years—and I know that in the province, if you speak the truth to Party members, they respect and appreciate you all the more. But in the village, [if you] don't grovel before them or [if] you reproach one of them, then he will set his entire village fraternity on you, and in the end, before you can turn around, you'll be on trial or slandered, and even if you're found to be in the right, you've already been covered in shame, you're an invalid.

Finally, in closing. Go incognito to the village assemblies, and you will hear the hirelings and careerists. In their speeches, according to my calculations, 3 percent of all the words they say are "your kulaks, kulaks," when, in fact, there's not a single one in the entire village.

It is already quite clear that the middle peasants have to accept these compliments at their expense. What do you make of all this? To me this is public hooliganism. And the newspapers? Read any one and in each you'll find: "Strike a blow against the kulaks"; "They gave the kulaks the forest"; "The chairman of the village soviet or chief forester is in cahoots with the kulaks," etc. But if you inquire of this village by mail, "How many kulaks have you?" they'll answer you, "Not one so far." Likewise, not one newspaper can make do without scoffing at religion. This is already not public hooliganism, but "hooliganism against the whole world." After all, it would be a whole lot better if the proletarian state would [serve as] an example of politeness. Is it really so impossible to choose another word in place of "kulak," an inoffensive word? For example, substitute the word "prosperous." In the village the difference between the prosperous and the poor is that he wears more becoming clothes and shoes, reads the newspaper, and isn't an inebriate. Of course, there are very poor peasants and widows. In short, I propose that the word "kulak" be taken out of use completely since it has begun to be considered a curse word . . .

So let's eliminate all the unrewarding abnormalities and conflicts that have a place in our Union family. This will mold everyone into one, single fighting host [*bogatyrskoe telo*] that won't be afraid of the foreign kulak.

Written two months before the Fifteenth Party Congress resolved to increase pressure on capitalist and kulak elements, Shokin's letter indicates that class relations in the village had already noticeably deteriorated. In addition to criticizing censorship, taxes, and an incomprehensible policy regarding productive peasants, Shokin also sharply distinguishes between venal local officials and well-meaning higher authorities, a political dualism that resembles traditional peasant faith in the benevolence of a supreme power holder. According to this worldview—dubbed naive monarchism by historians—local authority is corrupt and arbitrary, and true justice can be dispensed only by the tsar or his Soviet counterpart. This idée fixe is also present in the "declaration to the province and county" composed by the peasants of Stanovoe village, Ponyri district, Kursk province, and sent to *Krestianskaia gazeta* on 4 March 1928. Describing their troubles and hardships, they identify the local administration as "wreckers who prevent the building of the new life" and request that the "higher bosses" investigate and punish the guilty. "[Our] only request," they conclude, "is that this matter needs to be investigated so that no petty militiaman comes along, as usually happens, and conducts the inquiry by dancing to the tune of the village soviet chairman. At the very least we need [someone] from the provincial Party committee with a purely revolutionary spirit to call together the village assembly and right there hold an examination like the Last Judgment. If not this, if not [someone] from the provincial committee, then [it would be good] even if a commission passing by in an airplane would come down and do this so that every peasant would understand that Soviet power does not want wreckers to rule in its country, that it wants a purely proletarian spirit and a purely heartfelt cause."[39] In the 1920s and 1930s, the exploits of pilots received great publicity in the press, and their pioneering achievements were extolled as victories for socialism. A surprising effect of that ballyhoo may be seen here. One would be hard pressed to find an image that better combines peasant faith in higher authority and contemporary Soviet fascination with aviation than that of a flying commission swooping down to right the wrongs perpetrated in a provincial backwater.

Cut from the Same Cloth?

While peasants could and did employ a certain conception of authority as a stratagem to further their own ends, the higher leadership, too, sought to exploit peasant trust in order to deflect criticism and blame for problems onto lower-level officials.[40] Many in the leadership symbolically emphasized their closeness to the people by deliberately speaking in simple language and dressing in boots and *kosovorotki*—the traditional Russian blouse. Sometimes this made the desired impression, but not always. The "all-Ukrainian elder," the chairman of the Ukrainian Central Executive Committee, G. I. Petrovsky, a man who took pains to emphasize his worker past, must have been surprised when he read this letter of 19 June 1928 from Ya. Yu. Stepanov of Novo-Nikolaevka village, Pervomaisky region. Despite writing to the head of the republican government, Stepanov addresses Petrovsky as "citizen" and employs the familiar *ty* (you) and not the more respectful *Vy* (You).

· 91 ·

Letter to G. I. Petrovsky, chairman of the central executive committee of the Ukrainian republic, from the peasant Ya. Yu. Stepanov, 19 June 1928. RGAE, f. 396, op. 6, d. 28, ll. 64–65. Original manuscript.

Citizen Grigory Ivanovich Petrovsky!
You have assumed the authority of the all-Ukrainian elder [*starosta*]. This truly is a great achievement on your part [as you stand] before the worker and peasant. You should be proud of this achievement. Hardly anyone among the millions of [people] like me would be capable of such an achievement. But you, my friend, have probably already forgotten that you

are just like me, no? True, there is a difference between you and me—you are a worker and I am a peasant: that is the difference. I am saying that you have forgotten, and this, my friend, has become apparent because you have started to oppress the peasants so badly that it is just terrible. Aren't you ashamed of this? Think who it is you are oppressing. The ones who feed you, the builders of the state, and those who [carried] the Red coat of arms of our republic on their very backs and presented it to you. Think it over and you will have to say yes, we have to support the peasantry because this is the laboring mass without which no person, or even those microscopic animals and insects that only live on the ground, has existed. But if you think about it and say that this is unjust, then this will simply be a vile act on your part, one that nature itself has already determined to soon punish— namely, by famine. And this year it will be unavoidable. Although you, at any rate, will hardly bother about it then. The well-fed can't understand the hungry. You will be sure to get everything that you want, but the poor people, what will they do for bread? In 1921 and 1922, everything burned as if aflame. The people perished like flies in November. This was under your rule. Perhaps you have forgotten—remember!

I, as a matter of fact, think that you sit in the center, in Kharkov, and have absolutely no idea what is going on in the fields with the sowing and the peasants' landholdings. Are they being improved or ruined? Listen here and I'll tell you. I live closer to all this, and maybe you will understand and tell whoever's been writing in the newspaper that the peasant economy has increased 25 percent above the prewar level that he is a bold-faced liar. That's not how it is, and won't be any time soon. I can't hide anything from you now, I want to tell you everything. The peasant lives between a rock and a hard place; that is, the old is being destroyed, but the new has not yet taken hold. Maybe the economy can be improved by forcing contributions to peasant bonds, more than half of which haven't paid a thing, but I really can't say.[41] I only know one thing, that this is an interesting comedy—bonds for strengthening the peasant economy. Can you really improve the economy through these bonds? You are destroying it. For me to pay for all the bonds, I sold everything down to my last shirt. There is no money. Where can [I] buy it [the shirt] back? And now I'm hungry and practically naked. That's where the problem of improving the peasant economy stands.

But maybe that's how some of you want it? Maybe some of you want to wear the tsar's crown? Then be up front about it. Why hide it? I don't think anything should be hidden. After all, we're all equal. We bear everything on our backs—hunger and cold and almost all the deadly misfortunes of the revolution. And if that's the case, how come we don't live alike, you and me? Tell me, when will I and the millions of toilers like me be able to live freely, without oppression, as in a free country? You probably will say when there is socialism; that is what the government is working for. Is that right? This won't happen, my friend. And if that's what you're thinking of trying, then we'll all die from hunger. No one living on this earth, you included, will become an angel under socialism. I'll give you a good example and you can judge for yourself. Go home and say to your wife, "Well, wife, get your things together and turn them over to some commune or other, and you and the kids go there to work, and [you'll] get everything [you need] there—food and clothes, [although] of course [you'll have to stand] in a line." What will your wife say to that? And she is the ultimate judge. And you are the first, if I'm not mistaken, who stands up for socialism, all the while not wanting to work under socialism. Why? Here's why. Right now you are probably sitting there getting my letter at your desk, eating and drinking the very best [provisions] available in the USSR, and I'm far away from you, working myself to exhaustion. I've had a roasted potato with onions and am writing you this letter. Do you want to trade places? No, really? This, my friend, is one [example].

Here is another. You are clean and scrubbed, dressed to the nines with glasses on your nose. You have a broadcloth suit, rawhide boots, and a silk necktie around your neck, and I'm in a torn calico shirt, barefoot and practically naked next to your finery. I don't know if I can wait [for socialism]. Maybe you are equal with me in other ways. You may take me by the hand like a distant father and say, "Friend, let's go to socialism." True? With me alongside, you will be ashamed of how I dress. But we are people just like you. I have one heart, blood.

But this still is not all for socialism. In Kharkov, you have a nice apartment with plush furniture and fancy yourself the power of the people, and for this you receive an unlimited salary. But I live with my family in an old clay shack, we sleep in the rags we walk around in, whatever I earn I give away to Soviet power. But I will remain happy, as they say, for a piece of dried bread, and every year my economy is increasing—in minuses. True, over the last four years under the new economic policy the peasantry has risen. From which it is obvious that the state has also begun to rise. What are you doing this year? What are your expectations? After all, you have swindled the peasantry. Where can one place one's hope? Right now, in the villages, without exception, you have real enemies. They are still hungry, and you are establishing socialism. What will my wife say seeing your wife in her splendid attire, going around in a fancy automobile, carrying a dog on a chain in her arms while my [wife] is covered with dirt and buried by work? Do you really think they will live together under socialism? No chance. But they are equal, like you, the all-Ukrainian elder, and me, a simple peasant. So, let's stop accusing each other of unfairness.

Let's start to talk about the building of the republic. You have built Volkhovstroi and Dneprostroi and carried out electrification everywhere.[42] Everything is built for the cities, that is, for you. In ten years what have you done for the village? Absolutely nothing. You will say that we gave the peasants land as a gift. This is fair, but not entirely. It wasn't you who gave the peasants land, but the peasants who gave you power so that you would lead with the peasants and watch over them. True, you are taking care. For yourselves you are putting together the best circumstances and comforts, and from the peasants you take the last shirt. You guarantee yourselves the seven-hour workday but forgot about the peasant, [forgot] that he works a full eighteen hours. How can you have friends in the village. Can a bee be friends with the insect who takes all his honey? Never. If the peasant has something good, let's say a good hammer, motor, or even a tractor, the authorities take it and give it to the collectives so it can be ruined. Will this improve the economy and get the laboring peasant to work? No.

In conclusion, I say: Think about it, friend! You will probably be a little angry. Here's what should be done: change the law back to what it was under Comrade Lenin, a sensible fellow—that is, bring back the New Economic Policy. Otherwise, you will ruin all his hard work and are doomed to go back to the way things used to be, isn't that so? That's what's needed; otherwise, the entire new system of the young republic will go under.

STEPANOV, YA. YU.

In this letter the gap between official rhetoric and the reality of peasant life is starkly drawn. The awareness that the hardships weighing on the people were not being shared by the leaders only sharpened popular resentment of the new ruling elite, leading many to the conclusion that the leadership "talked socialist" but "lived bourgeois." Denunciations of the privileged life enjoyed by Party and government officials became more common as the people were called on to make greater sacrifices. Certainly, the visible gap between official discourse and everyday reality fueled these feelings of indignation, as seen in this letter from the peasant A. Grigoriev of Morshchakovo village, Tver province.

· 92 ·

Letter to *Krestianskaia gazeta* from the peasant A. Grigoriev on the exploitation of the peasantry for the benefit of government bureaucrats, 4 July 1928. RGAE, f. 396, op. 6, d. 114, ll. 747–748. Original manuscript.

Editorial board of *Krestianskaia gazeta,* may we put a question to the government of the USSR?

Only the government of the Union of the USSR benefits from the slogan "Proletarians of all countries, unite." In reality, especially for the peasants, there is no Soviet power. What's going on is not, as the rulers write, "the building of socialism," but [the building of] full-blown bureaucratism. The highest [level] of government itself is based on swindling—that's the opinion of everybody down below. The peasant is enslaved to the point where he would be better off if he agreed to be a serf and worked for a lord. At least then, one would feel that he is working and earning without all the deceptions.

For some reason, you write about the unemployed in other [countries] but are blind to the thousands of unemployed and the daily victims in the Union who show up at the labor exchanges because cities like Moscow, Leningrad, and others are full to bursting. And even though you are covering this up, the peasant knows all about it and sees how well Soviet power [is working]. Every peasant who reads a newspaper and read Comrade Yelichev's letter [Document 87], all of them as one man approved his righteous words. Now, in every village, there are rumors of an impending war. It is quite a torment for the peasant. If all power is based on bribes and deceit, then the peasant will never have a good life, but this is certainly good for the government.

Why were supporters of Trotsky, Kamenev, Zinoviev, and others kicked out of the Party? Because they are lousy at chicanery. You think this is right. Trotsky controlled every front, but everything he had is gone. No, he'll get everything back, but you will collapse. I say, end vodka [sales] so that the peasant isn't fleeced.

Please don't look on [me] like a counterrevolutionary. I am a pure proletarian who gets no support. Worse, they enslave [me] in a cooperative. Only those who have the money for dues are members. But the poor rarely have money. The death of Comrade Lenin was a shame. He died early, unable to carry this business through to the end. So, government comrades, in case of war, don't rely on the peasants too much. Not a single one says that he will defend Soviet power willingly. Even for ten rubles you can't buy bread. Where can one escape to? Our grain goes to feed England, France, and Germany while the peasants sit and go hungry for a week at a time.

Please, give serious attention to the peasants; otherwise, you will be ruined.

A. GRIGORIEV

Interestingly, the previous two letters—both written in mid-1928, shortly after the Party had reinstated forcible requisitioning and emergency measures to overcome acute grain shortages—point to the death of Lenin, in 1924, as a watershed event marking a worsening in the policies and behavior of his successors toward the peasant. That Grigoriev also bemoans the fall of Trotsky, Zinoviev, and Kamenev (none known as especially pro-peasant) suggests that invoking the good offices of absent leaders—regardless of their actual policy orientations—served the letter writers as a stick to be wielded against unpopular measures. Peasants may have believed that invoking prominent Bolsheviks made their charges all the more effective.

Rulers and Ruled

As before the revolution, large numbers of letters denounced the behavior of local authorities and the bureaucratization and arbitrariness of institutions that were now Soviet. The increasing practice of appointing Party and soviet officials from above exacerbated the situation. No longer answerable to those below them, communist officials now formed a separate caste, elevated not only above the citizenry at large but above rank-and-file Party members as well. In a show of superiority, Party bosses gradually began addressing their subordinates with the informal, second-person singular *ty* rather than the more respectful plural *Vy*. This attitude carried into their relations with noncommunists. "The Red experts are worse than the private owner," one factory worker declared in 1926. "They pass by without a greeting, but the owner says something and offers his hand."[43]

The caste solidarity of local officials often expressed itself in the formation of "family circles" or mutual-protection alliances between Party and soviet leaders in a given locality. Primarily designed to frustrate close control by Moscow over local affairs, these informal arrangements facilitated the creation of satrapies headed, usually, by the Party committee first secretary. Such alliances fostered an insularity that, in the NEP's ideologically lax atmosphere, encouraged illicit or corrupt behavior. A control commission reported that "drunkenness has infected all the responsible officials; it has spilled over into debauchery, scandal, and driving around with prostitutes. Drunkenness has also penetrated the Komsomol." The representative of the Agricultural Workers' Union (Vserabotzemles) in Kursk allegedly fired all the women who refused to have sexual relations with him. That this behavior often took place in plain sight and was commonly known did much to discredit the Party in the eyes of the general population. As one worker said, "Communists get drunk, the chiefs organize binges, [so] why shouldn't the worker drink?"[44]

Innumerable letters related instances of official corruption, as in the following from I. F. Goloborodko, a resident of Yekaterinoslav province.

· 93 ·

Letter to *Krestianskaia gazeta* from the peasant I. F. Goloborodko detailing the corrupt and dissolute behavior of local communists, 4 May 1925. RGAE, f. 396, op. 3, d. 263, l. 25. Original manuscript.

A *well-known supervisor*

Reading the lines of *Krestianskaia gazeta,* I had occasion to read the passages in Comrade Stalin's speech on the preparation of the Communist Party for the purge in 1925.[45] This has already taken place, and all those officials discovered [engaging] in inappropriate behavior have been removed. Many worthless elements were evicted, but many still remain in remote corners of the countryside and settlements, undetected by the center but in full view of the peasant. The peasant sees them all for what they are. He knows them because he has to deal with them. The peasant has to cough up a bribe and even dole out moonshine. But he is afraid to file a complaint against him [i.e., them], but, all in all, they, the bribe takers, are well aware of the peasants' opinion. And if there is some dirty business afoot, they [the peasants] know who's behind it.

I am a peasant from Lugovoi settlement, Novo-Mikhailovka village soviet, Ivanovo district, Melitopol region, Yekaterinoslav province, Kudinov [sovkhoz]. I will explain further.[46]

In the sovkhoz in Gofeld, Melitopol region, Yekaterinoslav province, there is a sovkhoz supervisor who's more concerned with moonshine than with the sovkhoz. The supervisor is Comrade Tikhon Grigorievich, with his wife, Aleksandra Sidorovna—the Shumiakins. They claim to be Party members, but they drink moonshine until they drop. Even in 1923. In 1924, the Shumiakins gave a pipe for a moonshine still to one of the renters, Ivan Gora, on the condition that Gora distill them some moonshine from his own grain in 1925. During Shrovetide, Comrade Shumiakin spent a three-day drunk at [the house of] his renter, Dikovets. On the night of the third day, he went to another farm [*khutor*], the Neifelds', drunk and with a revolver in his hands. He started trouble, foully cursing Cheredniko [and] Pavel Gordeenko and asking for moonshine. Gordeenko had no moonshine and turned him away. Looking into the commotion, his [Shumiakin's] son, a Komsomolist with a rifle, and the renter Ivan Krivorodko ran over to take him away, but he refused to go with them. He began to ask Gordeenko for girls. But on the Neifeld farm there are no girls. Shumiakin made a threatening gesture without consequence and went to the renter Trofim Grechko, where there had recently been a wedding, and took over the stove and slept until morning.[47] [. . .]

We renters ask the editorial board of *Krestianskaia gazeta* to take this worthless element away because we are fed up with how he, Shumiakin, addresses the renters: [as] kulaks, bandits, parasites, etc.

Signed, a visitor and house guest at Neifeld's farm, I. F. GOLOBORODKO.

As Party and state institutions grew and their staffs expanded, peasants began to say that the quality of the apparatus's personnel had declined since the years of revolution and civil war. One anonymous letter to *Krestianskaia gazeta* from August 1928 charged: "They've taken any trash into the Party. I see it all the time, and everything is worse. In 1917 the people were the best; there were some bad ones, but only a few, but now everything is worse. They are all careerists and drunks. If only it was better . . . White Guardists in punitive brigades tormented the poor, they murdered, but now [that] they crawl into the Party, they are also brothers. The Party member gets drunk and brawls. I know more than one who befriends poor women where they live. And the people don't understand some things and whisper a lot, [saying things] like: Party members know and torment the people so that they can live well and get fat . . ."[48]

As appointments from above became more widespread, elections lost their significance and, if they did occur, served only as formal rituals confirming the candidates chosen in advance on the basis of Party lists. I. Vasiliev, from Gdov county, Leningrad province, noted the growing interference of the county Party committee (*ukom*) in all business, however petty, and the suppression of any initiative: "The peasants are posing the question: will they now send a chairman for the [soviet] district executive committee from the city, or will we elect one ourselves? Even during an election they ask to have the [county] Party organizer removed from the district. 'If he stays,' [the peasants say,] 'elections or no, it doesn't matter, we'll have no independence.' "[49]

Appointed officials, or *nomenklatura,* became a special stratum of administrators, with their own interests, lifestyle, and mores, who in many ways resembled the old tsarist officials. That many of these individuals had adopted communist principles for strictly mercenary reasons was no secret. In the following letter, a village correspondent in the Tatar republic comments on the lack of ideological devotion that he observed among newer Party officials.

· 94 ·

Letter to *Krestianskaia gazeta* from the village correspondent Grigoriev on the opportunism and careerism of local communists, 7 March 1926. RGAE, f. 396, op. 4, d. 26, l. 172. Original manuscript.

Is it really socialism, and not capitalism, that's being transplanted?

They write a lot in the newspapers—and at various conferences, congresses, and assemblies they talk a lot—about socialist construction in the USSR. They write and say a lot about the leading organ—the Party of the communists, who, as the newspapers say, actively, ideologically, and successfully carry out the program of socialism in the midst of the rural masses. This is just idle talk, just filler for the pages of the newspapers.

In reality, the rural [situation] is far from what they say [it is] at conferences and congresses and in the newspapers. Under Soviet power, I have met only one principled and genuine communist in the countryside. This was in 1919, under military conditions, in Chistopol canton. The communist Georgy Ilich Kornilov, from Chistopol, was named district executive committee chairman of the Almetievo executive committee. This was someone who deserved to be called a communist. During Kornilov's brief stay in Almetievo district as chairman of the district executive committee he established order. The poor and even the middle peasants were quite happy with Kornilov, but not the well-off ones. The late Kornilov would enter [the house] of a rich man at will (as if it were his own house); he'd eat and even drink a cup [of tea] with a little honey. The master of the house was somewhat pleased that the chairman of the district executive committee, Comrade Kornilov, had dined with him and had a cup [of tea]. It followed, the rich man figured, that Comrade Kornilov would treat him leniently. Comrade Kornilov did quite the opposite: it's necessary to lay some sort of tax [on the rich] to get money for the district executive committee or at least for the needs of the poor. Kornilov said, "Yesterday I had dinner with such and such a rich man and saw what a nice setup he has. We must tax him." And he levied [the tax] and helped the poor, and the latter were very pleased with him. The rich, however, were very upset, especially if Kornilov had found out that a certain rich man was doing particularly well. But no one could say that Comrade Kornilov took anything for himself besides dinner, even from the rich. It just didn't happen. He lived according to the communist program. Kornilov had nothing besides the clothes on his back. It's a real shame for someone like Comrade Kornilov! Soon he was in the grave. Koshurin, Aleksandr, replaced Comrade Kornilov. He was also a communist, and he also passed away. But what a communist Koshurin was! During his life so many peasants endured his insults, and there was no end to it!

There are a lot of communists like Koshurin. In a short time Koshurin got quite rich. Despite the fact that he spent a lot of money to live luxuriously, he nevertheless became a rich man. Koshurin is a striking example. A lot of communist did likewise, and non-Party people hitched their wagons to them as well.

Not a few parents, the family elders, oppose the communists, but their kids are "committed Komsomolists." The parents are religious and have no faith in communism, but their kids are just the "opposite." It's all a put-on, done on purpose to nail down [a job] and get a paycheck. These committed comrades don't give a fig for socialist construction. I'll give you an example. In Almetievo, there's a medic (in a military company), P. Frantsuzov. He has no sympathy for communism whatsoever, but his son and daughter are "committed Komsomolists." The son of the local priest, Smelov, through [someone's] influence, also joined the Komsomol. There are other similar instances proving clearly that even though they pose as communists and are principled, they are working for capitalism and not socialism. I have not seen a single Party member who, out of principle, provided any benefit to the poor peasant, to society, to the cooperative, to the state. On the contrary, they are getting rich quick and want to get even richer off these people and institutions.

That's why it seems to me that it's not socialism but really capitalism that's being transplanted.

The peasant GRIGORIEV

The following anonymous letter to *Krestianskaia gazeta,* written in the name of the Donbass workers and peasants, also addresses the quality and ideological commitment of communist officials. Sent from an important mining and industrial region—the Donets basin in eastern Ukraine—it is an emotional, and often ungrammatical, indictment of Soviet bureaucratic degeneration and the new Party elite, which the writer likens to the old ruling class.

· 95 ·

An anonymous letter to *Krestianskaia gazeta* on the bureaucratic degeneration of the Communist Party, 15 August 1928. RGAE, f. 396, op. 6, d. 114, ll. 709–710. Original manuscript.

Dear editor! Don't hide this letter, show it to the people.

Russia, now the USSR, is called a proletarian country. It threw off the yoke of capital, and now power is in the hands of the toilers themselves. If we take a good look at the government itself, who's in charge? Our elected officials have been given power by the people, but they do the bidding of the center. The local officials are now so bureaucratized that life is impossible. [They levy] taxes on top of taxes [and show] no mercy for anyone. The village soviet has the authority to levy [taxes] as it wishes on its own discretion. The people are groaning, but they [the officials] don't even want to listen to their groans. The peasantry endures the punishment as though paying a tribute to someone. What is all this? The army was, is, and will be [drawn] from the peasants. Even now, who defends the city? The peasant. But they are squeezing the last [drop] of juice from the peasant. The peasant has no union. They don't allow him to have this power. Why? Why did the peasant lay down his life? For Soviet power, [which] has improved life for the worker, but his [the peasant's] elected officials suck his blood, they squeeze [him]. There is full-fledged bureaucratism everywhere in the Union. Criticism is widespread, but it is unspoken. All the filth has gotten into the Party. The road into the Party is wide indeed! That's what they want. They hold power now, [they're] at the helm, and the peasant-worker has already lost hope of fair elections. Assemblies of workers or peasants in the village vote [for] our cause, but the elected official is really a Party-drunkard.[50] The workers see that hardly anyone from the masses is promoted, only Party members. What class is this whose rule gets stronger every year, every day? They collect a good salary and live on the backs of the worker and the peasant. The worker gets fifty rubles; as for the peasant, he's a mole who crawls underground, he doesn't need light, and he doesn't see freedom.

What, after all, is the Communist Party? It is the people's party, the fighter for their power, but the people are left groaning so that the communists can live. He who was nothing before has become everything.[51]

[. . .] No, comrade leaders, for the worker-peasant, life is worse than a dog's; his life is a nightmare. The country is full of tears, full of dissatisfaction, enemies, conflicts. But the class [of local officials] celebrates and reports from every little corner of the USSR to headquarters—to Moscow— that everything is good. Moscow, Red, revolutionary, dear leaders of the whole world, give revolutionary attention to the moaning people. Moscow, issue a command to all the ends of the USSR and give the toilers more rights over the administration of the country because everything has perished. It is a shame. They fought for freedom, fought, fought like lions at the fronts [during wartime]. But most importantly, with the help of non-Party [people], purge the Party of trash, bloodsuckers. This slogan is needed, for if there is no purge, then the tsar and his system would be better!

I am writing from the depths of the masses. I see everything: express trains, cars [loaded] with champagne and whatever the soul desires, but the worker must go on working and not ask to eat. The tactic is obvious; if they were to say [anything], they would get the answer: you are politically unreliable.

What is happening? Moscow is aflame. Once again, all the counter [revolutionaries] must be destroyed. They think that a sick person has written [this]. No, not sick, but patience is exhausted, and I ask the comrades, the great leaders of our country, to take up the cause of Ilich [Lenin] and continue it. Don't abandon the administrative apparatus. Let the leaders wade into the depths of the population; have them visit the entire Union and, by sounding out the people, straighten out this miserable life altogether.

Dear editor, don't tear up this letter, but pass it on to the leaders. Let [them] hear our call: We the toilers are perishing from the old [prerevolutionary] government bureaucrats who have penetrated the Party. Dear editors, [here is] a hearty proletarian request: pass this worthless letter on to the dear leaders M. I. Kalinin, Rykov, and all the remaining brother-revolutionaries.

With a proletarian greeting, the workers and peasants of the Donbass. We are waiting for your control lever of freedom. Long live pressure on the bureaucrat.

Like previous letters writers, this one contrasts current corrupt practices with how things were done in Lenin's time, providing evidence that relatively early on, the founder of Bolshevism had begun to serve the Soviet public as an idealized symbol of principled, nonbureaucratic rule. The letter's desperate tone may be partly explained by the deteriorating relations between regime and peasant after the 1928 grain collection campaign, as well as by the letter's point of origin in the Donbass, a region that was much in the news at this time.

In March 1928, the press announced that a large contingent of engineers had been sabotaging mining operations in the Shakhty region of the Donbass. The resulting show trial of fifty-three noncommunist and foreign engineers received great publicity and initiated a backlash against the authority and privileges enjoyed by "bourgeois experts." The Shakhty Affair, now known to have been an OGPU invention, deepened mistrust and hostility toward technical personnel, especially those who had received their training under the old regime. In its wake, "specialist baiting" and charges of "wrecking" against engineers and industrial managers became more common.[52]

Written in reaction to the Shakhty trial, the following letter takes the lessons of the affair in a direction the higher authorities probably did not intend. It was composed by a village correspondent from Rotmistrovka village in the Shevchenko region of the Ukrainian republic.

· 96 ·

Letter to *Krestianskaia gazeta* from the village correspondent P. A. Korzhenkov on
the need to increase self-criticism and worker-peasant oversight in Soviet
institutions, 1928. RGAE, f. 396, op. 6, d. 114, ll. 748–750. Typed copy.

*More self-criticism: place all work under the control of the
worker-peasant mass!*

Comrade Rykov said that for the last ten years we in the Shakhty region
have been led about by the nose by a group of counterrevolutionaries.[53] I
find Comrade Rykov's answer quite laughable. Of course, I have not read
Comrade Rykov's entire speech, but I will say, where were the Party, the
trade-union forces, and the GPU [such] that for ten years they allowed us to
be led by the nose? In all likelihood they knew about the counterrevolutionary work. The workers complained, but they paid no attention to the complaints of the workers. In all likelihood there were complaints about the
Shakhty group of engineers even in the center, but there, no doubt, they only
came up with a resolution "to investigate." The investigating organs looked
into it, and the complaint was returned [with the conclusion] that the engineers checked out, and the center answered: The charge is not corroborated.
After all, it's as plain as day that there are red-tapists, bureaucrats, and, in
general, elements alien to Soviet power in the investigative organs who try
to discredit the government in the eyes of the workers and peasants. After
all, the Riazan provincial court affair, which the village correspondent
Shchelok first exposed, is factual, and this hero died a suicide victim [. . .]
Take the affair of Veli Ibraimov, the chairman of the Crimean Central Executive Committee, who conducted counterrevolutionary work with his
group.[54] This proves that there are elements alien to Soviet power in our
leadership who need to be swept out with the Red broom. Otherwise, we
will not eliminate our dependence on counterrevolutionary international
capital. This will only be accomplished when the peasant and worker are
not afraid to speak out freely about the work of this or that institution.
Otherwise, things will turn out badly with us. You see a mistake—for example, [there's] a peasant in the district executive committee, there's been
a crude mistake of slipshoddiness and total bureaucratism—then say something [about it] to the chief, that is, the chairman of the district executive
committee, [and] he'll give you a good slug, but the fact is, our villager paid
the full tax, he was a little late [in paying] and paid a fine, and then they hit
him with an additional fine, and he finds himself a delinquent taxpayer. He
goes over eight kilometers to the district [office] to inquire [about it]. They
tell him, yes, the fine against you is wrong, but pay it anyway, and then
write a declaration; we'll look it over and return the money according to the
declaration. This offends the peasant. They themselves see that it's wrong.
Sure, write another declaration, but they won't just correct the mistake and

end it. He wrote them their declaration in which he calls them bureaucrats and red-tapists. They say they'll give it formal consideration. Of course, he called the person who considers it a bureaucrat, and the head of the district executive committee places the matter under article 76 of the Criminal Code, and the entire business drags on, all over ten kopeks.[55] And the peasant tells me, they won't overlook the kopeks, so I won't let them off, either. So you see, a simple matter, but the peasant, a simple middle peasant, says, "There's no justice in Soviet Russia." All the same, I think there is justice, and these red-tapists who mock the peasant will be punished. I will write about the trial of this peasant when it occurs and will try to send you his declaration and the court's verdict. So maybe his declaration was written like a satire, a bit too venomous. So, what of it, I think the peasant is right. In other words, whoever works in the tax department should count ten times before he makes a single entry [in his account book]; then everything will be in order or, as they say: anything worth doing is worth doing right. And our peasant is afraid to speak out, often because he feels he is under pressure. This pressure needs to be removed—his intimidation— and the peasant allowed free criticism; otherwise, you won't eliminate red tape, bureaucratism, and counterrevolution. To accomplish this, the center needs to set up a department of investigative officials—traveling inspectors, that is: one hundred people who would be devoted to Soviet power, because the investigative organs we have still don't have the necessary standing. This is proven by the conduct of justice officials who conceal too much.

Village correspondent 764902/2, KORZHENKOV, P. A.

By the end of the 1920s, accumulated personal experience with Soviet institutions and the highly publicized exposure of cases of official crimes and misdemeanors were combining to discredit the government in many eyes. This letter sharply expresses that distrust. Korzhenkov, an obviously loyal Soviet citizen and village correspondent dedicated to exposing malfeasance, accepts the charges against the Shakhty engineers as fact but rejects as impossible the government's claim that no one in authority knew that Shakhty was riddled with saboteurs. After all, each citizen who dealt with Soviet institutions quickly learned that dishonesty and obstruction were the norms. Surely the government, with its wide sources of information and investigative powers, was aware of the dangers. Thus, Korzhenkov is led to the conclusion that the investigators themselves—the OGPU included—must be corrupted by anti-Soviet elements. In the late 1930s, under different circumstances, a similar rationale would lead to bloodletting.

A number of letters in this chapter express their authors' awareness that a wide chasm of power and privilege separated Soviet Russia's rulers, its elite (*verkhi*), from the people (*narod*). This sentiment, certainly not a new one in Russian history, dates back at least to the era of Peter the Great's Western-inspired renovations of Muscovite institutions. In Soviet times, however, this situation ran directly counter to regime rhetoric. Obvious social and political inequalities coupled with instances of official arrogance and corruption helped weaken the revolution's claim to have destroyed class privilege, and belied the Communist Party's declaration that it was exercising its dictatorship for the good of the working masses. That socialism was proving much more difficult to introduce than anticipated by either the regime or its rank-and-file supporters deepened frustration and caused both to seek out scapegoats. As the decade wound to a close, "wreckers" and "saboteurs" in the form of non-proletarian "alien" elements, bureaucrats, kulaks, and others came under increasing scrutiny and pressures. Like Korzhenkov, some citizens located the culprits in the upper reaches of the state itself. In 1929, the Party initiated a weeding out of just such elements in the form of a general purge of the state apparatus. In the course of the purge most government offices were cleared, temporarily at least, of old-regime holdovers. But suspicion and hostility fell not only on individuals whose commitment to socialism was suspect. More and more, both the elite and the people held the NEP responsible for delaying needed social transformation.

CHAPTER 6

Whither Socialism?

"And why not?" fired back the agitator, "must capitalists alone eat roast pig? It won't hurt you if you, too, taste it once in a while."

Laughter, loud and derisive, greeted his words.

"You don't believe it possible," continued the agitator, unperturbed.

"Who said we don't?" sneered the old man with the long shirt. "It must be possible since you Communists are saying it. When we have the *kolhoz* we shall have a real heaven on earth! No wonder you are against the priests. You don't want to wait for a heaven in the hereafter. Our *babas* will be wearing silks and diamonds, and maybe we shall have servants bringing us tea and pastry to our beds. Haw, haw, haw! I served once as a lackey in a landlord's household. I know how rich people live."

"Everything is possible, grandfather, if we all pool our resources and our powers together," replied the visitor.

More laughter and more derisive comment.

"You Communists are good at making promises. If you would only be half as good at fulfilling them."

"What promises have we failed to fulfill?"

"All, all of them," a number of voices exclaimed.

"*Nu,* be reasonable, citizens. There is a limit to jesting."

"Well, you have promised us a world revolution. Where is it?"

"Yes, where, where?"

"The German workmen were going to send us textiles."

"The American workmen were going to send us machines."

"From every country workmen were going to send us things—that's what you've been promising us."

"*Ekh,* people, quit talking nonsense," shouted someone in the back of the room, "this world revolution's got stuck in the mud on our Russian roads."

A howl of laughter broke loose in which the agitator himself joined.

"It will come yet, this revolution," he said after the roar had subsided.

"Like the devil it will! Other people have more brains than we have."

—MAURICE HINDUS, *Red Bread,* 1931

By the mid-1920s, the Soviet economy appeared far healthier than it had been only a few years earlier. Successful harvests in 1925 and 1926 and the return to prewar production levels in most industries by the latter year seemed to vindicate the Party's strategy of economic concessions to agriculture coupled with modest investments in industry. Not all difficulties had been overcome: grain production had yet to attain prewar levels, and procuring sufficient quantities of grain at prices favorable to the state remained a ceaseless struggle; some key industries, like metals, had not recovered as fully as others; and continued overall progress depended precariously on an unlikely future of successive good harvests. These problems notwithstanding, there could be no denying that the NEP and the reestablishment of the market had laid the foundation for economic recovery. But there was also no escaping the fact that contradictions and ambiguities beset the revival that the NEP had engendered. Rising unemployment and difficult living conditions plagued the working class; peasants complained that taxes were too high, manufactured goods too scarce, and prices paid for their produce too low; the communist rank and file and their supporters, especially the young, chafed at the continued indulgence shown to non-proletarians who worked for the state or who were openly prospering; and a growing segment of the leadership voiced concern over the pace of recovery, the potential for economic expansion, and the relative strength of the private economic sector. Despite successes, therefore, the NEP compact was coming under close scrutiny as the decade came to a close, and people increasingly asked, "Where is Soviet socialism?"

Elements of the NEP were coming under reconsideration at the highest political levels by the middle of the decade. With agricultural recovery and industrial restoration, Party leaders recognized that the nation had entered a new stage in its economic development. Having repaired the existing industrial plant, the leadership now looked forward to its expansion. In December 1925, the Fourteenth Party Congress expressed this new disposition when it committed the Soviet Union to attaining economic self-sufficiency by developing a heavy-industrial base that would convert the country "from an importer to a producer of machinery." The congress, moreover, elevated this undertaking beyond the economic by identifying it with socialist construction and the continued survival of the USSR in conditions of a hostile "capitalist encirclement."[1] In linking industrialization, military necessity, and socialist transformation the congress provided the blueprint for a change in the status quo that threatened the NEP emphasis on balanced growth, or "equilibrium," between agriculture and industry. Emphasis on industrial expansion also gave concrete expression to the Stalinist assertion that through sacrifice and heroism socialism was attainable in one country. No one yet spoke of abandoning the NEP, but under these new circumstances the primary question was not whether to increase the pace of industrialization but just how far to depress the accelerator.

The high-level debates over economic policy took on increased political significance in the wake of Lenin's demise. Lenin's first stroke, in

May 1922, had all but removed him from day-to-day activity. Except for a notable period between October 1922 and his second stroke in March 1923, he remained politically ineffective until his death in January 1924. Without him the apparent unity of the Politburo broke down, and his colleagues entered into a lengthy power struggle marked by bitter factional fighting and shifting alliances. First, Zinoviev, Kamenev, and Stalin formed a bloc to isolate Trotsky and his supporters. In mid-1925, after Trotsky's defeat, Zinoviev split with Stalin ostensibly over the latter's support for Bukharin's program of further economic concessions to the peasantry. Like Trotsky before them, Zinoviev and his ally Kamenev proved no match for Stalin organizationally or tactically, and they suffered an ignominious defeat at the Fourteenth Party Congress in December 1925. In a final, defiant assault on the Central Committee majority, Trotsky and Zinoviev overcame their mutual antipathy and, together with Kamenev, formed a United Opposition in April 1926 around a platform calling for increased peasant taxation and greater industrial investment. In the course of the struggle, the United Opposition advanced many trenchant criticisms of internal Party practices and current economic policy—its call to step up industrialization proved particularly prescient—but its challenge miscarried. In November 1927, Trotsky and Zinoviev were expelled from the Party, and their followers were purged from important Party and government posts.[2]

Even as the Central Committee majority upheld the NEP, the decision to promote industrial expansion taken at the Fourteenth Congress moved it further in the direction of the economic program advocated by its opponents, and raised a host of complicated questions relating to the distribution of scarce investment resources. These questions received a good deal of attention in the years leading up to the adoption of the First Five-Year Plan in April 1929. The discussions revealed that dissatisfaction with the slow pace of industrial growth under the NEP was not limited to the opposition. A cohort of industrial partisans now occupied important state positions, and with the support of provincial officials eager to develop their territories, they advanced a program that gradually overpowered resistance from economic moderates. While Party leaders remained outwardly committed to the NEP, between 1926 and 1928 investment favoring heavy industry increased significantly, and grandiose construction projects like the Dnepr Hydroelectric Station (Dneprostroi) and the Turkestan-Siberian Railroad (Turksib) were launched.[3]

Just how far industrial expansion could be pushed before it undermined the equilibrium between agriculture and industry fostered by the NEP remained a troubling unknown. Policy makers' concerns that too much emphasis on heavy industry would worsen the consumer goods shortage and precipitate another scissors crisis that could break the smychka were constant. Naturally, the availability of investment capital worked to restrict the rate of industrial growth. As before, the sources of such capital were limited. Cost-cutting measures had already been introduced in 1921, when previously nationalized enterprises were removed from the state budget and made to operate on a profit-loss basis. As the decade progressed, enterprise directors and "scientific managers" experimented with various "rationalization" schemes intended to increase productivity while further reducing labor costs. With the decision to step up industrialization, the high costs of the state and economic bureaucracies also came under scrutiny. In June 1926, the government initiated a "regime of economy" campaign to reduce administrative costs in all state and economic organizations. The Party and the government bolstered this campaign in the summer of 1927 by instituting a mandatory 20 percent across-the-board reduction in administrative expenditures for the coming year.

While logical under the circumstances, efforts to marshal every spare kopek imposed great hardships and suffering and did much to discredit the NEP among affected workers, clerical personnel, and their trade unions. In the first NEP years, workers' living standards had ebbed and flowed. Wage gains achieved between 1921 and 1924 were offset by inflation and the need to purchase scarce consumer items on the expensive private market. Although working-class life was never easy, as the economic situation improved after 1925 workers saw benefits in the form of stable prices and rising real wages. By 1926, average real wages had returned to prewar levels, and improvements continued into 1928. Even those workers whose wages fell short of the average now enjoyed a standard of living higher than it was before the revolution. Yet workers had reason for concern. Between 1923 and 1929 the national unemployment rate more than doubled, and in Moscow total unemployment surpassed 20 percent by the end of 1927. Life was particularly hard for less-skilled workers who labored at below-average wages. Families requiring more than one breadwinner to make ends meet were not uncommon. In the bigger cities the influx of migrants seeking work not only contributed to unemployment but created a serious housing shortage that fueled worker discontent. In Moscow, workers' housing was dirty, dilapidated, and unhealthy, as well as hard to come by and cramped—allowing 5.6 meters per person on average in 1927. As the decade progressed, the prominence and increasingly privileged lifestyle of a prosperous NEP moneyed class consisting of private traders, well-placed state officials, and educated technical experts directed workers' discontent over their living conditions into class channels. By decade's end, workers and the press were openly venting their anger at the material advantages enjoyed by "NEP-men" and the "new bourgeoisie."[4]

Though first and foremost directed at agricultural recovery, the peasants, too, had cause to be dissatisfied with the NEP. Successful harvests and the private purchasing network combined with tax cuts to make the countryside cash-rich, yet manufactured goods remained in deficit and failed to meet peasant demand. Cooperative stores that were intended to keep the villages supplied with finished goods continued to be poorly stocked, largely with goods of inferior quality, and the alternative—buying from private intermediaries—was costly. Other economic measures were also applied contradictorily or incoherently. This was most apparent in regard to state prices for agricultural products: they were unpredictable and sometimes changed in the course of a single growing season. Bureaucratic manipulations of prices usually worsened the situation and raised serious questions about the viable coexistence of economic planning and the market. Contrary to intentions, the low prices offered for grain had the least effect on, or benefited precisely, those groups the regime considered a threat—wealthier peasants, who could hold grain in reserve until prices improved, and private traders, who were positioned to outbid state collectors. The less well-off, assuming they did not abandon agriculture altogether, had to choose between accepting state prices, consuming or feeding their livestock more of what they produced, or switching to crops that promised a better short-run return.

In addition, despite a reduction in the agricultural tax for the 1925–1926 economic year, throughout the NEP period peasants of all strata complained that they were too heavily taxed. In fact, following the reduction, total receipts had fallen precipitately. The state collected just 252 million rubles in 1925/1926 compared with 326 million the previous year. The cut resulted in a significant decrease in the percentage of peasant income going to the tax and bolstered the peasants' purchasing power. In 1926, the opposition charged that the lower tax rates favored the kulaks and better-off peasants and made the issue a point of contention in its conflict with the majority. A more progressive tax raising the rates on the better-off while exempting the poorest 25 percent of the peasantry was applied in 1926/1927 and resulted in a large rise in overall receipts. Unlike before, the new tax was assessed not on the amount of land cultivated but on total income. The OGPU reported that poor and middle peasants who made their living solely from agriculture supported the new tax, but individuals from these groups who engaged in off-farm work (*otkhod*) or had other non-farm-related income were unhappy, fearing that the tax would be assessed on their gross earnings with no allowance for the cost of maintaining their families. The better-off strata denounced the new assessments wholesale and in several regions responded by reducing their livestock holdings and sown areas, by ceasing to rent land and purchase agricultural machinery, and even by cutting down fruit orchards.[5] In 1928, after the Fifteenth Party Congress determined to apply more pressure on "capitalist elements" in the city and the village, the kulaks' fiscal burden increased still further; the maximum marginal rate was raised from 25 to 30 percent, and a surcharge was levied on the wealthiest peasants.[6]

The direct agricultural tax was not the only fiscal burden borne by the peasantry. Along with the direct tax, peasants also paid, as before the revolution, a host of indirect excise taxes on already expensive consumer items like sugar, tea, tobacco, and, most important in terms of revenue to the state, vodka. The weight of these taxes fell most heavily on the poor. Generally, village soviets did not have their own budgets and received a tiny share of the agricultural tax to cover their expenditures. When the need arose to collect funds for matters of local interest, the village assembly could vote to levy a tax on its members in the form of the *samooblozhenie*, or "self-assessment." The self-assessment was arrived at democratically and distributed equally on a per household or per head ("eater") basis, but since it took no account of income disparities between households it was, like the excise taxes, regressive.[7]

The actual collecting of taxes presented its own difficulties and also fell disproportionately on the less well-off. During the tax collection campaign at the end of 1926, after the reduction in the direct tax had gone into effect, poorer peasants complained that the total of their other obligatory payments now exceeded that of the agricultural tax itself, and they simply lacked the means to pay it. The OGPU reported that the simultaneous collection of all obligations and arrears was proving especially burdensome on poor peasants and weaker middle peasants. The better-off and those considered kulaks, on the other hand, were finding ways to avoid assessments on their full wealth. Local efforts to collect from tax delinquents by inventorying and auctioning off their property or taking them to court seemed to specifically target poor and middle peasants. In some locales, poor peasants reportedly sold off the last of their livestock to settle their tax debts. The better-off, however, managed to evade punishment. Among examples of the illegalities (*samoupravstva*) that accompanied tax collection, the OGPU provided the case of one district executive committee chairman who went from house to house brandishing a pistol while demanding the immediate payment of the tax. The inequalities in tax payments between the rich and poor allegedly caused better-off peasants in Chita to jest, "Soviet power has classified us among the proletariat."[8]

As seen in the previous chapter, village inhabitants looked on the rural communists as aliens. They resisted, often violently, those Party initiatives that they deemed contrary to their interests or patently unjust. Yet, whatever hostility they may have harbored toward communists, peasants did not reject the socialist message out of hand. Certainly, the perilous nature of rural life bred in the muzhik a suspicion of change and a justifiable disdain for utopianism. Nature was pitiless, and the struggle to survive it a razor's-edge proposition. But the Party agitator in the epigraph is not simply derided for promising paradise on earth but also called to account for failing to deliver it. For many peasants and workers, particularly those whose situations had improved little under the NEP, only socialism with its promises of justice, equality, and material improvement could vindicate the sacrifices made since 1917. Thus, the question stood: Why has socialism not yet arrived?

Repeatedly, in the letters below, the correspondents point accusing fingers at certain features of the NEP for delaying the introduction of socialism. High taxes and high unemployment are blamed for perpetuating poverty and for hindering the full development of the country's productive potentialities. The greed and venality of those whose primary concern is individual gain is frequently contrasted with the universal benefits that will derive from the collective organization of labor. Opprobrium is heaped on NEPmen, priests, and those who profess communist principles but in reality are out to feather their own nests. And in a manner that brings together Russian historical experience and socialist idealism, special contempt is reserved for bureaucrats.

Aleksandr Herzen, reflecting on his youthful exile to the provinces in 1835, wrote, "One of the saddest consequences of the revolution effected by Peter the Great is the development of the official class in Russia. These *chinovniki* are an artificial, ill-educated, and hungry class, incapable of anything except office-work, and ignorant of everything except official papers. They form a kind of lay clergy, officiating in the law-courts and police-offices, and sucking the blood of the nation with thousands of dirty, greedy mouths."[9] Similar sentiments are expressed in several letters below suggesting that for much of the country little had changed in the ninety years since Herzen's banishment. Above all else, an administrative post was, and remained, a sinecure for collecting bribes and favors. Significantly, the complaints in our letters not only bemoan corruption but the very existence of a class of salaried officials. To many who wrote, salaried officialdom drained the budget of funds that could be put to more productive uses and elevated recipients above those who labored in the factories or the fields, that is, above the actual creators of wealth. This leads one anonymous letter writer to conclude—without reference to the Paris Commune or Lenin's *State and Revolution*—that "under socialism no one should get a salary" (Document 111).

Embedded in this inimical attitude toward bureaucrats are two of the elements that underlay peasant notions of the just society—that one's own labor should determine one's material well-being and that social differences should be kept to a minimum. For generations, the repartitional commune had served as the institutional guarantor of both these propositions through the redistribution of strip farmland in response to demographic change in the peasant household. Mutual assistance in hard times and labor cooperation in the face of crises were also "socialistic" features of communal life. In the early twentieth century, Russia's rulers discovered to their chagrin that peasants could not easily be weaned from the commune. The tsarist effort to weaken the commune after the 1905 Revolution in order to strengthen private landholding achieved promising but limited success. During the civil war, the Bolshevik attempt to undermine the commune in order to facilitate collectivization and raise productivity simply failed. In the wake of each upheaval, the peasantry as a whole remained committed to the equalization, mutual support, and collective activity practiced within the commune.

The repartitional commune had evolved in the eighteenth century as a means to facilitate tax collection. In Russian conditions of scarcity and subsistence agriculture it had the added benefit, from the peasants' point of view, of curbing the acquisitive activity of individual peasants that might threaten the stability and survival of the community. Nineteenth-century Russian radicals devoted much attention to whether peasant communalism could serve as a launching pad for socialism. Orthodox Marxists rejected the idea out of hand. For them, the peasants were petty-bourgeois property owners who wanted no more than to possess the land they worked. The peasants' socialism, such as it was, was an outgrowth of their primitiveness and was doomed. Based on the English historical experience, Marxists expected that under the pressures of capitalist development, the commune would disintegrate, splitting the peasant class into a small stratum of agricultural entrepreneurs and the pauperized, landless majority they would exploit as laborers. For their Populist (*narodnik*) opponents, on the other hand, the matter was not so clear-cut. As peasant socialists, the narodniks emphasized less the inevitability of capitalism's iron economic laws and more the collective social relations that communal agriculture engendered. The Socialist-Revolutionaries—the narodniks' political descendants—enshrined this faith in the socialist possibilities of the commune in their 1905 party program.[10] At the time of the October Revolution, Lenin, the most astute Marxist student of the peasantry, was shrewd enough to co-opt the

Socialist-Revolutionary program and not to call for the dismantling of the commune.

As Figes and Kolonitskii have emphasized in their study of language and symbols in the revolution, the peasants little understood the programmatic variety of socialism that made its way from the cities to the country via newspapers, handbills, and revolutionary agitators in 1917. The non-Russian words and ideas of this new discourse—"class struggle" and "bourgeoisie," for example—were alien to the peasant lexicon and mindset. The resulting confusions and malapropisms that these urban concepts engendered could indeed be comical. But, as these and other scholars have also shown, Russian peasants were quite skillful in adapting alien ideas in new ways to express their own values and revolutionary objectives.[11] Certainly, there was much in the fundamental ideals of socialism that peasants could find agreeable. The glorification of labor and opposition to exploitation, the emphasis on the collective, the striving for equality, and the calls for universal brotherhood and peace were principles and goals toward which the peasantry was predisposed.

By the mid- and late-1920s, after several years' of continuous exposure to Bolshevik agitation, much of it carried out by returning Red Army veterans, the peasants were certainly familiar with the specifics and socialist underpinnings of the Party program. War communism had taught them not to trust the Bolsheviks or their agents, and they were more than a little jaded about the Bolsheviks' ability to fulfill their more spectacular promises, a feeling captured in this chapter's epigraph. Nevertheless, as is evident in many letters below, even where they expressed hostility toward the Party, rejected particular elements of the Bolshevik program, such as the collective farm, or dismissed utopian dreaming about world transformation, peasants found much virtue in socialism's core values and continued to hope for their realization.

In the village, "socialism," like many ideas that came from outside, proved to be a protean and plastic concept. Peasants were able to define socialism as that which benefited the village by raising its moral and cultural level, improving agricultural production, and strengthening the economic position of the individual farmer. To the extent that Soviet policy furthered these ends, socialism was being constructed. Low prices for agricultural produce, high taxes, expensive manufactured goods, shortages of mechanical equipment and fertilizers, on the other hand, delayed or were inimical to socialism. Socialism also promised equality and control over those issues deemed critical to village life. In this regard, peasants found it easy to integrate certain village traditions into their conceptions of socialism. Periodic land redistribution based on a household's needs and working capacities guaranteed a degree of economic equality by preventing disparities in wealth between households from becoming too great. Likewise, village decision-making institutions, like the assembly, embodied a basic democracy while also providing mutual assistance in times of distress, ensured freedom of action, and protected against interference from outsiders. Such practices and institutions were deeply rooted in village culture and peasant psychology and complemented the peasants' understanding of socialism.

At a minimum, socialism carried the promise of material progress. One did not have to be a kulak to find Bukharin's admonition "Get rich!" agreeable. But individual enrichment could not come at the expense of the whole and had to be a result of honest labor. The socialist formula "He who does not work, neither shall he eat" encapsulated a fundamental peasant belief that applied not only to the former ruling class but to Soviet officials and other peasants as well. In the letters below, bureaucrats, kulaks, and the poor—who were often considered lazy or unproductive—are repeatedly portrayed as unfairly benefiting from privileges, influence, or taxes taken from the hardworking. Many peasants found this contrary both to the spirit and practice of socialism, which may help explain the relatively weaker class solidarity shown during dekulakization in 1930 as opposed to that exhibited earlier during war communism.

In addition to socialism's material benefits, its advocates have always asserted that socialism would provide the basis for a more just society, and almost any discussion of socialism turns on questions of ethics and morality. Many of the opinions that came from below fall into this tradition. In reading these letters we sense that their authors, too, understood socialism as a higher form of social organization, one they associated with their own definition of the good life. In several letters below, the peasants imbue the idea of justice with notions of love and goodness that may derive from Orthodox Christianity or the religious writings of Leo Tolstoy. For this very reason, even letter writers sympathetic to socialism despaired that it could soon be implanted in Russia. Questions of economy and politics aside, disrespect for the law, alcoholism, violent behavior, and a host of other personal vices convinced more than one writer that before Russia would see socialism its population would have to undergo a moral regeneration. For these writers, socialism was as much a question of personal responsibility as of property or class relations.

This is not to suggest that all peasants were equally enthusiastic about socialism or sanguine about its potential benefits. Peasants from different economic strata expressed different views on the Party, the collective organization of agriculture, and other related issues. Certainly, as some commentators note, for the mass of peasants socialism remained a vague, poorly understood concept (Documents 106 and 110). Others, agreeing with Maurice Hindus's peasant that the brainless Russian muzhik had indeed been duped, adopted a cynical approach to what they considered socialist claptrap. Regardless of how well peasants understood socialism or what they thought of its workability, though, socialism provided them with a very attractive myth of a more just social order and an idyllic future, exactly the sort of myth that has animated peasant protest movements since the High Middle Ages.[12] In the Soviet case, the socialist myth also happened to be the founding myth of the state itself. This enabled peasants to invoke socialism to protest their condition by adopting the regime's own language and stated values, a tactic that ran much less risk than confrontation or outright rebellion. To this extent, then, and to the extent that the content of socialism could be shaped to conform to their own interests, many peasants were prepared to accept socialism as a worthy goal and to identify themselves with it.

Documents

World Revolution

Occurring as it did during the First World War, when the lower classes of all countries were being sent to a seemingly senseless slaughter by their political elites, it is not surprising that Russian workers and soldiers, the latter overwhelmingly drafted from the peasantry, viewed their revolution as a blow struck not only against their ruling class but against exploiting classes everywhere. The fraternization that occurred on the eastern front in the wake of the October Revolution seemed to vindicate the socialist dictum that workers had no country and that their real enemy was the bourgeoisie of each nation. Socialist theory, its popular literature, even its anthems, all proclaimed that the coming revolution would be global. At first, Bolsheviks themselves could not conceive of socialist revolution on any but an international scale. From the very start, then, those who supported it were inclined to view their revolution as having worldwide ramifications.

Even as the European revolutionary wave ebbed, the idea that socialism would be victorious internationally or not at all remained common currency. Throughout the 1920s, fears that "international capital" was preparing to launch a counterrevolutionary crusade against Soviet Russia, repeating the foreign interventions that followed the revolution, remained very real both within and without the Kremlin. Russian hopes that foreign workers and peasants would quickly overthrow their capitalists and landowners arose as much from a desire for Soviet security as for the liberation of their fellow laborers. T. G. Bezugly from the village of Novo-Andreevko in Yekaterinoslav province, evidently submitted a lengthy treatise to *Krestianskaia gazeta,* but only a fragment has been uncovered. Here is how he interpreted the coming world revolution in a section written on the eve of the seventh anniversary of the October Revolution.

· 97 ·

Excerpt from a letter to *Krestianskaia gazeta* from the peasant T. G. Bezugly on communist political successes beyond the borders of the USSR, 23 October 1924. RGAE, f. 396, op. 2, d. 18, ll. 327–328. Original manuscript.

The coming revolution in the West and the East.

In the previous chapters, I came to the conclusion that the bourgeois states of the West and the East were heading down the slope of social revolution. Having examined the means of production and exchange established by the bourgeoisie, I note that gangrene has enveloped the entire bourgeois system of the West and the East. I see that any sort of scientific and humane foundations [for this system] are completely absent. The bourgeoisie's mindless wasting of the fruits of worker-peasant labor, [and their] craving for profits has attained unprecedented proportions. In all the *Reichstags* and parliaments of the bourgeois states, I see that the elections and reelections are bringing the bourgeoisie many failures. In the reelections, the workers and peasants go over to the communists at every opportunity, and as a peasant, I salute with delight the coming of the day when the cry "Down with the bourgeoisie!" resounds from every tongue among the toiling masses of the West and the East. And with the same unanimity that the Soviet republics of Russia and Ukraine were proclaimed on 25 October, the workers and peasants of the West and the East are now following the example of the latest system [of government], namely the USSR, taking it upon themselves to immediately study the machinery of the socialist system of the USSR, [which is governed] according to the will of the workers and peasants under the leadership of the Communist Party. I see the bourgeois way of life of the West and the East is nullifying its own existence.

Much of Bezugly's letter repeats clichés and formulas that could be found daily in the press and serves as a good example of an individual adopting the terms of official discourse. By providing a ready-made intellectual apparatus that enabled a coherent understanding of complex domestic and international developments, such comprehensive analyses were, no doubt, highly attractive, and all the more when coupled with the reassurance that the proletarian cause was prevailing worldwide and would soon end the Soviet Union's dangerous isolation.

International solidarity and revolutionary optimism also pervade this combative letter written in August 1924 by A. Fatin, a villager from Briansk province. On turning twenty-one, workers and peasants had to fulfill a period of military service in regular Red Army or territorial militia units. Fatin evidently reached service age in 1923, hence his reference to the "1902 draftees."

· 98 ·

Letter to *Krestianskaia gazeta* from the village correspondent Andrei Fatin on the strength of international working class solidarity and the young generation's determination to defend the Soviet Union, 12 August 1924. RGAE, f. 396, op. 2, d. 18, ll. 40–41. Original manuscript.

We are not afraid of capitalism.

On reading *Krestianskaia gazeta,* no. 38, which carried the piece on the foreign bourgeoisie's war preparations, the elders of our village of Fedorovka, Dubrovka district, Bezhitsa county, became quite terrified by the unparalleled miracle that the foreign bourgeoisie is performing in devising ways to wipe out workers and peasants. The young, however, especially the draftees of 1902, only smiled and said, "Let them invent. This doesn't frighten us. All these inventions will fall on the head of the bourgeois himself because we are sure that the foreign workers and peasants will not open fire on their brothers, the workers and peasants of the USSR. Right now, the foreign workers and peasants are in the tight grip of bourgeois capitalism. But when the bourgeoisie arms its workers and peasants, then they will loosen the claws of the bourgeoisie and free its victims, who have been held in bondage and oppression for a full one hundred years. We, the 1902 draftees from the village of Fedorovka, will be ready at a minute's notice to stand in the ranks of our invincible worker-peasant Soviet Red Army, and we vow to lay down our lives to liberate the people from capitalism. [. . .] It won't always be that the bourgeoisie will be able to defeat us with their poisonous rockets of death, 114-kilogram bombs, and long-range guns. The Soviet government should expect from us, the 1902 draftees, steadfast, manly fighters against the bourgeoisie for the Soviet government and its forward vanguard, the Russian Communist Party, and a strong worker-peasant alliance, [that we grow] stronger with each step! Long live the victory of the proletariat over capitalism! Long live the Soviet Red Army and its commanders! Long live the final and decisive repulse of the bourgeoisie! Long live our revolutionary leaders of the proletariat M. I. Kalinin, L. D. Trotsky, L. B. Kamenev, and the others. Long live the memory of Vladimir Ilich Lenin. Long live our valiant vanguard, the Russian Communist Party! The only hope of liberating the workers and peasants from the clutches of capital rests on you, friend.

Correspondent FATIN, ANDREI.

This village correspondent dares the technologically superior Western powers to make war on the USSR and risk inciting their own workers and peasants to revolution. By contrast, the following letter, addressed to scientists, is imbued with the horrors of modern war. It was written to *Krestianskaia gazeta* in March 1926 by a self-proclaimed "freethinking citizen."

· 99 ·

Letter to *Krestianskaia gazeta* from an unknown peasant calling on scientists to refuse further participation in military research and development, 17 March 1926. RGAE, f. 396, op. 4, d. 198, ll. 52–55. Original manuscript.

To the men of science

Take all measures to disarm all the people. Replace the word "war" with the word "peace." Instead of exchanging gas and bombs, exchange love and benevolence. Evil, gas, and bombs, and tanks bring much unhappiness, especially to the common people. Who will be dominated by these horrors? The lower class, of course. Who will swallow the asphyxiating gases and inhale the gunpowder? They will. And who will wind up orphans and cripples? Again, the lower class. The peasantry, which feels the effect of war intensely, awaits the time of complete disarmament.

How is it that the lower class, especially the peasantry (let's say of every country), knows that this specified evil brings harm, particularly to the lower class? Why do they fight? Very simple. The dark people are constantly occupied by their work and rarely touch politics. But men of science weave their intricate net so that he [the peasant] should go and kill other peasants and workers and they him.

In order to destroy all this evil, men of science need to become aware of this. They should expose all this evil to the light and unmask those harmful elements who benefit from this dirty business. And we, the lower class, have to sort out the harmful sciences from the useful. For the good of the people, [we must] elevate the science of love and universality to a place of honor. But to harmful science [we must] say directly: "Shame on you! Cultivated beasts, it is time you reconsider and come to your senses."[13]

Healthy science should work to foster peace, and not to annihilate everyone, so that all may live happily. Friendship makes a happy life; not hostility but the kinship of freethinking citizens.

I am a poor, uneducated, natural-born peasant. After a great deal of thinking, I have come to the conclusion that the duty of science, first of all, is not to be a shark and discover all the evil doings of the sharks. It would be useful to disseminate this across the border, where there are so many of them.

To the men of science

Evil cannot extinguish evil, because a force can never destroy itself. For this one needs a different kind of force. Good. Love can stop evil at the root. This is an incontrovertible fact. Love and good have no enemy. No matter how much evil and hatred may grow, only friendship, love, and good can defeat evil. Take a good look around and think how much one person can be an enemy to another, creating such horrors—asphyxiating gases, tanks, fighter planes, and long-range guns—and all this is the work of people who consider themselves intelligent and educated. But, as a peasant at the plow, from the ignorant and remote countryside, I say to the educated people, "You are trained animals, you are even closer to [the level of] predatory beasts than ignorant people who completely lack any sense of morality." Learned people are building these obscenities for the sake of their own welfare—just to get more money. These intelligent people are only worth a farthing. Soon the time will come [when] this honor will turn to shame. You must remember that there is no end to evil and that force begets force, courage [begets] courage, and so it goes since time immemorial.

I could write more, but it's probably pointless. I got *Krestianskaia gazeta* for 16 February of this year [containing] the article "The Danger of a New War Has Not Passed." This is the only answer to the sharks of the whole world for these terrible things.

A freethinking citizen.

Advances in military technology reinforced popular mistrust toward educated "men of science" who worked tirelessly to perfect the means of mass destruction and who, like the title character in Aleksei Tolstoy's popular science-fiction novel, *The Death Ray of Engineer Garin,* create evil weapons. The letter writer's outrage is inflamed by the knowledge that scientists, who should be working to improve the lot of humanity, continue to develop weapons that will only bring death and suffering to the common folk. This class awareness is bolstered by faith in the power of Christian virtue to overcome the evils of technology. In this homily of love, brotherhood, peace, and friendship, the writer reveals the affinity of peasant religious values with the fraternal spirit that animated popular belief in an internationalism of working people.

Imagining Socialism

Obvious social inequalities and the restrictions on fundamental freedoms caused many citizens to question whether the USSR was actually advancing along the path to socialism during the NEP years. The following letter from early 1926 addresses this theme in a particularly plaintive way. One note of interest: The author sets out to equate the Communist Party leadership with the patriarchal rule that he claims arose in primitive societies. In referring to the Party leadership, he employs the Russian term *partiatki,* a neologism, evidently of his own making, which might be unsatisfactorily translated as "partycrats." He contrasts partiatki with patriarch (the Russian term for which he misspells). To retain this wordplay and the flavor of the original, we have not translated *partiatki.*

· 100 ·

Letter to *Krestianskaia gazeta* from the Moscow resident M. V. Lobkov decrying the emergence of a privileged Communist Party elite and the restrictions placed on free speech, early 1926. RGAE, f. 396, op. 3, d. 368, ll. 138–139. Original manuscript.

The questions. Can communism be achieved—that is, an equal or almost equal life for all citizens of the republic? Yes, communism can come about, but not international, not all-Union [communism], but the working class and the peasantry of Soviet Russia don't need international communism as much as the rest of the world needs it. Only this doesn't mean that all principled communists, including the people's commissars, must be new modern partiatki. That is, they shouldn't be set apart from the ranks of the simple workers and peasants because of their education and intellectual development the way the first patriarchs were set apart from the primitive communards. Don't give them 192 rubles a month, [which they received] not only when there was famine along the Volga and people died of hunger but even [receive] now. Never set [them] apart from the masses; otherwise, I have the right to call any people's commissar a new patriarch because I know they are becoming like the primitive patriarchs. The question must be put to the entire peasantry and working class: How do they want to pay for the labor of all the non-peasants and the average non-workers? [Should they] enjoy more of life's benefits or the same as themselves? He who fears equality also fears communism because without equality there is no communism. And if it is forbidden to equalize wages, then it is because the communists themselves oppose it. But if they wanted it, then the workers and peasants would begin to dance for joy, and the workers and peasants of the whole world would be delighted.

I cannot express all [my] thoughts in detail because [I am] hungry, unemployed, and covered with lice; in short, I envy the poet Yesenin, who recently hanged himself. I'll probably meet Yesenin before [I'll see] the beauty of nature's spring. Maybe there is only a week left to me in this life, but out of love for the laboring peasants and simple workers, I will tell [everything]. Let them all find out what the new oppositions said at the Fourteenth Congress and familiarize themselves as well with the discussion of Comrade Trotsky, and this will reveal the difference between the Trotskyists and Zinovievists.[14] I have also heard from simple, chatty peasants and workers all over that Trotsky was right. That if the All-Union Communist Party had heeded his advice, then life would have gotten better for the simple peasants and workers. Now, regarding the new discussion, once again they are speaking in the same spirit. [They are saying,] "If only the Party would do as Kamenev and Zinoviev and their supporters advise it to do." In a word, every worker and peasant feels that life could be better than it is now if full freedom of speech were established. Words, so long as they are not crude, won't hurt the communists. And even if it hurts the communist to hear crude words, not to hear them is worse still. Every peasant and worker is afraid to say or write what he dislikes about Soviet power; that's why the workers don't express their true feelings. It seems that freedom of speech and the press are almost as restricted as under the tsar, and if you speak the truth against the government—that is, [if you express] your unhappiness and grievances, then again not [not clear in original]. Before they imprisoned revolutionaries for freedom of speech, but even the revolutionary can easily [not clear in original]. If I can escape Yesenin's fate, then gradually I will, in spite of the lice, the hunger, and the like [not clear in original] come.

LOBKOV, MIKHAIL VASILIEVICH, Moscow,
Orlikov Lane, first boarding house.

Through the fog of despondency, the author in his own, often confused way makes a number of cogent points relating to the gulf now separating theory from reality. Equality and freedom—in this case, the equalization of wages and freedom of expression, two virtues identified with socialism—were, in his view, imperiled. Although Marxist theory promised that under socialism remuneration based on work performed would ultimately be supplanted by the satisfaction of all one's needs regardless of one's work, wage differentials were a fact of life in the Soviet Union. As one of the many NEP unemployed, Lobkov had even more cause to despair that this aspect of the socialist future continued to remain elusive. That the "new patriarchs" who sat atop the wage hierarchy also saw fit to suppress criticism of the current state of affairs added to his sense of hopelessness and his contemplation of suicide as a way out of the Dostoevskian squalor that enveloped him. The letter provides a glaring example of the morbid state of mind that Party leaders dubbed "yeseninism" following the poet Sergei Yesenin's suicide at the end of 1925.[15]

The evident disparities of wealth and privilege characteristic of NEP Russia encouraged people to compare it unflatteringly with the prerevolutionary past. "What is the difference between primitive and today's people?" asked one Moscow worker in 1926. The answer: "In my opinion, there is none because now, as before, there are rulers and slaves, just in different guises." Another worker from Moscow province noted, "They said that [under socialism] everything would be equal; in fact, as it was before, so it is now: some live well and others poorly." The OGPU also recorded the following questions posed by workers: "Communists are not gentry. Then why do their wives go around in hats and they have a maid?" "Tell me, what is the difference between a minister and a people's commissar? For example, Lunacharsky's wife has diamond rings on her hands and gold around her neck. Where did they get it? Answer me." "Before they exploited us, but everyone had shoes and clothes, but now everyone is hungry with nothing to do."[16]

Such observations were not limited to domestic affairs. The popular view of the international situation often conflicted with official versions, as can be seen in these sharply drawn questions from workers: "How can America's wealth and the well-being of [its] people be explained? After all, it's a capitalist country, and capital only oppresses the people. Why don't the Americans envy our socialism?" "In France and Germany there is freedom of speech and the press, but not here. So where is socialism, there where there is liberty and freedom of the press or here in the dictatorship of the All-Union Communist Party?" Aware of these and similar attitudes, the leadership tried to limit contacts between average citizens and foreign visitors to stage-managed encounters, a practice that elicited the following question: "Why don't our workers in production ever see anyone from the visiting delegations of foreign workers?"[17]

Contrary to the principle of international working-class cooperation, the hardships of daily life in the Soviet Union were also invoked to argue against economic alliances and supplying aid to workers of other countries: "Foreign concessionaires are sucking the juice from our workers." "You are sending aid to the English workers [during the 1926 general strike] while your own are dying of hunger." "They've sold oil to America for five years, and we ourselves are without fuel." One unemployed worker, upon reading newspaper reports on aid sent in support of the Chinese revolution, declared, "What the hell is this for? We spend money on them, are teaching them, and our own people are dying of hunger." Doubts were also expressed as to the ultimate efficacy of the revolution: "If there had not been a revolution, our country would be richer than America."[18]

What must intrigue the student of the period, however, is that as widespread as these opinions surely must have been, they never completely discouraged the popular belief that a new society shorn of flaws could still be constructed from the old. For many of our writers, it was precisely the socialist possibilities opened up by the 1917 revolution that justified this belief. For these individuals, the revolution remained a powerful symbol signifying justice and equality, a symbol, moreover, that could be invoked in protest against the current state of affairs and against the Bolshevik regime itself.

According to the peasant F. A. Martian, from Dmitrovka village near the Rubanka station on the Southern Railroad in Belorussia, successful communist construction necessitated a strong alliance between workers and peasants and the establishment of a broad cooperative network. In the excerpt from his letter reproduced below, he turns his attention to the role of technology.

· 101 ·

Excerpt from a letter to *Krestianskaia gazeta* from the peasant F. A. Martian explaining the proper way to build communism, October 1926. RGAE, f. 396, op. 4, d. 24, ll. 232–233. Original manuscript.

The poster "Lenin is the banner, Leninism is the weapon, and world revolution is the mission."

I happened to see the contents of this poster at a city tea room, and I was quite amazed to see that world revolution is [our] mission. This is one thing we don't need. Revolution is a rapid movement, and it can't happen without bloodletting. This will make it much more difficult and almost impossible for us to reach communism. We don't need the mission of world revolution. We need to establish the following mission: 1) to smash capital and 2) to build communism. To do this, we need to make a socialist revolution in one country—that has already been accomplished. To smash capital there and achieve communism will only occur when each person is an exploiter—not an exploiter of man, but of machines. To do this, we can utilize capital and the international bourgeoisie. We will build communism through capital and the bourgeoisie. There is no other way. We can make use of capital and the bourgeoisie, and it will be easy. I was more than a little surprised that so much time has already passed since the socialist revolution. Power is in the hands of the toilers, but so very little has been done. We already could be knocking at the gates of communism.

Peasant F. MARTIAN

Please put my article in the newspaper.

Although Martian rejects the Marxist-Leninist position on international revolution, he, knowingly or not, comes very close to Lenin's own view on the need to exploit the culture and technology of the bourgeoisie in order to build socialism. His equating of the proliferation of machinery with communism also reminds the newspaper reader of the mechanized utopianism promoted by A. K. Gastev, the worker-poet who, as head of the Central Institute of Labor (TsIT), spearheaded the Soviet Taylorist movement in the 1920s.[19]

In this letter from mid-1925, Ya. M. Rudikov from Saltykovo village, Kursk province, expresses his impatience over the delay in introducing socialism.

· 102 ·

Letter to *Krestianskaia gazeta* from the peasant Ya. M. Rudikov calling on state leaders to move the economy in a more socialist direction, mid-1925. RGAE, f. 396, op. 3, d. 352, l. 12. Original manuscript.

Comrade editor! Why has our Bolshevik All-Union Communist Party circumvented the communist program and pushed socialism so far off? Everyone says that we still haven't grown into it, but the fact is that with the transition to the New Economic Policy— NEP—state capital, in my opinion, has already sufficiently increased, and now, in my opinion, would be the time to move on to socialism, if only partially. The New Economic Policy boosts state capital, but the workers and peasants only get horrible unemployment and hunger, and this is not to the good. Without work all the peasants just wander around this way and that way, there is nothing anywhere, they're down to their last ounce of their bread. In our village of Saltykovo, Burkma is putting up seven derricks.[20] They are drilling for magnetic ore, and [it is] very valuable. It seems to me that right now we have to start to build mines and factories. They say our state cannot build all this by its own means right away and that our country has either to turn to the capitalist countries for support for a loan or to offer concessions for several years to the capitalists of different countries. It is my belief that this is not necessary. As I see things now, in our village of Saltykovo there are five hundred households and a population of more than three thousand. Of these, fifteen hundred worker-souls have absolutely nothing to do. One hundred households have horses, while four hundred households have no horses, but all the same, land is rented out on *ispol*[21] to those with horses, [and those without horses] walk around like fools with nothing to do. But the slogan is "He who does not work shall not eat." But what's the way out? There is an exit. To make the transition to the socialist path our Bolshevik Russian Communist Party needs to do the following: dispatch five tractors and trucks to our village of Saltykovo, put the land of those without horses to one side, and work [it] all with state equipment—there are enough people [for this]. And since all the citizens' land will be put under cultivation, they can all turn to state work, building railroads [and] constructing and working the mines and factories for refining ore. We're not going to achieve socialism through kommunas and sovkhozes, collectives and trusts. I have investigated life on the Kolmykov kommuna near Stary Oskol. The young do the plowing, the young feed the livestock, and the adults walk around in single file to no purpose; they just have nothing to do. Now again, on one side of us is the Kilkhen sovkhoz in the village of Korobkovo. I looked in here. The sovkhoz also gives all land to the peasants' on ispol. Half the harvest goes to the sovkhoz and half to the peasants. There is nothing to be learned here. The peasants work now as they did before, and the administration lives behind closed doors as if at the side of God. It is well paid and eats and rides around in a lordly manner. It's possible that the former manager-specialists came from the Whites, since before 1923 they were nowhere to be seen. But it's the specialists who run the sovkhoz! We have illiterate women, and they can pass out the land on ispol better than these specialists and get half for themselves and half for those who did the work. So it's obvious [who] benefits from all this is and what we are accomplishing. Everything ends up in [someone's] pockets. And so I'm thinking this is the way to ensure a good life, a socialist life. We can set up an artel of surveyors, and an artel of joiners, and an artel of stone masons, an artel to make bricks and to fire them, to fire lime, an artel of shoemakers, an artel of tailors, and workers, and provide work. The people have no work! Germany

could do this, but we can't!? This is an embarrassment, all the more so in a
worker-peasant state. Right here everything can be set up: storage sheds,
schools, hospitals, even baths and gardens, and model field cultivation. This
needs to be proposed in an actual resolution for the entire RSFSR, and then
there won't be so much unemployment and hunger, and everything will be
worked out according to a plan, and they won't walk around one behind
the other from nothing to do, and the specialists will be put to work and not
just get a salary.

Comrade editor, if I had the means, I would go to the center, to Comrade
Kalinin, and talk with him about this, [about] how to build this life. Now
we can [do it]. It's clear to me that before, all this was impossible, but now
it's possible.

Please print [this letter].

A wealth of ideas from below on how best to establish socialism was
made available to the political leadership. Like some in the Communist
Party leadership, Rudikov had concluded that the NEP should be abandoned and industrial expansion should be undertaken. Unlike many
communists, however, he saw nothing good in collectivized agriculture.
But he did call on the state to support agricultural mechanization. Peasants thus freed from their labors could then form artels—a prerevolutionary form of cooperative labor organization—and join the industrial
work-force in collective fashion. In this way, Rudikov's plan looks both
forward and backward, cleverly combining technological advance with
traditional Russian egalitarian labor practices. His evidently sincere
wish to discuss the future of socialism in the USSR face-to-face with the
head of state also speaks to the regime's success in portraying Kalinin,
the former peasant, sympathetically, as a national village elder accessible to all.

In like fashion, P. F. Ponomaryov, from Durnovskaia stanitsa in Sta-
lingrad province offers his ideas on ending unemployment and establishing genuine communism in a letter from early 1927.

· 103 ·

Letter to *Krestianskaia gazeta* from the peasant P. F. Ponomaryov denouncing the
personal acquisitiveness of communist bureaucrats and presenting his plan to
transfer the economy to communal foundations, 16 February 1927. RGAE, f. 396,
op. 5, d. 30, ch. 1, ll. 538–539. Original manuscript.

For practical socialism

From the very start of the October Revolution, our Communist Party has been propagandizing among the toilers for a transition, for a new communist society. How much paper and how much money has the government spent to maintain these propagandists? But in the end, paper is only paper, and these propagandists who eat up from fifty to one hundred rubles [a month] are always ready to establish communism on paper. By their salary all the clerk-communists want to improve their own farm, if they have one; if there isn't one, then they improve their personal or family situation through their wages. But on paper? On paper [they are] for a new communist society! But what kind of communism is it when every clerk and non-clerk tries to bolster his economy, to stuff his pocket? Is this really how they are going to create a communist society? No! This strengthens private property. I'll give you an example: Ever since 25 October 1917, we had a communist here, Ivan Molofeevich Stepanov. He fought for Soviet power and held Party and social positions. He also stood up for communism, and as a result, he got rich and became a Red merchant [*kraskup*]. He trades in our stanitsa. Now he opposes communism and is for private property. And he is not alone; there are many like [him]. What does the peasant see in this? He sees that every communist looks to his personal well-being but that real communism interests him [only] on paper. We are all aware that in our country there is a mass of laborers, a mass of the unemployed. How can you make a free and not an involuntary toiler of a laborer? How can you give work to the unemployed? Here's how! So that our army of laborers should not expend its strength lifting up the private economy, which is the enemy of communist society, so that the unemployed should be guaranteed work, work that benefits communism: To do this, all the clerk-communists must be required to create a kommuna on a district scale, even though it will be filled with professionals, and to draw into this kommuna all of the district's laborers. The unemployed will be the first paid members, getting money for their service so they can buy the materials they need with this money. This way all the laborers and unemployed will physically create a communal economy. But the communist-clerk, as long as he still is a clerk, will completely invest his entire service pay in this communal economy. If he doesn't get a post, he gets no pay, and he must participate in the physical labor of the kommuna. In case of a transfer from a post in one district to another, he must register in the kommuna of that place and do as described above. Communists—peasants, tailors, cobblers, joiners, house painters, and the like—must also be attracted into these communes. Only by this practical communism will we get the desired results. Only by this practical communism will we turn NEP Russia into communist Russia.

Village correspondent P. PONOMARYOV

In the following letter, the peasant S. P. Romanov from Nizhnaia Aki-movka village in Briansk province, demands that the NEP be replaced by a highly centralized system of cooperatives. Like Ponomaryov, he also expresses his hostility toward bureaucrats ("hangers-on") who are not committed to socialist construction.

· 104 ·

Letter to *Krestianskaia gazeta* from the peasant S. P. Romanov calling for an end to the NEP and a fundamental reorganization of cooperatives, May 1929. RGAE, f. 396, op. 7, d. 14. ll. 169–170. Original manuscript.

To *Krestianskaia gazeta*. (For discussion by the government and the worker and peasant toilers.)

It is necessary to create a united front to struggle against nep.

It is the united Soviet land that alone defends the interests of the toilers. Our government is trying to raise up industry and agriculture, but the toilers of the USSR help out very little in this development, and this serves as a brake on our achievement. In his lifetime, Lenin said: "In the near future, we must overtake the industry and agriculture of the biggest capitalist countries: America, England, France, and others." Comrade Lenin is right about this. This can and must be achieved, but under these conditions. If the toilers give their advice [about] how to achieve this quickly, the financial component of the country still needs to be raised, agriculture must be amalgamated in collective farms, and handicrafts in artels, [in order] to create a unified cooperative union. Three currents have appeared in our All-Union Communist Party. These currents are: Trotskyism, the right deviation, and our genuine Communist Party. For my part here's what I see: many of our leaders are turning away and have already left us, the toilers, and are trying to provide a life only for the rich while they pressure and crush the proletariat. To our beloved leaders who have remained and to the party of steel, I convey the wish [that they] take the offensive (a purge of the party) in the hope [that] we non-Party toilers, the most reliable, will replace a hundred of those hangers-on with thousands of good [people], who will have to pass through the finest sieve so that no alien gets through [not clear in original]. In developing our country, in both industry and agriculture, cooperation must play a large role, for it beats back every private trader and wrecker-speculator, but it needs to be rebuilt in a new way. Before offering proposals for [its] reconstruction, I wish to make note of all the shortcomings of cooperation. Our present cooperation is unable to get goods to where they are needed. [. . .] How should it [cooperation] be rebuilt? Here's how: along the lines of the trade unions or as our Union administration is organized. All cooperatives should be amalgamated into a single whole, leaving aside handicrafts, which should be like an assistant to our production, but there should be a connection between these two in finances and goods when there is a shortage. In discussing my proposal, they will surely say that this will wind up engulfed in red tape, but this is a radical error. If a cooperative branch does not have an item it needs, then the district administration can, on its own, transfer [it] from the district factories or department of trade to that locality where it is needed, and if it's not available in the district, then [it will be] in the Central Executive Committee cooperatives, which should have all the information on goods. But in the case of a shortage of a certain product or good, this administration or organization can distribute everything correctly. In the share books one's property, [social] origin, deprivation of rights, etc., should be indicated. Through such an administration, we will attain the full cooperativization of the population and in this way do away with the NEP.

Non-Party middle peasant ROMANOV, SERGEI PAVLOVICH

The previous two letters provide some evidence that as the 1920s closed, popular hostility to the NEP was on the rise. In the writers' view the NEP was no longer furthering economic recovery but instead obstructing the implementation of needed socialist measures. The letters also convey a sharp sense of class resentment toward bureaucrats and nonproletarians, who are identified as the principal beneficiaries of the NEP.

The overwhelming majority of letters addressing the theme of socialism point to the absence of collectivism and comradely mutual assistance as the main obstacle on the path to the new society. Given the peasantry's communal traditions, it is perhaps natural that precisely these features should be seen as defining socialism. Using peculiar metaphorical imagery, the peasant D. K. Bolshakov identifies collectivism with equality. Bolshakov lived in the village of Viazovitsa, Ivanovo-Voznesensk.

· 105 ·

Letter to *Krestianskaia gazeta* from the peasant D. K. Bolshakov on the meaning of socialism, 17 July 1926. RGAE, f. 396, op. 4, d. 24, ll. 38–41. Original manuscript.

On the building of socialism

Comrades, we and you have written a lot here about what is bad in the country of the Soviets, [about how] the leaders get big salaries, etc. On the whole, one hears a lot of criticism from the peasantry, all sorts of curses, but not a single peasant will say nor offer his thoughts about how, in his opinion, we can quickly attain socialism. Indeed, I think that among the considerable mass of peasants there are intelligent heads that sometimes think about how, say, socialism should be built in the country of soviets. Maybe among this mass there are sensible people who could offer a good proposal or an opinion about this. After all, in spirit and nature the peasantry, as a whole, is cheerful and healthy. It is nurtured among the forests, fields, and meadows, and it could be strong-minded if Soviet power reeducates and cultivates it. But I, in my turn, according to my prejudice, in my simple peasant mind, and with no consideration for the findings of scholars, am thinking as follows: right now we are all still not equal, and everyone has not been educated the same, [but] when everyone will be at the same [level] of development and education, when one will not mentally lag behind another, when everyone stands in the same material situation—except for the leaders, who must be placated with a better material situation—until ideas have penetrated them, then the mass of the people is a stagnant swamp in which there is still no well-defined channel, but it is still a big swamp divided by all sorts of little streams that don't cross one another but go back and forth. When these little streams are drawn into one river that quickly and boldly flows forward, then it doesn't need any leadership, then each drop will not overflow the shores. Then the engineers and professors will not take more for their work than the peasant takes. When a tractor rumbles along a peasant's field which is beyond the strength of an individual peasant [to work], only then will the peasant see an eight-hour working day and realize that he did not participate in the revolution in vain. Not every peasant is aware of socialism. They don't even know the word "socialism." But the problem needs to be placed before the peasantry directly, that socialism is when people will have no property, when they will begin to lead a life like ants that is in general accord.

Bolshakov, D. K.

Please print this in *Krestianskaia gazeta.*

Many letters from workers and peasants to Soviet editorial boards or political leaders begin with an apology for their writers' lack of formal education and clumsy communication skills. In most cases, these expressions of humility reflected the genuine feelings of inferiority the semiliterate felt before the educated. By the same token, displays of self-deprecation and deference may also have been part of a conscious strategy of self-defense. Subordinate groups often employ such a strategy when challenging an existing order.[22] Adopting the proper self-effacing tone,

Bolshakov calls for an end to the inequalities that are undermining Soviet socialism. Until society is organized ant-like, on an equal and collective basis, until social life flows in one direction, there can be no socialism.

Not everyone was so enamored of the collective life, however, or found analogies to the animal world attractive. The following commentary on socialism and collective agriculture dates from March 1925. It is taken from an unsigned letter sent from a Siberian village.

· 106 ·

Letter to *Krestianskaia gazeta* from an unknown resident of Altai province opposed to socialism, 6 March 1925. RGAE, f. 396, op. 3, d. 100, l. 26. Original manuscript.

About socialism. It is pointless, dear comrades, to concern yourselves over how to reach socialism quickly because it can never be reached, because in the kommuna and artel, life is possible only when a man doesn't know his wife, or a wife her husband and children. You'll live there like cattle and just as laboriously. This is clear from the fact that those close to us and our kommunas have all scattered, and those who are living [there] do so not because they support the kommuna but because there is no place to hide—he comes to the village, but they don't give him any land. There are some [people] who live in a huddle, but they are disconnected from one another, and if they are together, then it's always a [matter of] big fights and arguments. So everywhere there are gates, but socialism is fit for cattle, not people.

In the following letter, the peasant Ye. I. Safronov from Selishche village, Smolensk province explains why peasants must abandon religion and dislodge parasitical village priests if they wish to enjoy the benefits of socialism.

· 107 ·

An antireligious letter to *Krestianskaia gazeta* from the peasant Ye. I. Safronov, August 1926. RGAE, f. 396, op. 4, d. 24, ll. 12–13. Original manuscript.

What the workers and peasants should be after.

Comrade peasants and workers! I want to place [an article] in *Krestianskaia gazeta*: "What Must Workers and Peasants Aspire To." Namely this: [they] must emerge from darkness, not listen to priestly deception, and destroy it. But exactly how to destroy it? It's very easy. We need to do exactly that which was stated before—to emerge from darkness we have to study and then study some more according to the bequest of our dear leader Vladimir Ilich Lenin. And we workers and peasants should fulfill the bequest of the dear leader. In the villages now we have many Soviet schools. Unlike schools under the tsar, our schools in the USSR are separate from the church. Before, we, the young generation, had some unseen God drummed into our heads. Under the tsar, if a student failed to memorize God's law, then he did not pass the class examination. But if we begin to study in our Soviet school, only then will we emerge from darkness and priestly deception [by those] who for ages have lived without working and feed on our hard-earned crumbs, for which we get calloused hands. But they, the bloodsuckers, have eaten nothing but the best: honey, fatback, eggs, meat, white bread, etc. And it's the ones who didn't work who devoured it all, while we peasants knock ourselves out day and night and come home hungry and cold and [find] nothing to eat. But the priest-landlords grew bellies without working, and didn't fear that God would punish them for this. And for God, we gave away our whole life. And so the comrade workers and peasants must fight this bloodsucking fraud, for everyone should work and labor, as they say, to live by his own labor.

But, comrade workers and peasants, this is not our only aspiration; we still need to aspire to socialism. But a lot of people don't know what socialism is. That's because earlier the priest-landlords were afraid that then there would be no reason to come to a priest; they were afraid just as the devil fears incense. Socialism is, for example, to unite, in place of associations,[23] [in] a universal peasant and worker life; all peasants and workers really need to unite, all peasants need to obliterate all boundaries, all peasants and workers need to work just as one, as they say, [and] then we will be a harmonious host. Only then will we have rich lives; then we will have our reading hut and our clubs, and as they say, there will be a fraternal host.

Composed by YEGOR IVANOV SAFRONOV

Safronov asserts that the true meaning of socialism has been hidden from the people by those pursuing their selfish interests behind a religious facade. Safronov then offers his own definition of socialism: workers and peasants organizing life collectively as a "harmonious" (*druzhnaia*) or "fraternal" (*bratskaia*) "host" (*rat*). The definition, which is clearly inspired by religious imagery, reinforces the idea that for the worker and the peasant genuine communion can be found only in the brotherhood of those who labor.

Not everyone who adopted a skeptical attitude toward religion was prepared to accept socialism in its stead, however. As one worker put it, "Who should the worker have faith in, the priest or the communist? The first promised us 'a heavenly kingdom' and the second an earthly paradise, but in the end you don't see one or the other."[24]

The Lazy and the Industrious

For the peasantry, as for the Party leadership, the complex demands of Soviet agriculture made it difficult to agree on a common path to socialism. Disparities in wealth, means of production, abilities, and the determination to succeed divided the peasant class. The state tried to exploit these differences and limit the influence of wealthier peasants through tax policies and legal measures that favored the poor at the expense of the better-off. As intended, these steps also exacerbated intra-class economic and social tensions. Successful, hardworking peasants especially resented policies that they believed encouraged idleness. This line of thought is expressed in the letter of I. P. Ogurtsovsky, who identifies himself as a middle peasant from Padora village in Pskov province.

· 108 ·

Letter to *Krestianskaia gazeta* from the peasant I. P. Ogurtsovsky explaining the harm that agricultural taxes inflicted on productive farmers, 17 March 1927. RGAE, f. 396, op. 5, d. 30, ch. 1, ll. 257–261. Original manuscript.

Even though I have a feeling, comrade editor, that this article of mine will lie in the editorial basket and not be printed, all the same I want to offer my thoughts for a discussion by the *Krestianskaia gazeta* editorial board and its readers. I am a peasant—a middle peasant by social standing. I am engaged in agriculture and consider this the highest of all pursuits and attach great importance to it because the entire welfare of the republic rests on it. A look at the pitiful external and internal condition of the peasants' holdings makes one heartsick. By this one may judge the material power of our Soviet state . . . Aware of the importance of raising the cultural level of agriculture, I am trying with all my might, with the aid of the science of agronomy, to enhance my holding in order to make myself an example for others. I acquired the knowledge to improve agriculture through the *Krestianskaia gazeta* publishing house. Thankfully, the *Krestianskaia gazeta* editorial board also helped me. Last year it sent me four rutabaga seeds with instructions for their cultivation. I put a lot of effort into this and the rutabagas grew wonderfully well, each weighing 6.8 kilograms. I took them to the agricultural exhibition along with the potatoes and Swedish oats ordered from Moscow from an advertisement in *Krestianskaia gazeta,* and a colt, too. For this, as a so-called model farmer, I got a "Pruzhinka" harrow[25] as an award for conscientious labor. I took it home . . . Thank you *Krestianskaia gazeta* . . . and to agronomy. From being a middle peasant soon I will become rich, and then I'll take on the shameful name "kulak" and will not wear bast sandals. I have taken one step closer to socialism by fulfilling Ilich's request: "Improve your holding, and in this way you will participate in the building of socialism." It's a pity that not everyone understands this and does not work at improving their holding, fearing that insurmountable obstacle that stands in the path of advanced development—the unified agricultural income tax levied for the rejuvenation of the poor. [To put it] in another, more candid way, looking at the problem in depth—[it is a tax] for the breeding and reproduction of poverty and the training [of people] in "laziness," in an indifferent and wasteful attitude to economic achievement. When he thinks of the agricultural income tax every laborer loses heart because the revenue of [his] holding will barely, if at all, exceed the minimum for subsistence. [. . .]

When compiling the draft of the income tax assessment that exempts from the tax the farms having only the subsistence minimum—of which there are many in the USSR . . . "and [which] are increasing"—they didn't take into account the natural-scientific proofs that man's thinking, that is, [the thinking] of the caveman, began to develop in the Ice Age, when the conditions of life of our beast-like grandfathers had deteriorated. Nature had stopped coddling [them] and in this way spurred them in the direction of intellectual development in the sense of the struggle for existence. And on the soil of this struggle their states of mind have reached modernity . . . Likewise, the unified agricultural land tax on a desiatina of land, taking into consideration its quality, might awaken the poorest people to reasonable, vital, cultural activity in the sphere of raising the revenues from agriculture by not giving indulgences to loafers and those who grab norms of land and get no use out of it and don't pay the state tax on it to boot. Then the material might of the Union would be strengthened more quickly, and the desired goal of socialism would be reached more quickly. All-Russian [Central] Executive Committee, heed the sharp cries of the laboring mass and fulfill the glory of socialist construction!

I. P. Ogurtsovsky

For Ogurtsovsky the way to socialism is through hard work and struggle, and the main obstacle on this path is a tax policy that rewards the loafer and discourages production.

In the next letter, Afanasy K. of Lugovoi settlement, Melitopol region, Ukrainian republic, also questions the wisdom of the state's support for the poor peasant and its punitive approach to the well-off.

· 109 ·

Letter to *Krestianskaia gazeta* from the peasant Afanasy K. calling for an end to the division of the peasantry into rich, middle, and poor categories, 22 February 1927. RGAE, f. 396, op. 5, d. 30, ch. 1, l. 267.

Where is our freedom?

Here it is, already the tenth year of the revolution, ten years since the peasants threw off the landlords' yoke. This had to be so that the land of Soviets could become a country of fraternity and peace, could forget hostility and grief, and so that everyone could think and live freely. But this hasn't happened; for some reason they are dividing us.

I think that citizens living in the territory of the USSR should not be divided up; each should have identical rights. But this is also not the case; they are dividing us.

You read the newspaper; There they write: Hardly anyone gets a pat on the head. Instead [they write that] you are a kulak, you a middle peasant, and you a poor peasant. Why do we still have this division, can the state really be interested in keeping us in three classes, is it really impossible that everyone should be equal?! I think it's possible, but the state is not doing it; it loves the poor peasant more than the well-off because even in the newspapers they call [it] a "proletarian state," but why is it not possible to call [it] a "people's republic"? How long will it be called the "proletarian state," and will it really get poorer and poorer and never think of getting rich? It probably can't be any different because [the thinking is,] I am a poor peasant, I have all the rights, but down with the well-off [peasant], because you are a dangerous element. [We] must still deprive [the well-off] of free speech or else watch out, he will crawl into the Soviet and will prevent the poor from building the state, but on whom does it [the state] stand, the poor peasant or the well-off? No. What would happen if we were all poor peasants? In the end, they'd just destroy their economy, and that's all. Then what would support the state, the poor peasants who now lie on the stove and spit at the ceiling because the state gives him [*sic*] credit? Won't he still be a poor peasant then? But I am well-off. I am always working, day and night without a break. I worry about paying the state tax and in general try to stay in the state's good graces, but it, on the contrary, because I am rich, takes away my right of speech because I am a dangerous element.

And they still want to build socialism. By this division of the peasants it's impossible to even talk about socialism, to build socialism. The whole country must be made aware of this. It's not necessary to divide the peasants into classes. Then we will achieve our goals. You want to live well? Give the well-off peasants full freedom, and then there won't be poor peasants in the land of Soviets, and it won't be called a "proletarian state" but a "people's republic." But if the well-off peasants are deprived of rights, then they won't even think of being state builders. Our country is poor, and under these conditions it will be the first conqueror of "Labor" (period).[26]

I ask the *Krestianskaia gazeta* editorial board to print this letter although I know that they won't print it. If I had written that my neighbor was an exploiter, it would be printed right away, but letters such [as these] are hardly ever [printed].

AFANASY K.

Successful peasants like Afanasy K. who found themselves on the sharp end of the Bolsheviks' class-splitting stick invoked the stereotype of the lazy muzhik spitting at the ceiling to discredit the Party's poor-peasant bias. For all prejudices and resentments in the letter , however, it expresses certain truths. Kulaks and well-off seredniaks knew how important their output was to the state's well-being. Common sense dictated that any attempt to create agricultural abundance on the backs of the least productive farmers was doomed to failure.

In the next letter, M. F. Kholin, from Kriusha village, Nizhny Novgorod province, asserts that socialism has no popular support among hard-working peasants and is simply a gift to idlers.

· 110 ·

Letter to *Krestianskaia gazeta* from the peasant M. Kholin calling for a general and open discussion of socialism among the population, 15 February 1927. RGAE, f. 396, op. 5, d. 30, chs. 2, ll. 771–772.

Lately a system has come into being whereby each, more or less, important legislative proposal of the government is disseminated through the press for wide discussion by the people. However, the most important problem, one that concerns life and death—the building of socialism—not only is not brought up for broad discussion but up to now no one in the village has said anything intelligent or convincing about its character. On the whole, the population only has a vague idea about it, but it senses with its heart that socialism is opposed to its (the village's) nature, and not without reason, [for] when you talk to local residents about socialism in general, then they unanimously disown it and see it simply as fertile soil for the growth of lazy people, drunkards, parasites, and other similar types who currently abound in fairly large numbers and who are lumped under a single yardstick—the poor. In this way they are getting the first place at the table of the republic, and for that which offends the middle peasant they praise Soviet power.

Therefore, to get down to building socialism, the population considers it necessary that this problem also be offered for broad discussion by the people. And only when it is proven that life in the realm of socialism is better than life in the realm of capitalism, only then will it get down to the building of socialism itself. But building it at the present time, without having the authorization of the people, is premature, since the mass of the population is opposed to socialism. That this mood is highly negative is certainly known to *Krestianskaia gazeta*.

About this I thought it necessary to inform *Krestianskaia gazeta*.

With respect, M. KHOLIN

Please do not make my name public.

KHOLIN

Rather than embodying the collective ideal of the peasant commune, socialism is "opposed to the nature of the village" and a boon only for those allergic to work, or so maintains this writer. Because it remains an abstraction, the author also asserts that most peasants have only a vague understanding of the meaning of socialism. That the peasants lack a tangible sense of socialism is the subject of a brief letter from Pavlovsk district, Voronezh province.

· 111 ·

Letter to *Krestianskaia gazeta* from an unknown peasant speculating on the characteristics of the socialist system, 1 February 1926.
RGAE, f. 396, op. 4, d. 27, l. 294.

At an assembly of citizens who enjoy carrying on discussions about the Soviet system, the question What sort of system will there be under socialism? frequently comes up. Some say that under socialism the peasants will work as if in a factory or a plant—everyone together—and then they will get what they need. But others say that under socialism everyone will be rewarded according to his output. If, for example, I gain for the state one hundred puds of grain, then I will get [something] for it. Whoever [comes up with] more, receives more. But everyone comes to the same conclusion— that there will be no socialism for some time, if at all, because everyone in the government gets a large salary, and under socialism no one should get a salary, since for them there is no point in building socialism. In regard to this problem, at the request of the citizens, I ask *Krestianskaia gazeta* to print an article in the newspaper that will explain to us in detail what sort of system will exist under socialism.

Searching for Socialism

The regime's efforts to set the terms of popular discourse achieved imperfect success in the 1920s, for NEP reality continuously undermined the claims of socialist achievement put forth by Party propagandists. Simple observation of their surroundings led people to conclusions about Soviet socialism that could be quite unsettling from the regime's point of view, as seen in this letter from A. T. Melnichenko of Bereslavsky settlement, Zinoviev region, Ukrainian republic.

· 112 ·

Letter to *Krestianskaia gazeta* from the peasant A. T. Melnichenko speculating on the fate of socialism in the USSR, February 1927. RGAE, f. 396, op. 5, d. 30, ch. 2, ll. 1076–1077. Original manuscript.

To the *Krestianskaia gazeta* editorial board. I ask the editorial board to print my letter in the newspaper.

In search of socialism

First off, I apologize. I am poorly educated and not a good editorial specialist; nevertheless, I want to say that I can understand our political-state system. I am the very smallest cog in the state administration, since the people are the power in our republic and I live in this republic. The fact of the matter is that we citizens, the dark mass, have come across a few things that we really can't understand. It's quite possible that this will not be of any significance or to the point, but all the same, I think that I can be excused, [for] otherwise I won't be able to put [my thoughts] together. Anyway, this is how we ignorant peasants understand it. I am familiar with the four letters of the name of our state—USSR—and will examine individually what each letter stands for, why it is so named, and will offer my conclusion on each letter. The letter "U" stands for "Union." This is because there are several republics in the Union. This is correct. One letter "S" signifies "Soviet." This is because in these republics power belongs to the soviets, and this is self-explanatory. The next "S" means "socialist." This I cannot define, and a lot of other peasants also cannot understand it. As a result, we, along with millions of peasants and together with the representatives of the government, maybe we ignorant peasants are deeply mistaken, but all the same, I cannot figure out where this socialism is. I see that a poorly clad peasant in worn-out shoes and a torn hat enters a store and buys 0.2 kilograms of sugar and 0.4 kilograms of salt and barely has enough money for the bill, and I think, there is no socialism in this, and I conclude no, socialism is not kept here, because he is poorly clad here, socialism is chilly, and the goods are such that socialism cannot live, it is not here. But then another citizen enters, well dressed, in a good sheepskin coat, shoes, galoshes, and buys 4.5 kilos of sugar, 2.3 kilos of plums, 2.3 kilos of cherries, 1.4 kilos of sultanas, 4.5 kilos of fish, 0.4 kilos of halva, and a chunk of toilet soap, and I think maybe socialism is here and conclude no, there is no socialism here; although it is warm in these clothes, [although] one can stuff oneself on these goods and die, no, it's not here. [. . .]

Then I see that in the city unemployed day laborers are sitting—is this where I have found socialism? And I come to the conclusion that no, under these conditions poorly clothed, half starved, this also isn't where socialism stays. And then the owner of a rich creamery comes running up and says, "I have an order for three men to come and clean my well. Hurry up. I'll go for a driver right now because in an hour I have to leave for the region[al center]." And I think that maybe socialism is here and come to the conclusion that no, it can't live in this hectic life, it runs away and disappears, it is not here.

And so, I didn't find that which applies to the word "socialist," and I think to myself, maybe a typographical error has crept in. Instead of a "C" they wrote "S." Instead of "Union of Soviet Capitalist Republics," they wrote "Union of Soviet Socialist Republics." I think that's what happened: it's a mistake. However, [depending on] what people smarter than me say, maybe I will come around to their opinion.

But it's difficult for the editorial board to let this [letter] slip in [the newspaper], in the sense that it is ungrammatically edited and written and of a purely peasant character and education.

Because in the newspaper the large part of the articles are written and edited by well-schooled authors, but the writings of unlettered peasants are well known; either they are not to the point or went too far into painful [detail], in particular [about] this non-Russian word "socialism," but, however,

I ask the editorial board to put my letter in an issue; and literally, don't remove a single word.

The end.

Melnichenko's unusual account is rooted in his personal experience. His effort to make concrete the slippery, abstract, and "non-Russian" concept of socialism avoids the usual propagandistic clichés and, had it been published, probably would have struck deep chords of recognition with the peasant readership. He is not as guileless as he pretends, however. His letter, though idiosyncratic, is not especially ungrammatical, and his dissection of "USSR" is a clever device, leading to his subversive conclusion that capitalism, not socialism, is thriving in the land of soviets. Under the guise of a naive and untutored peasant, he holds a mirror up to the existing system and uses the reflection to bring into question its central, self-professed principles.

The peasant S. T. Myskin-Zelenov from the village Sonin Lug, Orel province, draws on a broader store of knowledge and does not limit his discussion to the material possibilities of socialism.

· 113 ·

Letter to *Krestianskaia gazeta* from the peasant S. T. Myskin-Zelenov on the Soviet Union's socialist possibilities, 9 January 1927. RGAE, f. 396, op. 5, d. 30, ch. 2, ll. 1016–1017. Original manuscript.

To the *Krestianskaia gazeta* editorial board.

Are we heading to socialism, and when [will we get there]? The soviet republics of our Union occupy one-sixth of the globe; this is a vast space that allows them to carry on their own distinctive life. Because of an abundance of raw materials and many minerals, we can be independent from the capitalist states in everything. The features referred to provide the basis for socialism independent of other countries, and we will achieve full socialism as soon as the popular masses adopt the predisposition to recognize the truth and eliminate the egoism (self-love) inherited from the past. Looked at in the world context, our revolution is unique. Why? Because nowhere were the popular masses so enslaved, both economically and politically, as under tsarist Russia. But then the moment arrived. The spark of oppression under which the popular masses [suffered] was ignited, its ignition hastened by the war. The state boiler of Russia exploded, the lid blew, and the top tumbled to the bottom. I want to say that in the American or European states it won't happen like this, because there, the popular masses enjoy different living conditions. America will reach socialism on other rails—namely, through a highly refined [system of] education and the achievements of an amazing technology. Even though they write that the working class there is in the clutches of capital, one reads that on the contrary, machines are employed there in all branches [of industry] and that the workers operate them. The working class lives like our bourgeoisie and enjoys every possible modern comfort and luxury. As early as 1904, I read that in America the artels of different urban professions had united on cooperative bases; they ran farms and in their free time they performed physical labor and were not averse to getting a university education as doctors, etc. I also had occasion to read that during the Russo-Japanese War and, I heard this from prisoners, that at that time in Japan not only homes but even stores [had] doors without locks, and this is because strict laws or the upbringing of the popular masses has so ingrained mutual respect in the people as a whole. But with us, even at the present time, despite having attained the elevated rank of equal citizenship—out of which emerges self-administration (authority)— what is there? Hooliganism is growing all over more than anywhere else. This is an unhappy page of our citizenship. Sure, they say, this is a remnant of tsarism. But when will we overcome this? That is the question of the future. And when will the people, as a mass, understand the truth of mutual fraternity, love, and the sanctity of laws—of socialism? Because of our background, we are far from this. But at the state level, we will go forward. Among the people of the Union republics there are many sorts of people who hold up the banner of justice and the knowledge of mankind's immutable truth. Whoever doesn't know the sacred truth should try to perfect himself.

SERGEI TIMOFEEVICH MYSKIN-ZELENOV

The author does not deny the possibility of building socialism, but in his opinion the path will be a long search for truth and moral perfection. Despite having successfully overthrown the abusive old order, the people of the country are still not ready to act in accordance with these virtues. The brief pronouncement of a Moscow worker contains a more succinct but nearly identical thought: "Socialism will take on life only when it naturally replaces capital, but for us it is premature."[27]

Unhappiness with the condition of the countryside at the end of the decade comes through in many letters, although their authors are often unable to identify exactly why conditions are so unsatisfactory. The Siberian peasant F. Z. Dubrovin, from Suslovskoe village, Barnaul region, Altai province, expresses his discontent and confusion over the current state of affairs in this letter from April 1927.

· 114 ·

Letter to *Krestianskaia gazeta* from the peasant F. Z. Dubrovin on village
hooliganism and peasant disfranchisement, April 1927.
RGAE, f. 396, op. 5, d. 30, ch. 1, l. 401. Original manuscript.

I am taking the liberty of describing Siberian life to the comrade editor of
Krestianskaia gazeta. Under Soviet power life is not right at all. I do not
know what this is due to; either it is the leadership or it is just how the
people have turned out. Every year there is arson and hooliganism. They
burn grain stacks, they burn threshing barns full of grain, different houses,
buildings. They have already burned two windmills. Yesterday, 24 April, on
Easter, a mill with double millstones that could grind five hundred puds [of
grain] a day burned down. But we know who sets most fires. The big fire
setters are the loafers, the hooligans. What can be done now? To whom can
we turn, and where can we find relief from such lowlifes? Before our elders
would say the proverb "God is high and tsar is far away." But now there is
no God and no tsar, and the bosses do not want to take any action. Now all
one can do is wait to die. I had to hear from the elders that it will get worse,
but I don't know how. In our village of Suslovskoe we have nineteen disenfranchised persons. For what? Because they have a middle-level economy
and work day and night. These people are the real builders of the state.
They provide grain, meat, butter, leather, sheepskins, fiber, wool. And they
carry the label of alleged bourgeois. Please excuse me, I am poorly educated.
I am citizen FYODOR ZOTEEVICH DUBROVIN.

In the following letter, dated 18 November 1927, the peasant I.F., of
Berezovtsa village, Kursk province, indicts the leadership for perpetuating rural inequality. However valid the charge, in this case his wrath
falls on the wrong head. Demonstrating that many peasants had only a
tenuous grasp of who did what in the Soviet government, the addressee,
V. V. Kuibyshev, was not, nor ever had been, commissar for agriculture.
As a Politburo member, Kuibyshev had a recognizable name, but at the
time the letter was written, he served as chairman of the Supreme Council of the National Economy (Vesenkha), which oversaw Soviet industry.

· 115 ·

Letter to V. V. Kuibyshev, chairman of the Supreme Council of the National
Economy, from the peasant I.F. on the sorry state of agriculture in his village,
18 November 1927. RGAE, f. 396, op. 5, d. 30, ch. 1, l. 380. Original manuscript.

I am taking the liberty of pointing out to our worker-peasant government several errors and shortcomings in the administration of the people . . .

Comrade Kuibyshev! You are the commissar closest to peasant and agricultural affairs. You should examine and attend to what pleases and doesn't please the peasants about their government. I, when reading *Krestianskaia gazeta,* especially "Miting," [note that] you only see that everything is beneficial and good with us, as if there are no problems. But to my great regret, no, [this is not the case]. I will say that on the ten-year anniversary [of Soviet power], there is a lot that is better than before the war, but for the most part, things are worse than before the war. In our village agriculture is dead to the root. Each soul gets two thousand square sazhen of land. What can one get from such a scrap of land? Before the war the majority of us peasants lived at the factories, but now more than half have returned home—[and] there is no space [for them]. This is a problem! Because of the lack of land, it's impossible to practice agriculture. What can be done? And that is why, out of the burden of poverty and unemployment, hooliganism, theft, and vagrancy have begun to appear among us. Near us, three kilometers away, live the peasants of Brusovoe village. Without fear of error, they can be called modern gentry because they got hold of the land of three gentry and about three hundred desiatinas of land belonging to wealthy peasants. And now they have nine thousand square sazhen of land per soul, plus to this they added a lot of the woods and gentry buildings to their own. Several hardworking men from our village rent land from them and have to pay thirty to thirty-five rubles for one sowing of one hectare. Comrade Kuibyshev! You are the responsible peasant commissar for agriculture! Where is your slogan "All land to the peasants"? [Where is] "Equality and brotherhood"? What are you doing in the center—are you just picking up capital, getting five hundred rubles a month, and fogging up the brains of the peasants with your articles on the growth of agriculture? Don't you hear the groans of the people of our village? And there have to be a lot of villages like this. In conclusion, I will say that those who live better now are those with power and get five thousand rubles a [not clear in original], the new gentry. Give us socialism immediately!

I.F.

Unlike several previous writers, this peasant attributes low productivity and the rise in crime and violence to material causes—the land shortage and the inequitable distribution of arable land that the revolution and subsequent land reorganization was supposed to remedy—rather than the moral failings of the poor. Like Dubrovin, he wonders what the country's leaders are doing to alleviate the people's burdens and, in this vein, cannot resist remarking on Kuibyshev's salary, implying that he is not earning it.

L. N. Bondarenko, a village correspondent who familiarized himself with the Fourth Comintern Congress's draft program, submitted a long letter to *Krestianskaia gazeta* in mid-1928. His observations speak for themselves, as does the bitterness with which he makes them. His conclusion as to the best means of attaining socialism could not have pleased anyone in authority. He wrote from Iuzhny settlement, Kharkov region, Ukraine.

· 116 ·

Letter to *Krestianskaia gazeta* from the village correspondent L. N. Bondarenko on socialist ideals and Soviet reality, 24 June 1928. RGAE, f. 396, op. 6, d. 114, ll. 649–652. Original manuscript.

"Draft program of the Communist International"

To whom is the draft program of the Communist International directed? Who, among sensible people, can understand this wordy jumble, this verbose ranting about capitalism, imperialism, socialism, and the like? Who is this baseless verbiage supposed to convince? Absolutely no one, because everything that the Communist International stands against we ourselves have in abundance—capitalism, imperialism, monarchism, and bureaucratism—everything except . . . socialism, which is only an object of glorification and which doesn't exist at all, because all the vices that bloomed like a many-petaled flower in the present imperialist and capitalist era are now flourishing and even growing, only on the surface they are covered over by socialism. Can a person close his eyes to this, or look at it through rose-colored glasses? Just look around you and you'll see this very abuse of power, the disobedience, and even contemptuous disregard for higher authorities on the part of the lower organs of power, wasteful expenditure, violence against the individual, bankruptcy, drunkenness, petty tyranny, chaos, debauchery, hooliganism, arrogance, judicial and administrative red tape, etc. In a word, it's as if a kind of dark power set itself the task of bringing troubles and unhappiness to the masses, to compromise authority, focusing all its energy on economic collapse in order to visit more anger, debauchery, discord, etc., on the masses and thereby discredit socialism and the revolution and Soviet power itself. [. . .]

Comrades! Can this horror really be called socialism? If socialism can coexist with such vices as injustice, violence, bribery, slavish arrogance, the intentional distortion of the truth for personal gain, protectionism, caste exclusiveness, exceeding one's authority, [applying] pressure by means of judicial red tape with the goal of revenge, etc., then either the theory of socialism is speculative with no practical application in real life, or those people who shout about the building of socialism are not capable of carrying it out (I'm inclined to think that the latter proposition is likely). or else, finally, many wreckers have been spawned like those at Shakhty.

We cry that capitalism is an evil that impoverishes the poor and enslaves the proletariat, we curse capitalism, guard against it like a disease, and at the same time we ourselves strive after it, we fawn over and grovel before capitalism, we worship it and, finally, on the basis of destroying private property, take all state property in our own hands. Under the guise of the proletarian dictatorship, of proletarian command, we have reduced the individual to complete economic dependence, have enslaved him, made him weak-willed, serf-like, and, in this way have introduced among the mass of the proletariat uncertainty, poverty, bitterness, obsequiousness, immorality, etc.

What will this hypocrisy lead to? Everyone has already guessed that they understand its real meaning, and this awareness is penetrating into the masses more and more. If the capitalists say that capital is needed as a powerful lever for the extraction from the bowels of nature the riches lying there, that capitalism is an unavoidable evil and an irresistible organizing force for the development of culture that pushes mankind along the path of progress, then they, at least, are not lying, not engaging in hypocrisy, and certainly in their way are right, since another, more powerful means for the development of mankind has yet to found. Capitalists, notwithstanding their egotistical callousness, their thick skulls, their cruelty, are all the more attractive for their candor because there is no divergence between their words and their deeds, no levity, nothing utopian, no unrealizable theories; [for them] there is only one [consideration], though quite unattractive: actual reality, practically conforming to life. [. . .]

Who would have thought that socialism amounted to such a total destruction of property, that it would be so wildly violent, that even the poor peasants, who had managed to acquire a portion of the estate lands for blood money, up to several hundred square sazhen, would have it taken from them by violence, and that it would be given to better-off people or to cooperative enterprises that operate in the speculative form of the kulaks who hide in the cooperatives? This sort of socialism exists nowhere.

Socialism is a grand idea, unifying all mankind without the least exception. Therefore, to take it up with dirty hands is forbidden. Meanwhile, dirty hands are encountered at every step. [. . .]

If we acknowledge the theory of socialism as correct, meeting all the demands of a socialist life, then it must be so in practice, and under socialism those vices enumerated above are unthinkable. But the most important thing is that under socialism there should be no violence against the individual; otherwise it is not socialism but, so to speak, pseudo-socialism. It is unthinkable to imagine a type of socialism that takes without limit but offers nothing. A socialism that takes without restraint and gives back just so much—that kind of socialism is a perversion of the truth, unnatural and immoral because it serves as a bad example, as an excuse for any crime, and will not [allow us] to advance far beyond [the use of] armed force in restraining criminals. Moral authority is the only reliable means of restraint. But on what can it be based when the government itself has started down the path of violence and has legalized this violence under the guise of fabricated theories—requisition, nationalization, socialization, etc.—whose extraordinarily loose interpretation gives free play to arbitrariness and bribery?

Be that as it may, our life is abnormal and not only doesn't correspond to socialism but is directly at odds with it and is becoming more and more intolerable. Look around you, open your drowsy eyes, and you will see that dangerous symptoms of moral decay are encountered at every turn: bitterness, unhappiness, hooliganism, stupidly casting a stone from behind a hiding place at a totally innocent person's head. Theft spreads like ulcers, and robbery raids terrorize the population. These are frightening symptoms of an ethical and moral sickness in no way socialist. Seeing [the use of] non-socialist force against the individual, the arrogant way poor peasants are treated by big or petty authorities who, in their offices, deal in authentically bureaucratic answers, people in the end lose the ability to distinguish morality from immorality, cease to respect one another, and thereby don't feel obliged to stand on ceremony, become dependent on whisky and the barrel of a revolver, and hiss: "[Your] money or death?"

Just try to convince the thief that his trade is immoral, [that it] destroys human worth and that it will not profit him because today you took what doesn't belong to you, but tomorrow they'll take yours, and to this he'll answer you that you are no better than he is, that you only accidentally managed to grab up what he didn't, that the difference is only in the form of the grabbing, but it remains essentially the same. Under socialism, such a reproach of a robber is doubly convincing because socialism promised to shelter everyone under its protection, but it has turned out otherwise: far from everyone. The overwhelming majority remain outside socialism and are in a quandary. Moreover, socialism here, while moving from monarchism, patently has deviated toward genuine monarchical exploitation only under the cover of socialism. Maybe they don't notice this at the top, but down below they see it clearly and understand. Maybe the question will arise, Who among us is doing the exploiting when we have no capitalists, no landowners, and there have remained only the poor with a slight difference in social-economic correlations? There is only one answer: The exploiter is the government, which has seized all state property in its hands and which handles it irresponsibly and extremely uneconomically on an incalculably big scale. As a result, for the masses of people it makes no difference who the exploiters are, the capitalist, the landowner, or the government itself; they are sick of all of them and hate them all. [. . .]

In order for socialism to exist, it is necessary to cease and renounce once and for all any [resort to] pressures on the part of the authorities and to allow life to flow freely, and then it will find the natural channel for socialism. For this it is necessary, most of all, to acknowledge private property in land, free trade, free industry and, it follows, free individual initiative. Realization of socialism by force, by means of ignorant "*chinovniks*" who seek personal profit, cannot ever exist—freaks only beget other freaks. The experience of ten years has shown that what is happening now is impracticable.

Village correspondent, Iuzhny settlement, Kharkov region,
L. N. BONDARENKO.

Bondarenko's analysis reveals him to be a keen observer of NEP society and a powerful writer. To say that he undertook to speak truth to power is to grossly understate the force of his words. He obviously understood the vocation of village correspondent as that of a people's tribune; he is zealously exposing the abuses that passed for business as usual. Like a Soviet Savonarola, he enumerates the hypocrisies, vices, and corruptions of "pseudo-socialism" and minces no words in blaming those in authority for bringing about the economic and moral degradation of the common people. In using the word *chinoviks* to describe government bureaucrats, he uses a prerevolutionary term derogatorily. (We may also suspect that he himself was on the receiving end of that clandestine stone to the head and, like other village correspondents, experienced violence firsthand.)

In his letter, Bondarenko articulates what many people probably felt but could not put into words or were simply reticent to express openly about the current situation and the defects in the system. He offers several possible explanations for socialism's tardy appearance: that socialism is a noble theory but inapplicable in real life; that those entrusted with the building of socialism—Party and state officials—are incompetent and cannot put it into practice; or, as evidenced by the Shakhty Affair, too many enemies and wreckers are at work undermining the USSR's socialist foundation. As a moralist, the author is inclined to blame people who are trying to build socialism dishonestly with "dirty hands." Whichever the case might be, no one could be optimistic about the Soviet Union's socialist prospects.

As the 1920s closed, the Communist Party leaders, like many of the peasant letter-writers, became increasingly impatient with the snail's-pace advance to socialism. Like Bondarenko, they identified a number of factors contributing to this sluggishness, and like many peasants, they seemed unsure of just how to break the impasse. Unlike most peasants, however, some in the leadership identified private agriculture and peasant self-interest as the greatest obstacles to socialist construction. Such an analysis all but precluded the option of renouncing pressure and allowing life to flow freely in a socialist direction, as Bondarenko suggests. Events were about to reinforce this analysis of the peasant problem and offer a solution, provided the leadership was bold enough and resolute enough to pursue it. The time for action had arrived.

CHAPTER 7

The Great Break

At the end of 1929, with the growth of the collective farms and state farms, the Soviet Government turned sharply from [the policy of restricting the kulaks] to the policy of eliminating the kulaks, of destroying them as a class . . . The kulaks were expropriated. They were expropriated just as the capitalists had been expropriated in the spheres of industry in 1918, with this difference, however, that the kulaks' means of production did not pass into the hands of the state, but into the hands of the peasants united in the collective farms.

This was a profound revolution, a leap from an old qualitative state of society to a new qualitative state, equivalent in its consequences to the revolution of October 1917.

The distinguishing feature of this revolution is that it was accomplished from above, on the initiative of the state, and directly supported from below by the millions of peasants, who were fighting to throw off kulak bondage and to live in freedom in the collective farms.

—*History of the Communist Party of the Soviet Union (Bolsheviks), Short Course,* 1939

In October 1927, peasant grain deliveries to official collection agencies began to drop off unexpectedly. Over the next several weeks, the situation deteriorated to the point of crisis as deliveries fell far short of projected targets. By the end of the year it became apparent that unless peasants provided more grain, the cities would face a very hungry winter, and the regime's industrialization plans would be placed in jeopardy. Unexpected in its magnitude and severity, the "grain crisis" and the leadership's response to it set the stage for a complete reversal in the regime's relationship with the peasantry, a reversal that abrogated the smychka and irreparably damaged the mechanisms by which the NEP operated. By January, the central Party leadership was pressing local officials to use whatever means necessary to seize peasant—not just kulak—grain stores. The result was a war-communism-type operation that alleviated the short-term crisis but, by alienating the peasantry, led to a decline in agricultural productivity. Over the next two years, confronted by peasant reductions in sown area, the Party found itself routinely employing emergency measures to get grain. During this interval, Stalin and his allies on the Politburo argued for accelerating the creation of collective and state farms in order to resolve, finally, the grain-supply problem. In this way, the 1927–1928 grain crisis initiated the unraveling of the NEP. Consequently, by the end of 1929 the regime was poised to begin its final assault on independent farming in the USSR.[1]

The grain crisis occurred at the end of a year that had seen the Soviet Union suffer a succession of foreign policy failures. In China, Chiang Kai-shek, head of the Nationalist Party, slaughtered thousands of Shanghai communists in April, abruptly destroying the Moscow-sanctioned alliance between the Chinese Communist Party and the Nationalists. In May, the conservative British government cut diplomatic relations with Moscow. Still angered by the support that Soviet trade unions had shown British workers during the 1926 general strike, the Tories deeply objected to Comintern activity generally and to Soviet actions in China in particular, which directly affected their overseas possessions. In reaction, they canceled trade agreements between the two countries and ordered a police raid on the Soviet trade mission in London. Then, in June, a Russian émigré in Warsaw assassinated the Soviet ambassador to Poland, worsening Soviet relations with a regime that Soviet leaders already suspected of having territorial designs on Soviet Ukraine and Belorussia. The OGPU's clumsy repression of former old-regime officials in response to the assassination undermined any international sympathy this incident may have garnered.[2]

Thus, in both the East and the West the USSR's fortunes seemed to be at low ebb, and this situation had domestic repercussions. Based on previously classified documents, V. P. Danilov has called into question whether Soviet leaders truly believed that these events foretold imminent war.[3] In any case, Stalin and others did point to the worsening international situation to support the case for higher industrial tempos as necessary for national defense, which, by extension, meant turning the terms of rural-urban exchange further against the peasant. Rumors of imminent war also led to panic buying of scarce consumer items and gave peasants another reason to hold onto their grain.

Many factors contributed to the grain collection crisis, including peasant fears of war, but, for all the weaknesses of Soviet agriculture, the crisis was not a result of insufficient production. While total grain production in 1927 was slightly down from the record level achieved the previous year, state and Party leaders did not expect this to have a detrimental effect on food supplies; although less abundant than in 1926, the 1927 harvest was comparable with that of 1925, which had been considered a good year. The root of the problem lay in Soviet industry's chronic inability to meet rural demand for consumer goods. In 1927, after several years of tax reductions and an increase in off-farm work, the problem became acute as peasants—especially the better off—had cash but little to spend it on. The goods shortage had been exacerbated by the Central Committee decision taken in February 1927 to lower retail prices on manufactured goods despite their scarcity. Additionally, in September, in anticipation of the harvest, planners had also slightly lowered the official price for grain, which was already well below the prices paid on the private market. With little to buy and no need for additional cash, those peasants who were able to do so held their grain until prices rose, as they usually did in winter. If and when peasants did need cash, they could, under the prevailing circumstances, sell meat, dairy, or industrial crops that were then fetching higher prices.[4]

When the center began to receive reports about dramatic declines in the amount of grain being collected in the fall, it was caught off guard. Collection agencies had received only two-thirds as much grain in October and less than half as much in both November and December as they had the previous year. Nonetheless, at the Fifteenth Communist Party Congress held in early December, the leadership concentrated its attention on terminating the left opposition once and for all, and none of the main speakers indicated that an agricultural crisis was in the offing. By the end of the month, however, the scope of the problem had been recognized, and pressure was brought to bear on local officials to increase procurements. Various steps were taken to prompt peasants to part with their reserves. These initial measures were economic in nature and did not overstep the NEP. Central directives—most notably the 24 December Politburo decree entitled "On Grain Procurements"—aimed at depleting peasant cash reserves and called for the immediate collection of arrears in taxes, credits, and insurance premiums. Shares in cooperative societies—which peasants were induced to purchase—were increased, as were local taxes. Moscow also ordered manufactured goods transferred to grain-growing regions from cities, towns, and other agricultural areas. Raising the price offered for grain, however, was expressly prohibited.[5]

By January, such measures were proving inadequate to make up for the shortfall, and the center raised the pressure on local organizations to get results. A secret Central Committee directive to Party organizations of 5 January defined the failure to meet procurement targets as a "gross violation of Party discipline." A directive of 14 January warned that, depending on the region, the "breakthrough" in procurements had to occur within six to twelve weeks—that is, before the spring thaw made roads impassable. To aid and spur on local officials in their collection work, thousands of Party members were dispatched to the grain-growing regions as plenipotentiaries (*upolnomochennye*). Prominent Party leaders hastened to critical localities. In January, Stalin himself traveled to the important grain centers of Western Siberia, where he made a number of speeches rebuking the local Party organizations for their timid response to the crisis. He accused them of being in bed—figuratively and literally—with the kulaks and demanded that they employ emergency measures to seize grain surpluses, including prosecuting kulaks deemed to be hoarding grain as "speculators." Article 107 of the RSFSR Criminal Code carried penalties of imprisonment and confiscation of property for speculating.[6]

Stalin's proposals did not go unchallenged. The chairman of the board of the Siberian Regional Bank, S. I. Zagumennyi, objected that Stalin had erased the distinction between speculating—that is, trafficking in scarce goods for a profit—and withholding grain from the market until prices improved. According to Zagumennyi, poor and middle peasants, who held the former practice in contempt, would not understand how the latter practice could be construed as criminal. He warned that rather than weakening kulak influence, such prosecutions would have the opposite effect and, in Zagumennyi's words, would also "increase . . . the value of grain in the eyes of the countryside itself, and hence [lead to] a subsequent decline in its supply in the market."[7]

Notwithstanding the reservations of local officials, the center's pressure proved irresistible. The first three months of 1928 witnessed the abandonment of NEP restraint toward the peasantry and the implementation of emergency or "extraordinary" (*chrezvychanie*) measures to take grain. Three-member "troikas" of Party officials oversaw the collection work in given regions. The activities of private traders were severely curtailed, and markets were shut down in an attempt to close off non-state grain sales. The OGPU reported that as of 8 February approximately 3,000 private traders had been arrested in the country as a whole; by the beginning of April the number had climbed to 6,542. Roadblocks à la war communism were set up to intercept grain being transported between villages, and class differences were exploited to facilitate grain collections. Under pressure to procure specified amounts of grain, local soviet and Party officials determined the amount of grain each household possessed and then decided how much should be handed over for collection. To encourage poor peasants to help uncover reserves, they were offered 25 percent of the grain seized from kulaks in the form of long-term "loans."[8]

To enliven the collective farm movement, the Fifteenth Party Congress had resolved to intensify the "offensive against the kulaks" and to limit the growth of rural capitalism. The specific measures proposed were all in keeping with the NEP and did not foretell a general assault against the kulaks. Stalin made clear in his Siberian speeches, however, that in the current crisis such moderation was passé. He insisted that each kulak farm held grain reserves of fifty thousand to sixty thousand puds, enough to overfulfill the plan and end the crisis if only local officials had the will to take it by invoking article 107. But collection officials quickly realized in the course of their work that the quantity of grain held by kulaks and better-off peasants was insufficient to meet the targets set by Moscow. The bulk of the grain was in the hands of the seredniaks, and these reserves would also have to be appropriated. Since most middle peasants had already sold their surpluses, this meant taking grain being held back for seed, fodder, and consumption. Officially, only peasants holding reserves of more than two thousand puds of grain—that is, individuals identified as "kulak hoarders"—were subject to having grain confiscated under article 107. But local officials set about seizing middle peasants' grain regardless, operating as they were under the threat of removal from office, the loss of their Party card, and criminal prosecution if they failed. Thus, a campaign officially directed at kulak hoarders in no time engulfed the entire peasantry in what one historian has called "a reign of terror."[9]

To get grain, requisition teams routinely employed illegal searches and seizures and violence toward and arrests of individuals and entire groups of peasants. In the Kuban region, the OGPU reported that the head of a middle-peasant family of fifteen with a son serving in the Red Army was arrested "for nondelivery of surpluses." In its possession the family had thirty-five puds of flour. In one Don region village, over the course of a week, a gang of fifteen to twenty people under the supervision of plenipotentiaries from the district and the regional executive committees, and "without any mandates," conducted a round-the-clock search for grain exclusively among middle and poor peasants. According to the OGPU, those conducting the search included a well-known "thief-recidivist" and a kulak. During the searches, thefts of the peasants' possessions occurred, and those who protested were arrested and held in the village barn. The operation turned up little grain. In another village in the same region, a plenipotentiary from the district Party committee ordered a middle peasant to deliver forty-five puds of grain to the collection point. When the peasant explained that he had only twenty-five puds of flour for a family of nine and could deliver five puds at most, a search was undertaken. The hunt uncovered a single hidden sack of flour. As retribution, the Party representative ordered the local troika to tie the peasant up until he handed over his entire flour reserve. Later it came out that the peasant's own son had stolen the sack of flour from him and had hidden it without his father's knowledge.[10]

Incidents such as these were played out many times over and inflamed peasant enmity toward the Party and the state. Indignation was just as likely among poor and middle peasants as among kulaks. In conversation, one poor peasant from the North Caucasus said the time had come "to flog the bosses" (*nachalstvo*) because the bakeries in the regional center—Armavir—had no bread, and soon it would be the same in the villages. He attributed the situation to Armenian and Jewish control of Armavir. Another bedniak charged that the communists were "taking the grain for themselves and [. . .] condemning the people to hunger." "Soviet power is strangling us and gives us nothing," declared one middle peasant. "Only the clerks [*sluzhashchie*] are living well." One poor peasant even compared the current troubles unfavorably to grain requisitioning during the civil war: "They dupe [*durachat*] us and do what they want. Under the [civil war] procurement policy [*prodrazverstka*] it was far better. Then it was clear to all that the Red Army needed grain. But now no one knows exactly what these grain requisitions are for. The state says it needs to strengthen the peasant economy, but in fact it does otherwise."[11]

As a result of the emergency measures, grain collections in the first three months of 1928 greatly exceeded those of the same period in the previous year. In light of this success, the Central Committee announced at its plenary session held in early April that the crisis had been resolved. It soon became apparent that such optimism was premature. Over the remainder of the month, collections declined drastically, forcing the Central Committee to convene in an emergency session on 24 April to discuss the problem. The secret Politburo directive sent to Party committees the next day accused collection organs of resting on their laurels and demanded that the collection quotas for May and June be completely fulfilled. To achieve this, the Politburo called on Party organizations to mobilize their forces, to "smash" grain speculators, to hold the collection apparatus strictly responsible for fulfilling its targets, and to apply article 107 to "kulaks and traders who maliciously speculate in large batches of grain."[12]

Although the directive called for an end to the violent "excesses" (*peregiby*) of the first stage of the campaign it is hard to see how this could have occurred in the prevailing crisis atmosphere. On the other hand, the renewal of emergency measures did succeed in inciting more forceful reactions from the peasants. Beginning in spring 1928, peasant opposition to the grain collection campaign became both more intensive and extensive. According to national figures compiled by the OGPU, peasant demonstrations (*vystupleniia*) increased from 11 in March to 36 in April, 185 in May, and 225 in July, after which they tapered off. Actions defined as terror rose steadily throughout the year from June (43) through December (203).[13]

The grain crisis marked the beginning of the end for the NEP. Stalin explained the new realities quite plainly at the Central Committee plenary session held in early July 1928, when, in an ominous echo of Preobrazhensky's arguments for rapid industrialization on the backs of the peasants, he characterized the gap between industrial and agricultural prices as a "tribute" the peasants were obliged to pay to finance industrialization. There could be no talk, therefore, of reducing the price gap in the near future. Not all Party leaders supported the recourse to emergency grain-collection measures or passively accepted the threat they posed to the smychka. In response, Bukharin, A. I. Rykov, the head of the government, and the trade-union council chairman, M. P. Tomsky, formed an alliance within the Politburo to combat Stalin and his supporters' apparent adoption of the left opposition's economic program. Denouncing the excessive violence employed in the grain procurement campaigns and warning against reckless rapidity in industrialization, the trio fought a bitter battle during the fall and winter against a hostile Central Committee majority. In the end, though, the arguments of the "right deviation" in defense of NEP postulates proved no match for Stalin's organizational strength or for the Party elite's frustrations with agriculture and enthusiasm for the high growth rates projected in the First Five-Year Plan draft. At the Sixteenth Party Conference in April 1929, the three leading "rightists" suffered condemnation and defeat.[14]

In 1928, the total harvest compared favorably to the previous year's, but the production of the two most important food grains—wheat and rye—dropped significantly. Bad weather in the grain-producing regions was largely responsible for this outcome, but forced collections and low official prices also discouraged grain cultivation. The decline in output meant that grain collections would once again be difficult. Because of a decrease in fodder and the better-off peasants' desire to escape higher taxes, 1928 also witnessed an overall decline in livestock numbers. The slaughter of animals led to a rise in the marketing of meat but to a decline in dairy products. In any event, agricultural production could not keep pace with the growth in demand arising from industrialization, and in the course of the winter, bread rationing was introduced first in Ukrainian cities and then, by the beginning of 1929, in all the cities of the USSR. In short order, other food items were included in the "card system."[15]

The shortages resulted in a dramatic rise in prices, especially for wheat. According to R. W. Davies, between September 1927 and September 1929 the unregulated price for wheat rose 289 percent in grain-surplus areas and 354 percent in grain-consuming areas. Official prices, on the other hand, rose only a fraction of these amounts over the same period, opening a substantial gap between what collection agencies paid and what the peasant could get from private traders. To procure grain, the various collection agencies were now forced to exceed officially sanctioned prices and enter into competition with each other. This, in turn, had a detrimental effect on the state's budget and financial plans.[16]

In May 1929, faced with even greater shortfalls in collections than in the previous year, but not wanting to resort yet again to outright confiscation by state and Party representatives, the Politburo approved the wide application of a grain procurement practice that had been successfully employed in Siberia that March. In the "Ural-Siberian method," as Stalin came to call it, class pressure from below, or "social influence," would be brought to bear against the village's wealthiest farmers to encourage them to surrender their grain. Party plenipotentiaries would first induce the village assembly to approve the assigned grain procurement plan. A committee composed of reliable peasant activists drawn from the poor and middle strata would then identify the local kulaks and set strict fulfillment quotas for them. This committee would also see to the actual grain collection. In a variant on collective responsibility, non-kulak households were held responsible for supplying the difference between the total village assessment and the grain thus taken from the kulaks.[17]

On paper this method had the advantage of utilizing both class principles and the authority of the village assembly. It also leaned heavily on peasant traditions like democratic decision making and collective responsibility. Since no state or collection agencies were directly involved, the process presented itself, and was presented by Party authorities, as a form of peasant self-taxation. In practice, the village assemblies approved village collection plans only with great reluctance, and Party representatives found it extremely difficult to split the peasants along class lines. Kulak resistance also stiffened in response to the new tactics. To undermine kulak recalcitrance, provincial Party officials introduced harsh penalties, later endorsed by Moscow, including fines of up to five times the amount of grain owed (the *piatikratka*) and the exclusion of kulaks from cooperative stores (the social boycott). Continued refusal to deliver grain could lead to the application of Criminal Code articles that called for imprisonment, terms of forced labor, and property confiscation, all of which prefigured dekulakization—the elimination of kulaks. Ultimately, the total grain collected during the 1928–1929 campaign proved to be slightly less than the amount collected the previous year. Most importantly, the procurement of food grains declined from 8.6 million tons in 1927/1928 to 6.9 million.[18]

Brimming with confidence, supporters of the Party's "general line" anticipated that the harvest and grain collections of 1929 would far exceed those of the previous year. During the summer, they stridently rejected the less sanguine projections advanced by statisticians and agricultural specialists working in the Central Statistical Administration (TsSU), whose caution earned them denunciations as "bourgeois," "Mensheviks," and "rightists."[19] To ensure maximum collections, organizational preparations were begun early. The division of labor between grain collection agencies was rationalized. Manufactured goods, to the extent of their availability, were transferred to grain-supply regions to encourage peasant exchange even when this meant leaving the towns short. The center also insisted that local officials establish monthly grain quotas for collective and independent farmers—which in practice served as lower limits—as soon as possible.

In addition to organizational preparations, Moscow set in motion the class, manpower, and punitive mechanisms at its disposal. Once again, the campaign targeted kulak and better-off farmers. Through the widespread use of the Ural-Siberian method—now bolstered by stiff penalties for nonfulfillment of quotas—Party leaders planned to rouse poor and middle peasants against the village upper crust. To reinforce the Party's scattered rural contingent and assist in collections, upwards of 200,000 Party members, workers, Komsomolists, and others decamped from their cities and towns to serve as rural plenipotentiaries. Just in case local organizations failed to bend every effort in the campaign, the Commissariat of Justice recommended that article 107, originally intended to combat speculation, then extended to penalize kulak grain hoarders, now be used to punish sluggish or fainthearted officials.

The grain collection campaign in the summer and fall of 1929 shattered the NEP. Preparations notwithstanding, central authorities could not provide sufficient quantities of finished goods to satisfy peasant demand, nor could collection agencies establish grain quotas for every village in a timely fashion. Attempts to use the Ural-Siberian method to split the peasantry along class lines by transferring the burden of village quotas onto kulaks failed, for peasant assemblies sought to divide the obligations equally among all village households. In response, collectors imposed high "strict quotas" on individual kulak households and assessed stringent penalties for nonpayment. Peasants reacted to the campaign by doing all they could to keep their grain from falling into the hands of state collectors. As the campaign unfolded, measures reserved for kulak households like the "strict quotas" and auctioning of possessions, were applied against middle and poor peasants as well. These practices served to encourage peasant solidarity, and with the expansion of arbitrary and administrative excesses, violent resistance increased. In the countryside, 1929 proved to be even more violent than 1928. The total number of demonstrations increased 84 percent (from 709 to 1307), only partly diminishing during the harvest months and rising thereafter. Acts of terror also became more frequent, increasing an enormous 785 percent (from 1,027 to 9,093) and fluctuating between a low of 247 in April and a high of 1,864 in October.[20]

The collection campaign succeeded in increasing reserve grain stocks and provided enough grain to allow the Soviet Union to resume exports, both of which were necessary for the success of the industrialization program. Unfortunately, success came at the expense of the cities and towns, which continued to experience food shortages even though alleviating their condition had provided the rationale for the collection campaign in the first place.[21] For the peasants, after suffering two successive campaigns of administrative and punitive procurements, there could no longer be any doubt that the smychka was dead or that their role in the worker-peasant alliance was subordinate. They could now expect nothing from the state but the further exaction of "tribute."

Documents

The Campaigns of 1928 and 1929

When, in December 1927, the Fifteenth Communist Party Congress decided to intensify the "offensive against the kulak." Stalin also asserted in his speech to the congress that the period of peaceful coexistence with the capitalist world was coming to an end and that imperialist powers like Great Britain were preparing to intervene militarily against the USSR. He spoke of the Soviet Union's need to delay such hostilities for as long as possible and advocated a strategy that included "buying off" capitalist aggressors through mutually beneficial trade agreements and partial payment of debts.[22]

In a letter of 13 March 1928, written at the height of the grain-procurement campaign, I. S. Chernov, from the settlement of Anastasievka, Aktiubinsk province, Kirghiz republic, responds to Stalin and explains how the Party's reaction to the grain collection crisis was undermining the Soviet Union's strategy to avoid war.

· 117 ·

Letter to *Krestianskaia gazeta* from the peasant I. S. Chernov on the detrimental effects of the Party's kulak policy, 13 March 1928. RGAE, f. 396, op. 6, d. 28, l. 11. Original manuscript.

———————————

What is going on? You don't understand.

As is evident from Comrade Stalin's report at the Fifteenth Congress, the strengthening of our government in Soviet Russia depends on whether we can successfully put off the war with the imperialists for a little longer. But how can it be delayed? "Yes we need to buy them off," Comrade Stalin says, but in order to buy them off we must have the wherewithal, not cash, but grain, livestock, and different kinds of raw materials. But where can it be gotten? We must raise the production of these items by all means. But with us, just the opposite is happening. If what is going on locally had not been ordered by the center, as is evident from the Fifteenth Congress resolution, then I would have said that it was designed to turn the peasantry against Soviet power and to undermine it. [In the congress resolution] it is stated directly: "We should not strengthen the village bourgeoisie's appetite for acquiring individual capital." What appetites, one may ask? Just what the hell are the appetites of the peasant? To sow more, to better develop the land in order to improve the harvest, to get more healthy livestock, and to better feed the old ox, and even to pay taxes and to let the young grow up. But now, with us in Kirghizia, appetites are being suppressed because of all this. Everywhere the local (ostensible) Party members are crying about the need to dekulakize "or else they'll get fat." They have imposed a sort of indemnity on peasant-laborers who are not lazy and who manage their affairs energetically and who, as a result, have grain and livestock. Besides the tax that peasants assess on themselves [*samooblozhenie*]—a tax of three hundred to four hundred [rubles]—they lay another six hundred rubles on him just because he has it. And instead of gratitude for the fact that he [the peasant] works and has made gains, the peasants get a reprimand and prison. Of course, they have already lost the appetite to work, to plow deeper, and to sow more. The same holds for cattle and cattle raising. Now the toiler values nothing; he's lost his appetite for everything. [. . .] The people—the well-off—are being ruined, and as for the poor, they're poor, and that's how it is. And looking at everything that is happening, you wonder what is going on. Everywhere the newspapers shout, "We must improve agriculture." But no improvements are visible, and destruction goes on before one's very eyes. Yet one reads in *Krestianskaia gazeta*, no. 5: "Bessarabia is under the yoke of Romania. They are levying excessive taxes without any legal basis for it." It's amazing how the editorial board saw the disorders in Romania, but in their own Russia they don't see them, although the editorial board can honestly print that acquisitive appetites must be suppressed. Off whom does the state live if not the well-off? He provides the market with grain, and cattle, and pays the tax. It turns out they haven't built a new hut, just destroyed the old one, and the poor are no better off. And those who can live independently without reaching their hands out to the authorities for help are brought to their knees. This is a bad business. We won't get to any kind of socialism this way, only to poverty, and then the imperialists will take us by

the scruff of the neck with their bare hands. Some of you think and say that
the projected plan is correct, but I say to you that it is quite wrong.

The peasant IVAN S. CHERNOV.

Because of the turmoil caused by the grain collection campaign, village
soviet elections scheduled for early 1928 were postponed until 1929. By
the time the elections did take place, the category of those deprived of
voting rights had been broadened to increase the relative strength of the
poor peasants. The changes helped aggravate intra-village class relations
and, as secretly reported by the OGPU, caused middle and better-off
peasants to sound off and in some cases take direct action against the
new voting regulations.

· 118 ·

Excerpts from an OGPU summary reporting peasant reactions to the 1929 soviet
election campaign. GARF, f. 1235, op. 141, d. 147, ll. 6–18. Typewritten copy.

[...] V. Khudonogov, and S. Koronotov, from the Solontsa settlement,
Nizhne-Udinsk district, [Tulun region, Siberian territory,] said: "Now we
have to elect our guys to the soviet, but no poor trash and no Komsomolists,
because they'll make life impossible."

[...] In Pakhomovo district, Tula province, Davydkovskoe village, two
wealthy [peasants] in a talk with their fellow villagers said: "The reelection
to the soviets is a puppet show that the government plays with the peasants.
No matter whom we elect, they won't be able to defend our interests anyway, and they'll skin us like squirrels. The state only helps the worker. But
this isn't fair, since the peasants, not the workers, made the revolution.
There's only a handful of workers, but the peasants are the masses. They
won't take anything from the worker, but they take the tax, the self-assessed
tax [*samooblozhenie*], and indirect taxes from the peasants. There's the
class difference between the muzhik and the baron who holds power. No,
there'll be no respite until we shake the comrades up. That's why we need a
second revolution."

[...] On 22 January [1929], in Orlovskoe village, Nizhne-Kolosovka
district, Tara region, [Siberian territory,] at eight o'clock in the morning,
those deprived of voting rights gathered at the home of Mazurin and then
marched down the street with flags, calling the rest of the peasants to join
them. The peasants banded together, and the whole population marched
with the flags under the leadership of those without voting rights.

[. . .] There are districts where agitation for the creation of a peasant union is being conducted. In Pirogovsky district, Krasnoiarsk region, a procurement official for the Belsk credit association, Pavel Nardasov, (a middle peasant), said in a conversation with the peasants: "We must organize a peasant union and present our demands to the state. There are worker organizations and they get theirs—higher pay. But the peasants are squeezed from all sides, and it's hard to live."

[. . .] The agitation for the creation of a peasant union very often winds up with the formation of an organization. Groups are created that even middle peasants join. For example, in Kuskun village, Mansky district, Krasnoiarsk region, a group of fourteen kulaks, three traders, and some middle peasants that conduct their own meetings was exposed, and at every [village] gathering it organizes demonstrations and does anything to try and break it up. For example, at a meeting on 11 December [1928] questions were raised [about] the report of the regional executive committee and the self-assessed tax . This group very actively carried on debates whose contents were entirely anti-Soviet.

[. . .] In Talovka village, Uiar district, Krasnoiarsk region, the Driannyis, Gusevs, and the Chernyis (all strong middle peasants) gave it to the poor peasants on the village soviet who had passed a decision at the poor-peasant assembly on the self-assessed tax: "They tried to raise their hands and feet in order to skin the middle peasant. Just you wait, don't come to us for grain, [or] you'll get squat [*shish*]; we'd rather feed the dogs." At an assembly in the same settlement, Starkov, a rich [peasant], shouted: "Muzhiks, just think how much you pay—you pay the tax, you pay insurance, etc. The poor are all loafers and spongers. They're glad that Soviet power wants to ruin the middle peasant. In the future, I'll only pay twenty rubles. I'll reduce my holding."

[. . .] The Achinsk regional executive committee reports that in Yastrebovo village, Achinsk district, [Siberian territory], unknown persons posted an appeal to the middle peasants in open places containing the following:

Middle peasants, take note! The worst vagabond has raised his head against us, the lowest riff-raff is crawling into the soviet, into the new soviet the poor are advancing the most worthless bastards who hate the middle peasant and are trying to seize the last crumb of bread. Hear are the descriptions of the bastards:

1. Pelageia Tereshkina —a blabbermouth, a tramp, a scandalmonger, now conducts espionage.[23]
2. Anna Petrova—a similar creature.
3. Falei Pliusnin—took part in the Seryozh uprising, but among us he wore the mask of a poor peasant.[24] His mask needs to be removed.
4. Ivan Ginbut; Vasily Saulevich; and Pyotr Tereshin—the prime enemies. Under their leadership the toilers were deprived of their voting rights.

5. Vladimir Gavrilovich; Vasily Kuzmin; and Semyon Gerilovich—also worthless elements.

Don't forget with whom you must carry on the struggle. At the assemblies all the middle peasants need to act in concert to repulse any bastard. Through the concerted support of one another, complete victory will come for the middle peasant. Don't allow a worthless creature into the new soviet. Literate ones write your slogans and pass [them out] or hang them in an open spot. Hold the lazy and the loafers up to shame.

Please don't tear this down."

From 1928, grain collections assumed the characteristics of military operations. As in war, state agents exhibited a cruel and pitiless attitude toward the peasant "enemy." In their ruthless zeal, grain collectors frequently employed sadistic means to achieve their aims. Though officially condemned as "excesses" (*peregiby*), these measures took shape in response to the center's characterization of the crisis as a kulak campaign against the state and found support among rural activists and the propertyless segments of the village, who were encouraged to strike at the more well-to-do. The following account, taken from an April 1929 political report of the Central House of Peasants (TsDK), illustrates how the social boycott could degenerate into grisly physical violence.

· 119 ·

Excerpt from an account of "excesses" committed in the course of the grain procurement campaign, April 1929. GARF, f. 1235, op. 141, d. 113, S. 91–122. Typewritten copy.

[. . .] In their declaration, twenty peasants from the village of Bashkirskoe, Kurgan region, Ural province, write:

"After the general assembly, they detained several citizens and denied them food and drink for three days. On 10 March, at a general assembly, they hung boards with insulting inscriptions on the necks of those who had been boycotted and spat in their eyes. After the reading-hut librarian, Vlasov, commanded, 'Hey, poor peasants, go over there with these here kulak bayonets,' they jabbed each one in the side and the face demanding that they bring out everything that had been reserved for food and grain seed, saying: 'Tell [us where] the wheat is. Turn over everything. We're not giving you anything to sow with anyway.' To save their lives, the citizens began to bring out everything that had been set aside for seed and food." The declaration of these peasants closes with a request that the government review the activities of the local authorities and issue an order that seed be distributed to them.

In his declaration on the boycott, Solovyov, the Siberian, from Petropavlovsk region, explains: "In the villages of Yavlenka, Aleksandrovka, and Ilinka in Petropavlovsk region, they boarded up the windows of the huts [*izbas*] of those citizens who had been declared boycotted. Then they took kerosene and matches and placed a guard by the door so that no one could get out of the house in order to feed or water the livestock. This went on for three days. Children were crying and animals howling."

Industrial enterprises sent worker brigades to help the plenipotentiaries with grain collection. Their members also took the opportunity to expound on the Party's new rural policy and to promote the formation of collective farms. In a letter to Stalin dated August 1929, N. D. Bogomolov, a worker at the Chernorechensk factory in Nizhny Novgorod, recounts the conditions and the mood of the peasants he encountered as a grain requisitioner and Party propagandist in one village.[25]

· 120 ·

Letter to Stalin from the worker and Communist Party activist N. D. Bogomolev on his grain-requisitioning and propaganda work in the countryside, 14 August 1929. RGAE, f. 8043, op. 11, d. 16, ll. 58–59. Typewritten copy.

To the General Secretary of the Central Committee of the VKP(b), Comrade Stalin.

Comrade Stalin, allow me to answer several difficult questions that have come up in my work inasmuch as I was dispatched from the Chernorechensk chemical factory by order of the Nizhny Novgorod VTsSPS [All-Union Central Council of Trade Unions] Organizational Bureau and by the administration of the Joint-Share Association "Soiuzkhleb" [to serve as] a member of a brigade responsible for the handing over of grain contracted from the peasants and the sovkhozes, the sowing and harvesting campaign, and explaining the tasks of the Party and government.[26] The Nizhny Novgorod "Soiuzkhleb" sent me to the Central Black-Earth region under the authority of the Voronezh "Soiuzkhleb." Voronezh sent me to the region, to the [capital] city of the region, Ostrogozhsk. They put me to work in a [grain] elevator at the [railroad] station of the city of Valuika, and I found myself in Vendelevka district; here they have a district executive committee. The

VKP(b) district committee ordered me to the village soviet to work in the village of Salovka. [. . .] Grain requisitioning has been carried out here; the most recent was concluded at the end of this June. They left the peasants thirty funts [of grain] per person a month. Most did not have enough grain to get them to the harvest. They had to buy it and sell their cattle, their working oxen. Grain, [when] I got here, was [selling for] twelve rubles a pud by the private traders in the bazaar. A family with thirteen to fourteen persons does not have enough grain, and there aren't any supplementary fats for these thirty funts. For the entire family there is one cow, and a large part [of the peasants] live without cows; there are no pigs, and they won't acquire them. Whoever has the means keeps two horses or two oxen for a large family of thirteen to fourteen people, and a cow. They count him a kulak, and from such a proprietor they take everything down to the last, and he is forced to buy grain for his family and sell his working cattle. Comrade Yurevich of the VKP(b) Central Committee was here during the grain requisitioning.[27] He stayed and observed the whole situation of the individual peasant: how a family of fifteen to sixteen persons has only one cow, [how] they live on bread alone, [how] they have no small livestock—pigs or sheep—and have no fats whatsoever. Although four-field rotation is in use in the village soviet, they don't sow hay, and suffer [from lack] of fodder. [. . .] But after him [Yurevich] a local [official] arrived and began to take everything and more in violation of the decree; he took 35 percent from the top group and cut everyone down to the same level. He left thirty funts [per household]. This was after Yurevich. Yurevich was here from 25 May to 25 June; then Kharlamov, the local [official], conducted another requisition from 25 June to 5 July. They took the grain, and there was not enough [left] to make it to harvest time. So they had to sell their cattle to survive. Now they grumble that they took our grain for eighty kopeks and we have to buy it back for twelve rubles.

There are Socialist-Revolutionaries in Salovka village who jump on every little thing and hold the poor and middle peasants in the palms of their hands. They oppose everything Soviet power does. They put forward the whole SR policy: how we have no kulaks, how everyone is equal, [how] we are all poor.[28] We lived under the tsar, [when] there was more of everything and less tax, [and] now they let you own cattle, but they take it away if you don't pay the tax. There are no boots, but if they would just allow private craftsmen—there is a village, Urozovo, where they used to make boots—if they were given the freedom they'd bury us [in boots].

There are examples of going to extremes [*peregiba palki*]. Here is a typical example from a [civil war] partisan who is speaking at a general assembly: "What did I fight for, Comrade Bogomolov? I am not saying this just because VKP(b) Central Committee member Comrade Yurevich was honest. The country needs grain, and he made a proper inventory for this locality. My father lay near death; he was injured during a fire—he fell from the roof in January 1929." And this partisan, suffering because of a fire, asked

Yurevich to leave grain [so he could raise the money] for the rebuilding [of the house]. He had twenty puds of grain for fourteen people. And this local Party member, Kharlamov, put everyone at a norm of thirty funts. Kharlamov came to him: "No discussions, deliver the surpluses." His father lay in the stable ill. "He'll lay there until autumn. It's all right. We'll build there with the new grain [from the next harvest]," [he said, and he took the grain]. Of course, he told me this during the assembly, and he threw his hat and swore to God. "Where is there any justice here?" I asked him, "You searched for it?" "I didn't find it here. [...]" Now, what do they have in the cooperative [store]? The salt's no good—a simple thing—it's not ground up but in clumps, fit only for cattle; your basic soap, none for more than a month now; soles for shoes—something that a peasant really needs—none. There are only three handkerchiefs and ten pairs of gray felt boots and half a shelf of vodka—that's your rural cooperative. Now I'm in the middle of the fall sowing campaign, harvesting and grain threshing. They won't allow any socialist competition here: "We don't need anything, please; we'll go on living as we've lived. Don't concern yourself about anything. You take the grain, and there's nothing: no fabrics and no shoes." And they don't do anything about it. They don't want to organize a collective "because we see with our own eyes how they work in the kolkhozes—it's just no good." [...] I suggested to them: "Get rid of your property. Let's organize a sovkhoz, [and] then you'll get paid." But nine kilometers from this village is the Vintoropol sovkhoz, which I have inspected. And what do they have there? There are 7,542 hectares of land, eight hundred workers, and they feed them a gruel [*kander*] of unhusked millet and water and give them one funt of bread. The sovkhoz is under the direction of the regional trust. There have already been six different commissions and one from the Central Committee. They did nothing. I got there and under my very eyes three hundred workers were let go because of the bad food. There are no quarters and no bath for the laborers. And I told the workers' commission [*rabochkom*] and Party cell that we can't make guarantees [about grain deliveries], that we don't have enough workers for the threshing and harvesting. I got there on 14 July 1929. There was a lot to do, and they were firing people. They said: "We are not to blame, we have a trust!" There are 2,117 sheep and 87 pigs that the trust slaughters for bacon. Another example: fifteen tractors just stood there—no spark plugs and no fan belts. The trust received a telegram to take care of it immediately. It doesn't give a damn, but this can't wait. This is not in a factory but on a farm—you can't waste a minute of time here. Here's the discontent with the sovkhoz: "We'll go hungry. [We] get paid ten rubles. It's not right, and you won't trick us into entering and organizing such a sovkhoz." That's why I beg you, Comrade Stalin, answer my questions. What is the best way to get started? I explained to them that it's impossible right off to supply fabrics and shoes, and to get the factories running, because now we are operating according to a strictly laid-out economic plan. If we get these two branches going, then the main levers of our national economy will be in place—heavy industry, which will produce machines for production. I gave them a simple example: "Say a peasant has 250 rubles in cash, a family of fifteen, no shoes, and no clothes. What will he buy?" He says, "A horse." "It's the same for the state: first it needs a motor to drive the national economy." But not all of them believe this. "There is so much leather, yet boots are expensive." "How much would you pay for shoes? A pud of grain for one pair of boots?" "No, I would give eight puds for boots." And they don't believe that we are following a strict plan of distribution of all industrial goods through the cooperative organizations. Shortages are widespread. But they don't believe it. So I beg you, [please say] what is the best way to deal with this and explain it in detail. Please, don't refuse to answer.

BOGOMOLOV, member of the VKP(b) since 1926.
Address: p/o Vedelevka, Ostrogozhsk region, Salovka village, worker brigade member NIKOLAI DMITRIEVICH BOGOMOLOV.

———————————

Confronted by the dismal reality of rural life in the wake of grain requisitioning—an embittered peasantry, official arbitrariness, shortages, and collective farm disorganization—even a committed communist propagandist quickly found the rote arguments for Party policy deficient. Bogomolov's appeal to the general secretary for counsel is not without poignancy, but then it is unlikely that even as skilled a dialectician as Stalin himself could have convinced the peasants that everything they had recently experienced had occurred for the best or was laying the foundation for a brighter future.

With the introduction of the NEP in 1921, the Party's active promotion of collectivized agriculture had fallen off considerably. Between 1926 and 1927, the central authorities took measures to reverse this course in order to revive the kolkhoz movement. But it was the grain crisis that provided the necessary stimulus to expand collective agriculture. Strong supporters of the kolkhoz, like Stalin and Molotov, saw it as the only means of ending the state's dependence on the kulaks for grain. On the eve of the crisis, in October 1927, approximately 286,100—out of a total twenty-five million—peasant households belonged to kolkhozes. Within a year this number had slightly more than doubled, and by mid-1929, six months before the onset of wholesale collectivization, there were over sixty thousand kolkhozes containing more than a million households in the country. The average size of kolkhozes remained small in these years (16.9 households in the RSFSR and 19.9 in Ukraine as of June 1929), as did their marketed share of produce (5.7

percent in 1928/1929), but this modest growth did indicate that peasants, especially the poorer, would join collectives voluntarily under the right circumstances.[29]

Moscow expected rural communists to set an example for the peasants by organizing and joining kolkhozes, but as K. Skorokhod from Poplavsky settlement, Utevka district, in Samara province, explains, the rural comrades' enthusiasm often flagged when confronted with the reality of collective farming.

· 121 ·

Excerpt from a letter to *Krestianskaia gazeta* from the peasant K. Skorokhod on Communist Party members' reluctance to work in collective farms, 27 December 1927. RGAE, f. 396, op. 6, d. 61, ll. 129–130. Original manuscript.

About the collective.

Starting in 1918 and continuing to the present time, all the editorial boards of the newspapers and journals say that non-Party peasant-toilers would be attracted to the kommuna, the artel, the collective, or some other kind of peasant unit for the joint cultivation of land, as would all the national groups and social estates [*soslovii*]. Communist orators are so fervent when they talk about this at peasant assemblies that spittle flies from their mouths. But for some reason these organizations are not succeeding. And that's because the peasants won't go along: it's just awful! The communist orators speak quite well and sweetly [say] that life and work in the artel will be easy. After all, as the proverb says, "Many hands make light work." It's true, but the point is that it's very easy to write, talk, and sing praises, but when you get down to it, we peasants can't see how artel communists live. Here are some examples of what was and is:

1. Some peasant communists persuaded several non-Party peasants [to join them in a collective]. They set aside good land for their collective and cultivated the land together as an artel. But, alas, the communists say they have sore backs and [tell the peasants], "You peasants, cultivate your land and our land as an artel, on collective principles, and we communists will go to work with our heads—serve [in the apparatus,] that is." And so all the communists of this collective wind up at the district executive committee, like bees around a honeycomb, [working] in the credit association or cooperative administration, or institutions like that.

2. In a village near our settlement several communists partitioned off some land for themselves and worked as a collective of former officials: the chairman of the district executive committee, the chairman of the district committee of the poor, [the former] commander of a machine-gun unit on

the Urals front. And what happened? They were all removed from the Party, and the former chairman of the district executive committee never set foot in his own collective. After the liquidation of the district, he left his wife and went off with some woman to another village and now runs a general store. The former chairman of the district executive committee of the poor, who had adorned himself during his stint in the district executive committee with two revolvers and four bombs, worked a few years in the collective and went around the collective in Chuvash bast sandals [*lapti*], but he quit, saying, "He who worked [before], let him work [now], and good luck [to him]," and left for Samara, where he now trades in a meat market. The rest are still in the collective, but they work as individuals. The former battery commander on the Urals front, after the liquidation of the front, concerned himself with the peasantry on the liberated land and liberty, and also went among the peasantry for several years in Russian bast sandals.[30] And just this year he quit the peasantry, locked up his house, and went to work in the village soviet as a secretary at a salary of eighteen rubles a month. And these are the comrades that approved the kommuna, the collective, and joint land cultivation in general, and spoke from the tribune, and in public no spit flies from their mouths, just bubbling froth, singing and praising and teaching the peasants how to cultivate the land jointly, on socialist principles, but they themselves don't want to work the land on these principles. Truly, the ancient literature (history) speaks the truth, that they talk and don't do; they tie up the heavy burdens and inconveniences and lay them on the shoulders of the people, but they don't want [to work], only to show them where to go. [. . .][31]

K. SKOROKHOD

The hypocrisy of local communists toward the kolkhoz described here—promoting its advantages while evading membership—was not unusual; as of June 1929, Party members made up only 4.3 percent of all working adult collective farmers.[32]

While local communists' reticence to enter the kolkhoz may have discredited collective farming for many peasants, there were peasants who, out of economic necessity or commitment to socialist principles, found collective farming attractive. One such supporter of the kolkhoz was the Red Army veteran K. F. Khersun from Kuniche village, Kryzhopole district, Tulchin region, Ukrainian republic.

· 122 ·

Letter to *Krestianskaia gazeta* from the peasant K. F. Khersun explaining his reasons for entering a collective farm, 27 February 1928. RGAE, f. 396, op. 6, d. 61, l. 7. Original manuscript.

Forward to socialism.

Dear comrades, in view of the fact that I am a subscriber and reader of your publication, I want, however briefly, to report to you on the fulfillment of those duties which are the mission of the Communist Party. Although my relatives say that I have disgraced the family name, it is in my interest to stand before each [politically] conscious citizen of the Soviet Union. That is why I moved from a single holding into a collective. For four straight years I tried to find a way out in order to quickly move to such a farm. It is four years since I was demobilized, and, as is well known, I stayed in the Red Army for a year and a half and learned a little something: [I learned] what our Soviet power is trying to accomplish and what the benefits will be if you conduct yourself in a cultured way. It is well known that our countryside remained oppressed after the tsarist system, and [how] to make the transition to a new life, to a new socialist life, was poorly understood [because of] unconsciousness, and, in addition, [there was] kulak agitation and conspiracy. But it is possible to hope that the rural activists will quickly eradicate [all this] and that socialism is ahead of us. I was demobilized in 1924. In my village a collective was organized. And I said to my wife, "Let's join it; it's our only way to transfer to this type of farm." And just in time, too: the horses had croaked, the pigs had bought it, and all that was left was one calf and a pair of sheep. But my wife [opposed] this tooth and nail: "Go to a collective and not know what's mine!" What was there to do short of divorce? I nagged and nagged [her] and tried to persuade [her], but it didn't help. I could see that it made no sense living here, and I suggested to her, "Let's move to Kherson province. We can live there without the collective, and better." My wife agreed! But when I got there, I could see that with this family and in this condition not only can't you raise up your economic position but, on the contrary, you go down. I returned and began to raise sugar beets and somehow got hold of one horse. I could see that with such a wife you couldn't enter a collective. I left her and headed for Kuban province for work. I stayed there four months among the bricklayers, but because of an illness, I came back home and worked at home that summer. Then another collective was organized. Seeing that I still had not gotten rich, my wife says: "Ah, well, let's go to the collective; there they no longer mow with a scythe, and they don't do the threshing with a flail, and they look healthier." And so, after four years I made it to a normal socialist farm and truly disgraced my family name. After me six families entered. And it is a fact that these farms provide assistance to the state and they themselves live the best.

In my next letter, I will tell about this collective, how they built up their holdings from nothing. Right now we are in the middle of land reorganization. And that idiot that let valuable time slip away . . . [word not clear in original] [has found] self-esteem and self-enrichment here. In conclusion, I wish you all the best. Long live socialism and its leadership!

Addendum:
I would like to get a large desk calendar. On 1 November 1927, I had subscribed to *Krestianskaia gazeta* for a year. I want to participate in the free radio lottery. I subscribed to the journal *Lapot* [Bast Sandal] for half a year beginning on 1 March.

Respectfully, the peasant KLIMENTII FEDOROVICH KHERSUN.

Please, if you print this letter, expand it and correct it.

KHERSUN

This letter offers evidence of the economic salvation that a segment of the peasantry believed could be found in the kolkhoz even if this meant cutting family and social ties. It may be safely assumed that Khersun's commitment to socialism had been strengthened by his Red Army service. Peasants overwhelmingly filled the ranks of the Red Army. Military service not only broadened the outlook of the peasant conscript but also systematically exposed him to the Bolshevik political and economic program. Among the millions of veterans who returned to the countryside after the civil war, and the thousands who were later regularly demobilized upon fulfilling their service requirements, numbered many who were less wedded to the traditional way of life, more receptive to change, and armed with a coherent vision of the future. By the mid-1920s, these Red Army veterans were playing prominent roles in rural administration, chairing more than half the rural soviets and 70 percent of the district executive committees.[33]

Officials noted that reports about the kolkhoz movement were listened to attentively and that peasants willingly offered accounts detailing their own experiences. The compiler of the Central House of Peasants report recorded the following statement given by a peasant, Ksenia Sergeevna Panaskina, from Olshanka village, Briansk region.

· 123 ·

The testimony of K. S. Panaskina, a poor peasant, describing the obstacles
she encountered in her efforts to organize a collective farm, April 1929.
GARF, f. 1235, op. 141, d. 113, S. 122. Typewritten copy.

We decided to organize a kolkhoz. All the poor peasants and day laborers expressed this desire. Even a few middle peasants joined. In all there
were fifty-four families. We burned with the desire to work collectively.
There is land for the kolkhoz in "Rubanikhi" and "Zelenoe boloto." But
they won't give us the land, and the Briansk regional land department won't
allow us to organize a kolkhoz. They have been delaying since 1927. For the
second time I went to Narkomzem on my own peasant kopek. No go. I
went to Kolkhoztsentr, and they referred me back to the district. I went to
the People's Commissariat of Worker-Peasant Inspection. Here they reassured me. They questioned the region about the red tape. They gave me a
document. Will something come of it? I don't know if there's any hope."
(She sniffles, and tears roll down. She leaves the stand wiping herself with
her sleeve.) . . .

By mid-1929, collectivization had made its most impressive gains in
traditional grain-supplying regions like the North Caucasus, the Lower
Volga, and Ukraine, providing evidence of a link between the grain crisis
and kolkhoz formation. But, as R. W. Davies has pointed out, these were
also areas of notable economic differentiation within the peasantry. The
poor in these areas, as opposed to the poor in areas where nonagricultural work was available, had every incentive to join collectives. By October 1929, some peasant households in the RSFSR and the Ukrainian
republic, 7.3 percent and 10.4 percent of all peasant households, respectively, had joined collectives. The number of middle peasants who joined
collective farms remained small, though. As one historian has noted, the
kolkhoz movement may have already achieved as much as possible "under the then existing circumstances."[34]
   Peasant fears that collective farming would destroy their independence and promote licentious behavior were commonplace and generated all sorts of rumors and misconceptions. According to Lynne Viola,
the varied rumors about collectivization—that it would lead to the "nationalization" of women or that it heralded the appearance of Antichrist—
served the peasants as a means of defense and resistance by "subverting
official dogmas of class struggle between rich and poor peasants and
asserting in their place a battle between the forces of good (the peasants)

and evil (the state)."[35] Many peasants certainly believed in the literalness of such rumors, which often resembled those from earlier times of turmoil. Most troubling to the authorities was the use that opponents of the kolkhoz could make of rumors by building on peasant fears, rational or otherwise, to create a united front against collectivization. Here, for example, is a letter from the Central Black-Earth region written in substandard Russian with many misspellings that combines a rumor of mass rape in the kolkhoz with a defense of the peasants' economic interests and a NEP-like program to raise agricultural production.

· 124 ·

Letter to *Krestianskaia gazeta* from Ivan Nikoliivich calling for changes in the policy toward the peasant, August 1929. RGAE, f. 396, op. 7, d. 14, ll. 284–287. Original manuscript.

In the first lines of my letter, I bow to you comrades [of the] *Krestianskaia gazeta* and to Comrade M. I. Kalinin and his assistants. Dear comrades, I ask you to think over what I am writing in this letter. Dear comrades, you have made the five-year plan for the development of the national economy and you must see it through, but see it through my way. Not as you thought, by destroying the kulak economy and all private free trade and even all the peasants' machines, animal husbandry, and grain cultivation, and you want to drive everyone into the kommunas. I think that from the kommunas and artels of collectives the state will earn nothing, only losses. From all the kulaks and individual farmers and the other peasant producers there was a large income to the state and the like. You consider the kulak a rich laborer, and the peasants spill their blood and give it to those poor loafers in the kommunas and die themselves from hunger, eating only spring sorrel. But if you drive everyone into the kommuna, everyone will be hungry and won't get calico [word not understandable in original], and the women and girls will be raped, and in every kommuna there will be a battle and a slaughter. And it will end up a brothel [with] many children. The individual farmers work hard, but you take everything from them down to their dirty, black skins.

And now, dear comrades, I beg you to heed my proposal. My proposal is to give the peasant his freedom and not to consider him a kulak if he has two cows and one horse or rents land from someone who doesn't want to work it. I also would say [to] allow all free trade and any handicraft production. Furthermore, I say to you comrades that to wipe out drunkenness, you need to provide books and to hand out two bottles of vodka every month. Also, I say that if you would stop pregnancy among women and girls in the Soviet population, I think that if there is some way to prevent pregnancy, then you need to give it to each and every woman in the population between the ages of seventeen and forty-five, and give it by force if need be, and issue a strong order so that every woman will not give birth for five years; after five years she can give birth, then, again, the woman must stop. And if this woman should give birth, [then put her in] prison for one year and her husband for six months. Our country is poor because we have propagated too many people and we must stop further propagation.

Dear comrades, when you carry out my program, then you will fulfill your plan. This letter is from VANYA NIKOLIIVICH, Orel region, Narym stanitsa, Uritskaia rural district, S. district

Write an answer in *Krestianskaia gazeta*. I have a full year's subscription. So good-bye, may life be healthy.

In the end, for all the rumors and gossip that surrounded the kolkhoz, there is ample evidence that the peasants clearly understood the rationale that lay at the heart of collectivization—sometimes even better than its advocates. In one note sent to *Krestianskaia gazeta,* the Komsomolist and village correspondent A. Prokhorov extols the "common good" (*dobrobyt*) that collective labor will introduce. He disapprovingly reported that the peasants, biding time and sitting on the protective earthen mound, the *zavalinka,* that typically ran along the exterior of rural dwellings, often discussed the collectives and repeated what he termed "lies." According to them, collectivization was necessary in order "to collect the grain in one heap so that it will be easier to take from the peasants, as in Siberia. And even more, so that it will be easier to send it to the state . . ."[36]

Collectivization

Experience organizing collective farms during the civil war convinced many Bolsheviks that peasants could not be coerced into abandoning the commune and their individual allotments. Throughout the 1920s, the Party held that collective farms would proliferate only when they could attract members voluntarily. This stage would be reached, the reasoning went, only when the technical and organizational superiority of collective farming became evident to the muzhik. Until collective farms could be sufficiently supplied with tractors and other mechanical equipment, until they became staffed with well-trained agronomists who could introduce modern agricultural techniques, until their superior productivity and material benefits could be demonstrated, the mass of peasants, particularly the middle peasants, would never be convinced to pool their resources, amalgamate their individual holdings, and practice agriculture on a collective basis.

Socialized agriculture's dim prospects resulted in its official neglect through the first half of the 1920s, but in 1926 serious interest in collective farming's potential resurfaced, and by 1927 the Party had taken steps intended to spur kolkhoz construction. The most important were the creation of an organizing agent in the RSFSR—the Kolkhoztsentr, the All-Russian Union of Agriculture Collectives—and an increase in state aid. As a result of these and other measures, the number of kolkhozes quickly expanded. The 38,139 kolkhozes in the USSR on 1 October 1928 represented an impressive twofold increase over the previous year's total. Yet the size and socialization of the existing farms left room for dissatisfaction. The new kolkhozes were smaller in area than those already in existence (an indication of the poverty of their members) and overwhelmingly of the TOZ type, in which some cultivation was practiced collectively, but animals and implements remained private possessions.[37] A TOZ or SOZ (*tovarishchestvo po sovmestnoi obrabotke zemli*)—association for the joint cultivation of land—was the simplest type of collective farm.

In the summer and early fall of 1929, agricultural collectivization made further headway. Between 1 June and 1 October the total number of households in collective farms nearly doubled from 1 million (3.9 percent of all peasant households) to 1.9 million (7.5 percent of households). Important grain-supplying regions, like the North Caucasus and the Middle and Lower Volga, boasted even higher figures. Based on this upsurge, Stalin made the case that the peasants—including the middle peasants—were entering the collective farms in great numbers, not as individuals, but together in the form of entire villages, districts, and even regions. In his famous article "The Year of the Great Break," published on the twelfth anniversary of the October Revolution, Stalin also made grandiose claims concerning the size and productive capacities of the socialized sector, asserting that collective and state farm marketing had rescued the country from the grain crisis. All these claims were exaggerated at best, but Stalin's depiction of mass peasant enthusiasm for kolkhoz construction and of the decisive role the socialist sector had played in overcoming chronic grain shortages provided the rationale for intensifying the collectivization process.[38]

At the November Central Committee plenary session, optimism was on display as proponents of collectivization, no longer hampered by Bukharin and Rykov's cautious approach to the peasantry, argued that reality had overtaken the Party's modest projections and that it was now possible to broaden and accelerate the collective farm movement. In an address to the plenum, V. M. Molotov derisively dismissed projects for a five-year collectivization plan, declaring that by the end of the 1930 spring-sowing campaign "the issue of the final collectivization of our countryside will be resolved in all of the decisive areas of the USSR."[39]

In the wake of the plenum's decision, the Politburo appointed a commission to draft a decree that would serve as a plan of action for the coming campaign. Chaired by *Krestianskaia gazeta*'s editor, Ya. A. Yakovlev, who now headed the recently created All-Union Commissariat of Agriculture, the commission, after much debate and under the close guidance of Stalin and Molotov, who were not members, published its final draft on 5 January.[40] The decree "On the Rate of Collectivization and State Assistance to Collective Farm Construction" reflected the views of kolkhoz enthusiasts. It announced that in view of recent progress, large-scale collective farming was poised to replace large-scale kulak farming and, more ominously, that the Party could now supplant the policy of restricting kulak economic activity with "a policy of liquidating kulaks as a class." No details or timetable was offered for the assault on the kulaks. In regard to collectivization, however, the decree specified an optimal pace ("within five years, instead of collectivizing 20 percent of the crop area as assigned by the five-year plan, we will be able to collectivize the vast majority of peasant farms") and envisioned the complete collectivization of the main grain-growing regions by spring 1931.

By redefining the immediate potentials of the collective farm movement, the November Central Committee plenum sounded the death knell for private farming in the USSR. In the aftermath of the plenum, Party organizations in key grain-supplying regions approved bold collectivization timetables that would have seemed recklessly ambitious just a few months earlier.[41] Suddenly, the essential transformation from "petty-bourgeois" to socialist agriculture presented itself as a matter of months, not years. In this tempestuous atmosphere, the responsible Party and state institutions adopted an aggressive, and at times frantic, approach to the collectivization process. Successes and failures were reported in military terms as advances and retreats on the collectivization "front." As in war, popular opinion was mobilized against the enemy, in this case, the kulak and the "rightist"—the latter a label that could now be applied to anyone who questioned the wisdom or possibility of charging full bore into the collectivized future.

The suddenness of the decision to engage in widespread collectivization and the lack of adequate preparation for such an enormous undertaking also encouraged military-type solutions to what were deep-seated organizational problems, not the least of which was a dearth of qualified personnel. To supplement rural cadres and supply a pool of collective farm organizers, the plenum instructed trade unions and the Kolkhozt-sentr to recruit and train no fewer than twenty-five thousand politically reliable industrial workers to fill the breach as local officials and kolkhoz administrators. Working-class resentment at rising prices and workers' acceptance of the regime's explanation that greedy peasants were to blame may be gauged by the fact that more than seventy thousand workers volunteered to take part in the socialist transformation of the countryside, a number culled by nearly two-thirds. Those selected as "25,000ers" represented a highly motivated, politically committed (about three-quarters were Party or Komsomol members) phalanx prepared to implement collectivization as the necessary solution to the food supply problems that had plagued the cities for the past two years.[42]

Their training completed, the 25,000ers joined a contingent of Party and government officials, OGPU troops, demobilized soldiers, factory workers, and others, numbering upward of 250,000, that poured into the countryside in the first months of 1930. Most of these individuals came from cities and towns and had little, if any, firsthand knowledge of agriculture or rural life. Lack of expertise did not dampen the enthusiasm of those participating in the campaign, however. Certain that they were introducing a superior, more productive form of farming that would ultimately benefit peasant and proletarian alike, collectivization brigades set about their task vigorously. The center's emphasis on speed and results combined with the brigades' class-war approach to encourage haste and ruthlessness and make a mockery of official instructions to avoid coercion.

Between 1 October 1929 and 1 January 1930, the percentage of collectivized peasant households in the USSR as a whole had increased from 7.5 to 18.1 percent. It jumped dramatically thereafter, reaching 52.7 percent by 20 February, and 56 percent by 1 March. Between 20 January and 1 March, the percentage of collectivized households in the Russian republic grew nearly 2.5 times (23.5 to 57.6), and in Ukraine the gains were far greater (15.4 to 62.8). Such rates of collectivization could not have been achieved in so compressed a time frame without the widespread application of force. As soon became apparent to the central leadership, lower-level officials were using all the means at their disposal to achieve maximal rates of collectivization—partly to demonstrate their zealousness to their superiors—and were, in fact, engaged in a competition to outdo the results achieved by their counterparts in other regions. The chaos caused by the widespread formation of "paper" kolkhozes placed the spring sowing in jeopardy, yet Moscow did little to enforce the "voluntary principle" in collective farm formation until March 2.[43]

Collective farm organization was not uniform and could vary significantly depending on the degree to which kolkhozniks engaged in collective cultivation and retained ownership of their livestock, tools, and other possessions. In its 5 January decree, the Central Committee designated the artel—in which livestock herds and farm buildings became collective property—as the most suitable form of organization under current conditions. Later, in "Dizzy with Success," Stalin singled out for special criticism "overzealous 'socializer[s]' " who had sought to skip this intermediate stage and jump immediately from less socialized forms of organization, like the TOZ, to the most socialized form—the kommuna—where even poultry was collectivized. Indeed, even peasants inclined to combine their holdings in a TOZ would just as soon scuttle an entire collective agreement as see their chattels and household gardens socialized. As early as January, the OGPU reported many cases from across the Union of mass withdrawals from kolkhozes over just this issue. In February 1930, an agricultural artel that had been organized in Mineralnovodsk district in autumn 1929 decided to reorganize as a kommuna. About one thousand farms from various villages agreed to the new arrangement. Some peasants refused, however, and they were excluded from the kommuna. To their dismay, they were unable to recover everything they had contributed to the artel, including spring and winter crops and fallow land. Complaints to the district executive committee and the regional land department proved fruitless. In frustration, the complainants dispatched envoys to Moscow, but representatives of the village soviet had them arrested. But even the artel could prove to be too socialized for some kolkhozniks, and this also led to conflicts among the peasants. According to one letter to *Pravda*, "in October 1929 in Gorkaia Balka village, Vorontsovo-Aleksandrovsky district, Tersky region, [North Caucasus territory], a TOZ was formed. In January 1930, it was decided to reorganize as an agricultural artel unifying 750 farms. A segment of the members of the TOZ (102 farms) did not agree and submitted an appeal for withdrawal. Notwithstanding the complaint, however, their property was not returned. A special commission worked [on this issue] with no result."[44]

Officials frequently attributed such disruptions to "kulak agitators"—that is, to sworn enemies of the kolkhoz movement. But, as one peasant explained, even those who entered collective farms voluntarily did not always do so for the purest of reasons.

· 125 ·

The real reason peasants voluntarily join collective farms according to one peasant, no date. RGAE, f. 7486s, op. 1, d. 100, ll. 41–42. Typewritten copy.

If Comrade Stalin should show up at the kolkhoz incognito and would ask about the spirit of the kolkhozniks: "Why, say, did you enter the kolkhoz?" almost everyone would answer for many reasons: 1) to get land where I want; 2) to leave the commune for the kolkhoz, where I can have more freedom and fewer obligations; 3) they promised us all sorts of advantages and help in the kolkhoz. To the question, "But by regulation, you must cultivate the land socially and property must be social," he answers, "Yes! According to the regulation it is so, but we only carry out social cultivation for appearances and only under pressure, but China is coming, all this is breaking down, and we'll keep the property."[45] This is the secret ideology of the kolkhoznik—property [ownership].

While the state focused its attention primarily on important grain-supplying regions, more remote, less economically significant areas also sought relief. The travails of a settlement located deep in the Urals forest are discussed in a letter to *Pravda* from a Party cell secretary, I. L. Zhuravlyov, in Ust-Khmelevka village soviet, Tagil region, Ural province.

· 126 ·

Letter to *Pravda* from I. L. Zhuravlyov, Party cell secretary of a village in Ural province, describing the condition of his village and proposing measures to improve it, no date. RGAE, f. 7486s, op. 37, d. 102, ll. 115–117. Typewritten copy.

To the Sixteenth Congress of the VKP(b)
On behalf of the general Party assembly and in accord with the cell bureau, we request that you answer by letter, even if only in three words.

A letter.

Dear comrades, our leaders of the USSR. I have not seen anything in the newspapers about our area and am turning to you for advice and help because our village soviet has been forgotten, but not the entire district; it's as if we are invisible to Soviet power, for which we shed our blood in the October days, and that's why we are suffering and are unhappy.

Tsarism abandoned us to the immense Ural forests twenty-five years ago. The roads [here] are awful.

Butchers bought up the forest for vodka and left nothing but stumps. They still stand; the big ones have yet to rot. We don't have a machine to grind them down, and there's none in the district executive committee either.

The population of the village soviet is 320 households. There are 0.3 hectares of arable land per head, and pasturelands are poor and scarce. The area around the forest is large, but there is no way to clear it. Clearing it out by hand requires 150 days a hectare. Wouldn't we be better off resettling [in order] to be put on workers' provisions? We get a wage, but they only give provisions to those who work. For 80 percent [of the people] there is not enough grain. The people never get a holiday. Based on the eight-hour day, every year they work [the equivalent] of two years. Even children under ten years of age help in the heavy work.

There's no point thinking about literacy. We have no telephone and travel by horse. There is no teacher or medic—they fled from hunger. In the cooperative [store] there is no food. The frost often destroys vegetables and grain. The bog surrounding the forest burns, and the people are worn out by fires. They even say, "Let them shoot us, but if they don't give us any grain, we won't go to put out the fire." A man must eat twenty to twenty-five kilos a month, but they give him six or eight kilos. They even say, "Smash all the rulers [*vlastiteli*]."

A kolkhoz was organized, but they fled, and only thirty households, which are divided up into thirteen settlements over a distance of twenty-five kilometers, are left. They see in the newspapers, and they believe it, that life in the kolkhozes is better, but all the same they don't see any results. [Agricultural] machines won't work, because the stumps obstruct them.

The poor peasant helped carry out grain requisitioning, but he remains hungry. They sent the grain to the district executive committee, and they haven't sent any back.

[Here] are the views of all the poor people in the entire region:

> Take five years to collectivize Tagil region, but don't prohibit those who want [to join the kolkhoz from joining]. Don't carry out grain requisitions, but establish a firm price: oats and rye, two rubles; wheat, three rubles a pud.
>
> Don't spoil grain [by making] vodka—it's better to eat it.
>
> Raise the wage for timber requisitioning.
>
> Working time in the forest should not exceed eight hours of labor.
>
> Provide machines to clear the land at the expense of the state and give non-repayable state aid for [land] clearance, but pay the tax from the first year [of production].
>
> Don't burn the roots around the forest but make it into pitch because this is a valuable material. Build more pitch factories in the Urals and find ways to sell pitch. Free up credit for construction.

Install telephones in every village soviet.

Build roads in every hamlet.

Add on bonuses for the extermination of predatory animals.

Allow the sale of church valuables to benefit the peasants.

Allow the free use of firewood where land clearance is going on.

There is no need to lower the agricultural tax, but only kulaks should pay it.

Supply the poor and workers with grain in order to calm them.

Win over and separate the poor from the kulak, not by lying to them in eloquent words, but materially.

Increase the courses for village soviet secretaries, reading-hut librarians, and political study.

Secretary of the Ust-Khmelevka VKP(b) cell, ZHURAVLYOV.

The letter expresses the hopes of the poorest of the poor that the socialist offensive would alleviate their miserable condition. Others took the new policy as an opportunity for self-aggrandizement. Such was the case with one, Ruf Korolyov, who formed the "Ruf Korolyov kolkhoz" just outside Diachkino village, Glubokovsky district, Shakhty region. According to a letter describing his activities, "Ruf organized a TOZ by attracting the nomadic peasantry from the central provinces, luring them to railroad stations." He acquired tools and a tractor by taking out high-interest loans. Then, through intimidation, he achieved a personal dictatorship not only over the members of the TOZ but over the neighboring citizens as well. Corrupted by his new-found power, the letter states, "he abandoned himself to drunkenness and unrestrained folly. The collective members were actually his serfs and Ruf was the lord."[46]

From the outset of the campaign, collectivization brigades employed a variety of tactics to force peasants into the kolkhozes. In a letter sent to the *Pravda* editorial board under the title "They drove the people to insanity," the office workers of Moscow factory No. 22, who had evidently been in the countryside at the height of the collectivization campaign, offered this description.

· 127 ·

Letter to *Pravda* from a group of Moscow office workers describing the method used to force one peasant to join a collective farm, 1930. RGAE, f. 7486s, op. 37, d. 102, l. 221. Typewritten copy.

Comrade Muratov declared: "If you don't enter the kolkhoz, then it's nothing to us to shoot ten men out of one hundred or burn you from all four sides; so that none of you flees tomorrow, we'll make a sacrifice of you." He banged his fist on the table and closed the assembly. On the second day after the assembly, Comrades Preobrazhensky, Diakonov, and Yamilin came to the peasant Mikhail Molofeev of this village, who has a holding, one horse, and one cow. They declared, "Let's go to the barn, old man." The old man dressed and left with them. His family, aware of the discussions at the evening assembly, figured that they had taken the old man to be shot. The old lady, that is, his wife, went running scared to another village, Bulgakovo, located six kilometers away, where she was brought to her senses, after which she remained half deaf, but Yekaterina Alekseevna of the same family fell into an unconscious state and lay down all day, after which she was seen by a doctor, who said: "If a second instance of this sort of fear should recur, she would definitely go mad . . ."

The peasant P. M. Simakov, from Manchazh village, Kungur region, Ural province, sent the following letter to *Pravda* with twenty-six signatures in addition to his own. In the opinion of the editors he had added these in his own hand.

· 128 ·

Letter to *Pravda* from P. M. Simakov, a peasant from Ural province, denouncing collectivization and threatening mass resistance, 1930. RGAE, f. 7486s, op. 1, d. 100, l. 60. Typewritten copy.

Right now total collectivization is occurring throughout all of Manchazh district. But whoever came up with this law deserves to be called brainless. Really, comrades, war hangs over our heads, every day White bands invade the Soviet Union, and still you, comrades, compel [us] to enter the kolkhoz by force. It's said that whoever doesn't enter the kolkhoz will lose his rights and won't be allowed to vote.

This is fundamentally wrong, comrades. In the kolkhoz, only bureaucrats can live, but not us peasants. If they forcibly drive all of us [into the kolkhoz], then we, as one man, will go to war with the Soviet Union and will win back the old rights. This is how they live in the kolkhozes. Take, for example, our kolkhoz, "The Red Ray," in Manchazh village, which was organized back in 1928. It has eighty households. From every household there they took away the horses, and they are no longer theirs. There was one case in which a citizen gave up his horses to the kolkhoz, a poor peasant with a family of seven of which two worked. After three days, he went to ask for a horse: "I'm going for straw." They told him: "Now the horses are no longer yours. You must forget about them." On another day he took the horse on his own and sold it. For selling his own horse they put him on trial. Or this: the kolkhoznik comes for rations and they say to him, "Why did you swallow up everything so fast? There are no rations for you today." And the kolkhoznik goes home in tears, and at home the kids are dying of hunger and howl like wolves. This is what Soviet power is building. But we are partisans who fought for Soviet power. We don't want this. We fought for freedom so that we would be free, but not for such torment of the people. We are prepared to go forward with rifles in hand. If we are to be oppressed and starved, we partisans and citizens of Manchazh village and all of Manchazh district don't want it. If they forcibly drive us, the poor peasants and partisans, into the kolkhoz, then we will torch every kolkhoz building, and all the kolkhoz wealth with the red rooster, and not just the nearby kolkhoz—we'll also go farther out where there are large kolkhozes and burn them.[47] Even though war hangs over our heads, you raise this havoc. Let those kolkhozes exist that the people entered voluntarily, but don't forcibly drive [people] into the kolkhoz. Give us the freedom that we won; otherwise, we'll burn the kolkhozes and go to war.

On 7 March, days after Stalin applied the brakes on this first phase of collectivization with his *Pravda* article "Dizzy with Success," OGPU Deputy Chairman Yagoda sent him a "top secret" report detailing "distortions and excesses" in the campaign, which Yagoda described as "typical for almost all the areas of the Union." Yagoda provided numerous illustrations of dekulakizations of poor and middle peasants, mass dekulakizations, property theft, and the brutal and sadistic treatment of peasants, regardless of their economic category, committed by collectivization brigades in various provinces. According to the report, brigade members operating in Zinoviev region, Ukrainian republic, raped two kulak women and beat up a sixty-five-year-old man. They then forced him to dance and sing while pouring water on him and sticking a dirty cigarette in his mouth. In another village in the same region, a Komsomol secretary "forced a middle peasant to pull the end of a noose that had been thrown around his neck." While he was choking, the secretary mockingly offered him a drink of water. In one village in the Central Black-Earth region "a group of 30 young people led by the secretary of the Komsomol cell came to the house of a middle peasant, took all his belongings, beat up the peasant's wife, locked her in a room, and pocketed small items."[48]

Yagoda was clearly concerned that such cruel and undisciplined behavior was undermining the collectivization campaign by stiffening peasant resistance while evoking sympathy for the kulaks among the local population and even among local communists and soviet officials. In one instance included in his report, a group of Komsomolists and the plenipotentiary from the district soviet beat up a kulak who had refused to hand over his gold and money. "After the beating the Komsomol members, along with the district executive committee's plenipotentiary, tied a rope around the kulak's neck, bound him to the shafts of a horse [cart], and dragged him this way through the village to the district center. The kulak's screams brought out a large crowd of peasants who voiced great outrage over this abuse." At a village meeting in the Central Black-Earth region called to vote on collectivization, the village soviet chairman declared, "Whoever is against collectivization is against Soviet power— raise your hands." In reaction to this intimidation, the peasants denounced the chairman as a "counterrevolutionary" and abandoned the meeting without voting. This "despite the fact," in Yagoda's words, "that the groundwork for establishing a collective farm had been fully prepared." Yagoda also reported that in the North Caucasus a Komsomol member had warned two kulaks of their impending arrest, allowing them to flee, and in another case, "a candidate member of the party and five Komsomol members categorically refused to take part in an operation to remove kulaks."[49]

Defense of religious prerogatives played an important role in ethnic actions against collectivization. In January, the OGPU reported that many demonstrations were occurring in the villages of the German Autonomous Province. In one case, the simultaneous attempt to close the village church, tax the Catholic clergy, and socialize "nonworking cattle" had ignited a demonstration under the slogan "For belief in God and against the collective farms." Although the report attributed these demonstrations to kulak provocations, it also admitted that the active participants were "exclusively" women, drawn largely from the poor peasantry with some middle-peasant support, and that kulaks did not participate "openly." In Tataria, the arrest of a mullah along with other "kulak–anti-Soviet elements" brought out seven to eight hundred demonstrators, some of whom were armed with pitchforks and clubs. The incident ended with the mullah's release and the forced flight of the criminal investigation unit, whose members the crowd warned, "Don't come back here or we'll kill you."[50]

The collectivization drive also exacerbated tensions between kolkhozniks and peasants who remained outside the collective farm. Conflicts between the groups could be violent. A letter to *Pravda* with fifteen signatories describes how a drunken gang from the "Ray of Ilich" kolkhoz, in Tiumen region, followed one of their drinking bouts by attacking individual landholders. One kolkhoznik armed himself with a knife. As they fell upon the farmers, the kolkhozniks shouted, "Oorah, give us rifles, we have to slaughter the individual farmers!" During the assault a single landholder was fatally stabbed in the chest, and another suffered multiple wounds from a beating administered with a hunk of iron. The single landholders had to band together to drive out the invaders.[51]

The anti-kulak hostility fostered by the regime proved impossible to keep within strict class boundaries. Middle peasants were also victimized. In a letter to *Pravda,* a seredniak expresses indignation at his situation.

· 129 ·

Letter to *Pravda* from a middle peasant denouncing the discriminatory policies practiced against middle peasants, 1930. RGAE, f. 7486s, op. 37, d. 102, l. 108.
Typewritten copy.

They don't give the middle peasant a chance. Can this be the real policy of Soviet power, is this how the dictatorship of the proletariat is to be understood? For example, on the kolkhoz, *he gets the worst work and, in any event, he certainly can't become a tractor driver.*[52]

The state apparatus is proletarianizing [*orabochivaetsia*], the militia is proletarianizing, but the middle peasant is not allowed to work in a plant or factory. It's no use talking about the vuzy or vtuzy [higher educational institutions]: 75–80 percent of the students there must be workers or their children, and the quota for the middle peasant is zero.

His (the middle peasant's) only role is to increase the wages of the workers.

IVAN KUKAN, village of Gorki

One method of inducing peasants to join collective farms was to strictly limit the sale of goods to them in cooperative stores and shops. A letter to *Pravda* from Riazhsk district, Moscow region, from early 1930 offers the following exchange between a peasant and a cooperative store clerk.

· 130 ·

A verbal exchange between a peasant and a cooperative store clerk contained in a letter to *Pravda,* early 1930. RGAE, f. 7486s, op. 1, d. 100, l. 165. Typewritten copy.

Some peasants were poking around in the cooperative [store] and ask:

"Could I get a couple ounces of tobacco?"

"And do you have any eggs?" the sales clerk asks.

"What eggs?" the stupefied peasant asks.

"They're well known, [they come] from chickens. We only give tobacco [in exchange] for eggs. Bring ten and we'll give you four ounces."

"Then I'll have a spool of thread . . ."

"Bring eggs!" repeated the implacable clerk, "then take the thread . . ."

"Then the hell with it. I'll buy a little something. Throw a funt of salt herring or a kilo of some other fish together."

"We only give [them] out for eggs or [under] contract to the kolkhozniks. The district consumer society has forbidden giving them out to individual [peasants]."

In the course of the collectivization drive, grain-requisitioning methods were extended to other types of agricultural produce and nonagricultural items. One letter even speaks of the procurement of women's braids, for which a special price list was established for washed and unwashed hair sorted by color.[53] "Strict quotas" (*tverdie zadaniia*) were applied to agricultural deliveries and income from handicrafts. This excerpt from a letter of a village commune located in the Uralsk region of the Kazakh autonomous republic to the provincial Party committee describes the effect this had on livestock.

· 131 ·

Excerpt from a letter from a village commune in Kazakhstan describing the negative effects of the requisitioning of meat and other items, no date. RGAE, f. 7486s, op. 1, d. 100, ll. 139–140. Typewritten copy.

The arbitrariness of the local authorities has so terrified the population that we are afraid to come out openly and reveal the disgraceful offenses. [. . .] The burden of the meat procurement fell on the middle and poor peasants because they [the procurement officials] collected not [only] the surplus but [even] the pitiful scraps that remained [. . .] The appropriated meat is stored so carelessly, especially in Zhangala district, that it is rotting and gives off a stench that keeps you from getting close [...] The burden of the seed fund also fell on middle and lower-middle holdings [...] In December 1929 and January 1930, the local government imposed a compulsory wool [collection] on the population. The population then sheared the wool from their sheep and camels despite the bitter cold, and this resulted in a significant reduction in the number of livestock, since the animals, deprived of their natural protection, could not endure the harsh winter and went down [...] Thanks to the excesses perpetrated by the local authorities in the procurement of grain, wool, and meat, the quantity of livestock fell by 75 percent [...]There were mass arrests and the Ural's isolator [prison] overflowed. The majority of those [arrested] were middle and poor peasants. Owing to the arbitrariness of individuals and the local authorities, many laborers were treated as kulaks and were deprived of voting rights and brought to trial on the basis of article 28. [...][54]

As early as 22 December 1929, the OGPU was reporting that the sale and slaughter of livestock (including horses, cows, pigs, and sheep) in a number of important regions had assumed a "mass character." On 10 June 1930, the chairman of the Council of People's Commissars of the Russian republic, S. I. Syrtsov, summed up the negative effects of meat procurement in the Central Black-Earth region in a secret report to V. M. Molotov: "In the region there has been a catastrophic reduction of the herd. Compared to 1928, large horned cattle have declined by 32 percent, pigs—72 percent, sheep—50 percent. Procurement proceeds at the expense of a reduction of the basic part of the herd, damaging the possibility of recovery in the near term ... 80–90 percent of the procurement must proceed at the expense of those with a single cow. In fact, procurement is becoming a confiscation of the last cow and will meet decisive opposition from the peasants and a new flood of young working cattle being thrown on the market. We must clearly bear in mind what requisitioning will be like with the participation of the militia ..."[55]

Dekulakization

In December 1929, Stalin shocked the Marxist agrarian scholars who had gathered in Moscow for their first national conference when he announced that the recent expansion in collective and state farm grain production had made it possible to replace the policy of "restricting the exploitative tendencies of the kulaks" with a policy of "eliminating the kulaks as a class." In his address, Stalin did not specify the methods by which kulak liquidation would be achieved, nor did he indicate the scope the operation would assume, but his use of terms like "offensive" and "smash" made clear his willingness to resolve the rural class struggle violently.[56]

At the time of Stalin's speech, central instructions on dekulakization had yet to be drawn up, nor would they be completed until the Politburo commission chaired by Molotov issued its top-secret decree "On Measures for the Liquidation of Kulak Farms in Districts of Wholesale Collectivization" on 30 January. In the interval, local organizers proceeded with kulak expropriation as they saw fit. In the Crimea, for example, lower Party organizations began dekulakization on January 20 without the participation of the village poor or local collective farmers. In a form of "naked dekulakization"—dekulakization by fiat—they engaged in surprise nighttime raids and searches that uncovered stashes of kulak gold and valuables. Other peasants took advantage of the attacks to steal and conceal kulak property and livestock. In one case, enterprising middle and poor peasants managed to hide away four thousand sheep.[57]

This type of arbitrary dekulakization was no aid to collectivization efforts, and its chaotic character deeply troubled Party leaders in Moscow. In fact, as Lynne Viola and V. P. Danilov show, it was the OGPU, rather than the Politburo, that assumed the primary role in defining and organizing the anti-kulak campaign. As early as 11 January, four days before the creation of the Politburo commission, the OGPU deputy chief, G. G. Yagoda, instructed his subordinates to begin planning for the "wholesale purging" of kulaks from the countryside. He specifically asked OGPU collegium member G. P. Boky how many more people could be accommodated by existing labor camps and where new camps could be built. So that a "crushing" (*sokrashitelnyi*) blow could be delivered against kulaks, Yagoda was especially keen that the OGPU should take control of the operation. On 24 January, he expressed strong displeasure that Moscow province officials were carrying out dekulakization "under our nose" without OGPU knowledge. Yagoda's efforts were successful. According to Viola and Danilov, the Politburo commission's deliberations and final decree owed much to the OGPU approach to the kulak.[58]

The 30 January Politburo decree laid out the steps to be taken to accomplish the liquidation of kulak farms, kulak exile and resettlement, and the confiscation and disposition of kulak property and to allocate personnel and plenipotentiary powers for the campaign. In districts of wholesale collectivization, the decree called for the expropriation of kulak "means of production, livestock, farm and residential buildings, processing enterprises, and fodder and seed reserves." Beyond this, the decree distinguished three kulak categories and meted out punishments accordingly. In addition to confiscations, kulaks deemed active counter-revolutionaries faced incarceration in concentration camps or the imposition of the death penalty. Richer, more influential kulaks not charged with counterrevolutionary crimes were to be exiled to remote regions. The remaining, third-category kulaks could expect resettlement to different locations within their home districts. To prevent unwarranted dekulakization from spilling over to the middle peasantry, the decree limited farm liquidations to no more than 3–5 percent of the total in the most important districts. Other potential supporters of the regime were also offered some degree of protection. Families of soldiers and army commanders enjoyed immunity from exile and confiscations, and collectivization brigades were instructed to take "an especially careful approach" toward kulak families having members in the factory workforce.

The decree called on the OGPU to "carry out repressive measures against the first and second categories [. . .] during the next four months" and set the numbers of those to be sent to concentration camps and exiled to remote areas during this period at 60,000 and 150,000, respectively. Specific figures were given for the nine most important agricultural regions (from the Lower Volga region, for example, 4,000–6,000 people were earmarked for concentration camps and 10,000–12,000 for exile) while it was left to the OGPU and provincial officials to establish the totals for the remaining regions. Given the scale of the operation, both the Party and the OGPU required reinforcements to conduct the campaign. The Central Committee promised to transfer twenty-five hundred Party members from the major industrial centers to boost staff at rural Party organizations. Funds were made available for the OGPU to hire an additional eight hundred individuals—preferably older, reserve Chekists—to serve as plenipotentiaries in remote areas, and to increase by one thousand the complement of OGPU infantry and cavalrymen. The Commissariat of Transport and the OGPU were given five days to devise a railway plan for transporting kulaks to the camps and places of exile.[59]

Conditions of kulak resettlement were designed to be particularly harsh. Exile locations had to be "uninhabited" or, at most, "sparsely inhabited." Exiled and resettled kulaks could retain "only the most essential household items, some elementary means of production in accordance with the character of their work in the new locality, and the minimum food reserves necessary for the initial period." They could carry just enough money to cover the costs of the journey and establishing themselves in their places of exile, the sum not to exceed five hundred rubles for a family.[60]

Finally, in Bolshevik fashion, the decree attempted to promote and weld together both the mass-spontaneous and the organized-conscious elements of dekulakization in an "organic bond." It insisted that dekulakization serve as a means of mobilizing the rural poor for collective farm construction and warned against administrative dekulakization by fiat. Decisions to confiscate property and exile kulaks had to involve the participation and approval of collective farmers, poor peasants, and landless laborers at general meetings. Special plenipotentiaries of the district soviet executive committees were to carry out the confiscations themselves at the head of dekulakization brigades made up of these same groups, whose participation was mandatory. To fulfill their role as the conscious vanguard, provincial and regional Party committees were instructed to oversee the process and provide the dekulakization brigades with "constant leadership."[61]

Despite these instructions, arbitrariness and spontaneity continued to overwhelm political consciousness. To the Politburo's dismay, Party organizations were devoting more time and resources to dekulakization than to collectivization. On the same day that the Politburo decree appeared, Stalin sent a coded directive to all Party organizations warning them not to abandon collective farm construction for dekulakization work. The next day, the leader of the Middle Volga Party committee received an urgent telegram signed by Stalin, Molotov, and L. M. Kaganovich rebuking him for haste and "naked dekulakization of the worst kind," which "has nothing in common with Party policy." On 3 February, the Central Committee Secretariat warned key regional and republic-level Party committees that their precipitate actions threatened the campaign with "disorder" and "lack of coordination." It instructed them to coordinate their dekulakization activities with the OGPU and to strictly adhere to the terms of the 30 January instructions.[62]

The following letter to *Pravda* from the summer of 1930 addresses the arbitrariness involved in classifying peasants as kulaks. It was sent from a collective farm in Zamot, located outside Dmitrov, in Moscow region, under the title "Nothing's to be gained by blaming the little guy."

· 132 ·

Unsigned letter to *Pravda* from a collective farm member on the dekulakization of non-kulaks, summer 1930. RGAE, f. 7486s, op. 37, d. 102, l. 228. Typewritten copy.

The violence perpetrated by the brigands has brought the Central Committee of the Party, especially Comrade Stalin, into complete mistrust. They call him *zimogor,* a scoundrel.[63] They said that he was the one whose head was spinning. It does no good to blame the little guy. On the matter of collectivization, it was necessary to go in all seriousness right to the top and send out in advance a resolution of the Central Committee of the VKP(b) on the struggle with the distortions of the Party line in the kolkhoz movement and precisely define there the concept of the kulak. In the district there are ten kulaks, but five hundred people (households) were dekulakized. The peasants here are not opposed to collectivization but are absolutely against working with such a gang of bandits, presented to us as "the poor."

Despite the Politburo's conception of dekulakization as an organized mass movement pitting the village poor and the socialist vanguard against the rural bourgeoisie, in practice officials found it difficult or impossible to control class hatreds once unleashed. The 30 January Politburo decree specified that dekulakization brigades make accurate inventories of confiscated kulak property and that village soviets retain the confiscated items until their distribution to collective farms or to state and cooperative agencies as payment of kulak debt. Unadulterated greed, personal animosities, and the desire to redress past wrongs, however, turned property confiscations into outright looting. Peasants facetiously referred to dekulakization brigade members as "junkmen" (*barakholshchiks*), and countless reports from the provinces described acts of brutality and petty theft perpetrated by the brigades. The opportunities for material gain also served to expand the campaign's targets, for dekulakization brigades often applied a liberal definition of who was a kulak. In a "highly secret" report to the all-Union commissar of agriculture, the first Party secretary of the Central Black-Earth region wrote, "As a rule, during dekulakization, the directive on the partial confiscation of the means of production from kulaks expelled from collective farms has not been followed. Almost everywhere they carried out total confiscations, and not only of the means of production but of all other possessions as well. Mass excesses were permitted that stripped middle peasants, communists, workers, teachers from the village intelligentsia, and the like. The poor usually tried to expand the dekulakized contingent, since, obviously, they had a material interest. We tried all sorts of ways to eliminate these excesses; nevertheless, we failed to avoid them. In the course of the coming months we still have to eliminate them."[64]

One dekulakization victim who was familiar with the various policy positions taken toward the peasantry in past Central Committee debates gave the following account of what had gone on in the villages of Siberia.

· 133 ·

Letter to *Pravda* from the dekulakization victim Kapustin describing the excesses he witnessed, 1930. RGAE, f. 7486s, op. 1, d. 100, l. 127. Typewritten copy.

Here's how dekulakization occurred: fifteen men come at night and gather everyone up. They carry off the sour berries, salted pickles, and even the meat from the pot. They tore off my last fur coat, which I had not handed over, and for this I was arrested then and there. The question is, is this dekulakization or robbery? In my opinion, they dekulakized kulaks even at the time when they were conducting a limited policy. But at the time of the liquidation of the kulaks as a class, they carried out a form of robbery of the laboring peasantry. Not for nothing are they frightened of the new slogan about liquidating the urban bourgeoisie. This would begin to entrap the most responsible officials, who have clothes and shoes in abundance. They have dekulakized those for whose sake the Soviet Union came into existence, those who gave the state essential raw materials and foodstuffs.

As for the right deviationists. In the newspapers they write that during collectivization there were, allegedly, leftist deviations, and it must be said in more than a few places the slogan "Liquidate the kulaks" is itself a left deviation. TsIK writes about the struggle with the left deviation, [but] these are only words; in actuality, [TsIK] practices a left deviation. The rights correctly identified the problem. I am not an advocate of the old system, but it is impossible to agree with the line put forth at the Fifteenth Party Congress. Many souls perished during the expulsion of the kulaks. In [temperatures of] forty degrees below zero, they took families on horseback to Tiumen and Tobolsk. In the city of Tobolsk alone, about three thousand people are buried. These are completely innocent victims. It resembles the time when Herod gave the order to kill infants younger than six months old. Let them consider me a kulak because I don't want to enter a kolkhoz. But in what way are children to blame here? They write a lot about self-criticism, but in fact it turns out that it is absolutely impossible to criticize TsIK. It's as if it is something sacred or the imperial crown. Comrades Bukharin, Rykov, Frumkin, and Tomsky are right.[65] They know the peasant's way of life and the peasant's ideology better than you do.

My current address: Tobolsk, Nemtsov St., No. 27, KAPUSTIN.

Rather than submit fatalistically to this cataclysmic onslaught, many peasants responded in ways that aimed to obstruct dekulakization and undermine collectivization: they dismantled their farms and abandoned agriculture altogether. In what official documents described as "squandering" (*razbazirovanie*) and "self-dekulakization," peasants rushed to sell off their livestock, tools, machinery, and other belongings before they fell into the hands of dekulakization brigades. So that their valuable farm machinery would not benefit the collective farms, and by extension the Soviet state that was making war upon them, they often deliberately damaged it in an act of resistance that Lynne Viola has analyzed in terms of peasant Luddism. Similarly, in several Ukrainian districts, incidents of peasants torching their own houses and barns—as well as nearby kolkhoz structures—were reported. According to official reports, peasant arsonists were after the insurance payments. After liquidating their holdings, large numbers of peasants fled to towns and cities, where they lied about or otherwise concealed their social origins in order to enter the ranks of the expanding industrial workforce. However, as Viola notes, contrary to peasant intentions, these acts of resistance more often than not convinced local officials that dekulakization had to be intensified, not abandoned.[66]

Despite the 30 January decree's instructions to limit dekulakization to the village elite and to treat other segments of the population with circumspection, non-kulaks and supporters of Soviet power were also hit by the wave. A note sent to *Pravda* from Tambov region highlights one such incident.

· 134 ·

Unsigned letter to *Pravda* regarding the dekulakization of Red Army veteran Ye. D. Chalikov, 1930. RGAE, f. 7486s, op. 1, d. 100, l. 48. Typewritten copy.

They were not ashamed to dekulakize Yermolai Dmitrievich Chalikov, a civil war participant on the eastern front. [He served] in the ranks of the Red Army continuously from 1919 to 1926. He has an Order of the Red Banner for distinction in battle.

In 1927 he was the chairman of the village soviet. Yermolai Dmitrievich has a middle peasant's landholding. He never employed hired or mechanized labor.

Chalikov tried to contact Voroshilov.[67] But for [taking] such a bold step, they threatened him with arrest and even other things. Besides Chalikov, other Red Army men have been dekulakized.

Despite instructions to the contrary, the dekulakization of middle peasants and families of serving Red Army soldiers was commonplace. Contemporary reports attributed these "excesses" to the "bungling" of local officials or the unrestrained acquisitiveness of the poor. In Salsk region in the North Caucasus, dekulakization victimized former Red partisans. According to one report, "In this case, they stole mercilessly. They dekulakized the father of a Red military cadet." In such instances, the targeted peasants faced "clean-sweep" dekulakizations or dekulakizations "to the last crumb" (*raskulachivanie pod metelku*), in which all their possessions were up for grabs.[68]

Just who faced dekulakization in a particular village was often the result of a complex web of intra-village relations with class, personal, and political strands. During the voting on dekulakization in one village, an individual named Gudz said of the middle peasants that there really was nothing to consider: "In this business there can be no middle. If we're cutting, then let's cut, and we'll slip anyone who wants to be in the middle the point of a bayonet and herd [them] together with the kulaks." Local leaders were not averse to using dekulakization to settle accounts with their critics. As one dekulakized peasant wrote, "I received a copy of the district election commission protocol stating that I would be transferred into the home guard and deprived of the right to vote. I had never traded, and neither had my father, my grandfather, or my great-grandfather. I had never used hired labor a single day, neither before the revolution nor after the revolution. I had never been in the White bands and had never been a servant of a cult, since I was a nonbeliever from childhood. Everyone here is saying that they took away my voting rights because I wrote in the newspapers and was a village correspondent." Even wounded pride and affairs of the heart made themselves felt. One letter from Belorussia states that Iosif Veriga and his mother, citizens of Tirshev village, Rositsa village soviet, were dekulakized and exiled because Iosif had rejected the amorous advances of the former village soviet chairwoman.[69]

As seen in the following letter, there is also testimony that not only peasants suffered dekulakization.

· 135 ·

Letter from A. N. Misiurov, Ural province, on the dekulakization of non-peasants, 1930. RGAE, f. 7486s, op. 37, d. 102, l. 78. Typewritten copy.

———

In Votkinsk district, Sarapul region, the local organizations, in particular the chairman of the city soviet, Comrade Marinin and a few others like him, went crazy with dekulakization and the confiscation of property. They dekulakized a large percentage of the union membership, factory workers, families of Red Army soldiers, and the soldiers themselves. All this happened

under the eyes of the local prosecutor and the Party organization. Up to
now, they have yet to punish Marinin, the leader of all these bungled antics.
On the contrary, from the city soviet they sent him into kolkhoz work.

The criminal investigation into the matter of the suicide of the Red Army
soldier Fionovsky established that the latter had been deprived of the right
of citizenship and the rank of Red Army soldier.

Everyone is aware of this, but they are afraid to speak openly because of
the enormous pressure.

A. N. MISIUROV, Votkinsk, Ural province

The socialist offensive in industry and agriculture cleared the way for
radical assaults on all aspects of Soviet social and cultural life. Since
1928, the antireligious movement and its ardent enthusiasts in the Komsomol and the League of the Militant Godless, had taken advantage of
the new atmosphere to pursue an increasingly aggressive line. During
dekulakization this movement struck the villages with particular vehe-
mence. Clergy were identified as kulak allies opposed to collectivization
and were arrested and exiled in great numbers. Churches were ransacked
and closed, and the sacred objects they housed were defiled. When not
completely dismantled, the buildings themselves were frequently put to
secular uses.

Believers, particularly women, responded with anger to these outrages even to the point of violently opposing the vandals. Sometimes this
direct action had the desired effect, as in this incident from Ukraine.

· 136 ·

Letter from I. S. Zemliansky describing how peasants in one Ukrainian village
defended their church, 1930. RGAE, f. 7486s, op. 37, d. 102, l. 223.
Typewritten copy.

The men and women banded together and then, as if to a fire, ran off to
the church in order to occupy the church and prevent any knocking down
or breaking up there. Several of the women challenged the authority of the
village soviet and savagely defended the church. The authorities from Piriatin went there to put down the assault on the church, but there were no arrests, because the Piriatin officials promised to honor the wish of the peasant
men and the women concerning the church. Having settled the worst of the
peasant assault, they left, satisfied. But all the peasants are enraged; they all

rush about and are all senseless because who knows what will happen next? So we are reporting this to you, the higher authorities, and [leave it to] your discretion to soothe and assuage the hearts of the peasants who are up in arms over their churches both in Maershchin and in Karavai, etc.

I. S. ZEMLIANSKY, Mikhailovka village,
Dubensk region, Piriatin district.

In other cases, the outcome of these confrontations could be ugly, as in this incident at the "1 May" kolkhoz near the Kuban Cossack capital, Krasnodar.

· 137 ·

Letter from F. F. Taran describing antireligious violence in a village near Krasnodar, 1930. RGAE, f. 7486s, op. 37, d. 102, l. 220. Typewritten copy.

The "1 May" SOZ. The chairman of the village soviet and the Krestishch soviet, which also took part, held a meeting at which the village soviet chairman proposed: "Destroy the church in Krestishch and I will give you the motor that they confiscate from the kulak." Because the chairman of the village soviet and other persons who opposed the union of the believers had gotten roaring drunk, they conducted the raid on the church. No fewer than one hundred women came running. The drunkards began to beat the women and to throw [them] through the fence, shouting, "Don't just smash [just] the 'holy objects' but the long hairs too." About fifteen women ended up seriously beaten. The women went to a doctor in Krasnodar for documentation, and the affair is headed to court.

Citizen of the Krestishch settlement, FEODOSY F. TARAN.

The attacks on village spiritual life and the desecration of churches through their transformation into clubs, warehouses, granaries, and other profane venues fueled the sort of apocalyptic rumors and revelations that peasants had reverted to in times of crisis for generations. "Letters from God" circulated, stating that "the unclean one has come down from the heavens," that "the profane live in the world," that "whoever does not believe will be punished by wind and hail," and that soon "brother will turn against brother and there will be a blood-letting." Like peasant Christianity, prophecies of the apocalypse combined pagan and biblical imagery and, not surprisingly, focused on the supposed satanic essence of the kolkhoz. Reports from the villages noted that "rumors are afoot about house spirits, wood goblins, and all types of legends and fables in connection with collectivization. Icons appeared. Pilgrims thronged. Women complained that those who entered the kolkhoz were selling their soul to Antichrist." According to another report, "They are speaking of a certain stone that fell from the sky, about it raining stones, about drought that was sent down [as punishment] for people's sins . . . they are saying that everyone will be driven into the kommunas, where a shameful, sexual, depraved life will begin."[70]

Partisans of collectivization interpreted these cataclysmic predictions as nothing less than sabotage, as weapons in a "religious counterrevolution" committed to disorganizing collective farm construction. A workers' brigade sent to the Ukrainian "collectivization front" by the Moscow food-processing plant Mosselprom reported the following in a letter to *Pravda*.[71]

· 138 ·

Letter to *Pravda* from a Mosselprom workers' brigade describing religion-based opposition to collective farms, 1930. RGAE, f. 7486s, op. 37, d. 102, ll. 56–55. Typewritten copy.

In certain districts and localities, various religious sects have hindered collectivization and the mobilization of resources.

In Gelmiazov district (a district of wholesale collectivization), Shevchenko region, Kalaberdy village, churchgoers carried out agitation saying that in the SOZ the chairmen would be those who had previously been landowners or that "in the Bible and the gospel it is said that the SOZ will exist for twenty-eight days, the artel for twenty-one days, and starting from 1 February, if we go to the kommuna, we will live for forty-two days more, and then 'the end of the world' will arrive."

Because of this agitation, there were several declarations of withdrawal from the kolkhoz. The mobilization of resources was also weakened there, although in one night the church council collected 130 rubles. In another village, Bubnova Slobodka, a group of kulaks and churchgoers (members of the church council) carried out the [following] kind of agitation:

1) they will take the horses to work the Jews' land;
2) five puds of hair will be taken from every woman;

3) every female household head will have to give eight puds of shirts for rags to the collective;

4) girls weighing four puds will be sent to China;

5) they are collecting money to expel the kulaks;

Pray to god. Don't enter the satanic empire of the kolkhoz.

On the basis of religious convictions, sixteen declarations of withdrawal from the kolkhoz were submitted.

For example, in the village of Yablon, Kanev district, they organized the "The New Path" artel. Religious counterrevolutionary propaganda was conveyed there both inside and outside the kolkhoz. The church elder, [who] is in the kolkhoz himself, agitates [among] the others to leave it. He collected 184 rubles for church necessities at the very time that the financial campaign for the month was going very badly. A segment of the middle peasants (the blind Prokopadr and others) were pulled into the orbit of this agitation.

Lutovsky agitates [among the peasants] that "a stamp will be affixed to those who enter the kolkhoz and children will be put in nurseries."

As for the Party officials, either there aren't any or, if there are any, then they are weak and feebly fight this kind of agitation.

Religious counterrevolutionaries must be isolated. This is the request of the greater part of the rural masses.

Some clergy, to secure their positions, sought accommodation with secular authority even to the point of evincing sympathy with the regime's goals. Turning for advice to the Communist daily, *Pravda,* one village priest adopted a submissive and obedient stance.

· 139 ·

Letter to *Pravda* from the parish priest A. Pokrovsky seeking clarification on the relationship between the church and kolkhozniks, 1930. RGAE, f. 7486s, op. 37, d. 102, l. 52. Typewritten copy.

I sincerely ask that you clear up the following matters for me: in our parish in Riagovo district, in the villages of Podporokhi and Navolok, there is a kolkhoz.[72] Can they [the kolkhozniks] ask me, as a priest, to come to them and maybe perform some ceremony or other? So as not to be held liable before Soviet power, I earnestly ask that you send me your instructions. May I perform this or that ceremony for them without disturbing the class line?

I think that in strict accordance with the class line, I cannot have contact with them.

Each kolkhoznik has set out on the path to communism and therefore should have absolutely nothing to do with religion. If this is so, then I request that you inform all the kolkhozniks through your organizers. Let them know that they are antireligious.

It would please me very much if they would renounce religion; then all the kulaks here would be completely weeded out, and otherwise it will be impossible to advance this [class] line. Send me your instruction clearly stating that as the priest of the Riagovo church, I am strictly forbidden to perform any ceremonies for the kolkhozniks of Riagovo district.

Priest A. POKROVSKY.

The impact of collectivization and the antireligious offensive on family relations is brought out in the following letter from a priest's wife.

· 140 ·

Letter from D. Sigaeva, a priest's wife, renouncing her former life and beliefs, 1930. RGAE, f. 7486s, op. 37, d. 102, l. 218. Typewritten copy.

Citizen Dominika Ya. Sigaeva, of the village of Nizhny Samaevki, Rybkinsky district, Mordovskaia province, Middle Volga Territory. I am the wife of a priest. I have six young children. [I have been living] on an allotment since 1922. There is a son. From the very day of the [land] partition he argued that the policy of the Party and Soviet power was correct, he demonstrated the inevitability of the demise of capitalism, etc. At first, I would not even speak with him. I damned and cursed him. In the end, I came to understand him. And from this day on, when, as a result of dekulakization, I have absolutely no property, I renounce [my] old unnecessary and harmful view once and for all. From the present day, I am divorcing my husband. I no longer need exploitation and coercion. Let deprivations and misfortunes come. My own labor and the labor of my children are and will be my highest ideal. From this day, I take for myself and my children my maiden name, Tumanova.

With a fervent and earnest voice, I call on all those who, until now, have not recognized the power of the new society, free from oppression and force, to follow my example.

D. SIGAEVA

Curiously, Sigaeva, or whoever wrote the letter in her name, signed with her married, not maiden, name despite her pledge. Such expressions of self-criticism, even one that hit all the right notes, did not always pass muster. The official report accompanying Sigaeva's letter questioned her sincerity and interpreted her appeal as a "new trick of the decomposing old world." Nevertheless, it was felt that the document "can and should be used for antireligious propaganda." Forswearing one's "alien social origins" or religiosity could not guarantee against social ostracism or political denunciations.

Police tribunals, deportations, executions, wanton destruction, theft masquerading as property confiscation, forced resettlement, religious persecution, and many other cruelties accompanied dekulakization. On 28 August 1930 the OGPU reported that as of 20 May, when the first round of deportations ceased, nearly 68,000 kulak families consisting of 348,734 individuals had already been deported from nine major agricultural regions as second-category kulaks. Deportations resumed in the fall, and based on incomplete figures that included more regions of the country, the OGPU reported in early December that 112,828 families (550,558 individuals) had been deported as second-category. According to the historian N. A. Ivinitsky, a total of 115,231 families (559,532 individuals) were expelled from their farms in 1930, and a further 265,795 (1,243,860 individuals) in 1931, for a combined total of 381,026 (1,803,392 individuals). Of these, 133,717 families were "resettled" within their home regions, and 247,309 were deported elsewhere. He also calculates that an additional 71,236 individuals were deported in 1932.[73]

Danilov and Viola cite the following information from the OGPU regarding first-category kulaks—those deemed counterrevolutionary and liable to be sentenced to execution or to terms in a concentration camp: From January to October 1930, arrests totaled 283,717. Not all were arrested as wealthy peasants, however. Of the 142,724 first-category arrests between 15 April and 1 October, less than one-third (45,599) were arrested solely on the basis of class; the remaining two-thirds were arrested for participating in "mass disturbances," a term that was probably elastic but which gives some sense of the scale of peasant resistance. As for punishment, in 1930 alone, OGPU three-man tribunals passed judgment on 179,620 people, sentencing nearly 19,000 (10.6 percent) to death and 99,319 (55.3 percent) to correctional-labor camps. Instead of farming, the latter found themselves put to "productive" use digging gold or coal, felling trees, or building railroads.[74]

The hundreds of thousands of dekulakized victims were transported to places of exile or correctional-labor camps located in the far north, Kazakhstan, or beyond the Ural Mountains into western and eastern

Siberia. Besides "kulaks" and their "henchman" (*podkulachki*), this mass of uprooted humanity included families of priests, former landlords, representatives of the rural intelligentsia, and other "aliens" and "enemy elements." Responsibility for conveying deportees from collection points to their assigned places of exile belonged to the OGPU transport department. According to guidelines issued by the head of the department on 2 February 1930, trains consisting of forty-four freight cars, each carrying forty individuals, would be utilized as the primary means of transportation. Each train was also to have eight additional cars to carry food supplies—up to 490 kilograms per family—tools, and personal effects; one final car housed the thirteen-man guard and their officers. Depending on point of origin and destination, journeys were projected to take from five to sixteen days. OGPU deputy chairman, Yagoda, in his order on dekulakization, also dated 2 February, required that deported kulak families also take with them axes, saws, spades, and carpentry tools—an indication of what awaited them.[75] How many and which of our correspondents may have undergone this ordeal is impossible to say.

Many of the sources discuss the situation of the exiles in their new locations. A. I. Korostin from Cherdyn district, Verkhne-Kamsk region, Ural province, wrote to *Pravda* describing the negative impact the relocated were having on the local population: "The exiled deportees who live in Chuvasheeva village, Cherdyn district, walk to the neighboring villages around Chuvasheeva to congregate 'for the sake of Christ' and are introducing panic: 'You,' they say, 'don't enter the kolkhozes, or things will go badly. They placed a brand and cut letters onto the foreheads of us who are Kuban [Cossacks].' The population is afraid to enter the kolkhoz, especially the women. We demand that they be removed from Cherdyn district."[76]

Collectivization victims helped swell the population of the camps and colonies comprising the GULAG, but precise figures on the number of camp inmates at this time are lacking. There is no doubt that the sheer volume of dekulakized victims placed a great strain on the resources of the camp system, however. M. S. Ivanushkin, a clerk in the Zeisk Corrective-Labor Colony (ITK), where convict labor was used to prospect for gold, wrote that as a result of the grain procurement campaign, dekulakization, and the resettlement of people from restricted zones to remote regions of Siberia, the camp buildings were overfilled. The new arrivals were hastily thrown to work. "They are driving this raw, embittered mass and have agitated the population. Winter is coming, there are no living quarters, and the people are completely worn out." He noted that many were beginning to flee because "there are not enough guards, weapons, or ammunition." In response to the situation, he called for a review of all cases, for part of the prisoners to be released from custody, and for the sick to be excused from prospecting. A group of village officials from Krasnoiarsk reported that "they arrested people one night and had them sit in a railway car for three to four days and sent them to Siberia like lambs . . . they divided [them] up based on appearance and then assigned [them to work], asking only, 'How old are you?' "[77]

Resistance

Collectivization and dekulakization with their attendant "excesses" were nothing less than a frontal attack on the economic, social, and cultural institutions that made up peasant life. In response, the peasants fought back to defend their homes and traditions. From the far ends of the country came reports of active and passive peasant resistance to the onslaught. Sometimes opposition to collectivization took the form of pitched battles. According to a report of one such confrontation in the Volga region, "A brigade arrived to dekulakize the entire village of Cherepakha, Serdobsky district. There was a fight. Eight people from the brigade were killed; how many peasants is unknown. It was the same in the Yenota-evka, Chernyi Yar, and Vladimir districts." Still, as noted in a report from Ukraine (Document 141), disturbances were often limited to individual villages, and mass rebellions occurred relatively rarely.[78]

Women often took the lead in resisting collectivization efforts through direct action. As Lynne Viola has demonstrated, officials sought to denigrate their actions, calling them *babi bunty,* or peasant-women's disorders, and declaring them the result of the peculiarities of female psychology (hysterical, irrational) in combination with regressive class instincts (petty-bourgeois). In fact, peasant women understood collectivization as a threat to family security and an assault on their entire way of life—in no way an unreasonable analysis—and responded accordingly. That the punitive organs may have been more restrained in dealing with women no doubt recommended the practice to peasants seeking any advantage in their unequal confrontations with state power. The women, for their part, facing a mortal threat to everything that held meaning for them, were not inclined to hold back. According to a letter describing one incident, "the women [*baby*] grabbed axes and stones to drive the kommuna members from their farms."[79]

In places, resistance to collectivization escalated into a genuine peasant war. The deputy chairman of the Ukrainian GPU, K. M. Karlson, describes the situation there in a memorandum to the All-Union Communist Party Central Committee.

· 141 ·

Excerpts from a report on peasant resistance to collectivization from K. M. Karlson, deputy chairman of the Ukrainian GPU, to the Central Committee of the VKP(b), 19 March 1930. RGASPI, f. 85, op. 1s, d. 118, ll. 43–49. Typewritten copy.

Since the second half of February of the current year, in a number of regions in Ukraine—Starobelsk, Proskurov, Dnepropetrovsk, Sumy—mass peasant disturbances have begun which have provoked distortions in the fundamental line of the Party on the part of the lower Party and soviet apparatus in the conduct of collectivization and the preparatory work for the spring agricultural sowing campaign.

In a very short period of time, the mass outbreaks have spread from the regions listed above into the neighboring regions; however, the sharpest forms of peasant disturbances have taken place in direct proximity to the border, in regions on or near the frontier. Between 8–10 March, eighteen regions, with a total number of 110 districts, were already in the grip of anti-Soviet disturbances; however, the districts were not completely engulfed, only individual villages. Especially threatened were the regions: Shepetovka, where thirteen districts were enveloped by disturbances; Berdichev—ten districts; Odessa—three districts; Tulchin—fifteen districts; Kiev—five districts; Vinnitsa—four districts; Proskurov—eight districts; and Dnepropetrovsk—eight districts.

The basic mass of peasant disturbances is bound up with the categorical demands for the return of socialized sowing material and agricultural inventory.

The mass actions are accompanied by willful dismantling and plundering of socialized property.

The slogans of the active participants in the mass actions, [the slogans] that usually sustain all the mass actions, amount to a concrete protest against the implementing of measures, especially those regarding collectivization.

On these grounds, the COMMON CHARACTER OF THE SLOGANS of the peasant's disturbances is notable: "DOWN WITH COLLECTIVIZATION . . . RETURN OUR GRAIN AND INVENTORY."

Along with this, in many places the mass actions occurred under the patently counterrevolutionary slogans "DOWN WITH SOVIET POWER . . . LONG LIVE FREE TRADE."

In those cases where the mass actions intensified and carried an especially sharp character, they were accompanied, in some places, by armed opposition to the local authorities.

The participants in the mass actions, armed with shovels, pitchforks, and axes, as well as a few hunting rifles, shotguns, and sawn-off guns, do not limit themselves to appropriations of socialized property but cross over, in individual cases, to serious, active operations threatening the village soviets, dealing savagely with [Party] activists [*aktiv*] and representatives of authority, often selecting the village elders for the "leadership" of the village in the grip of the action, and passing numerous counterrevolutionary resolutions against the basic measures of the Party. [...]

The memorandum also discusses in detail the GPU measures to suppress the peasant disturbances. The losses directly attributable to the mass disturbances from the beginning of the active peasant disturbances until 18 March for thirteen Ukrainian regions were reported as 107 deaths for GPU and 147 for the opposition. The memorandum's author felt that these figures were understated, "since these do not lend themselves to an accurate accounting."[80]

But this is not all. Karlson goes on to state that "by way of liquidating hidden counterrevolutionary organizations and groups, restoring revolutionary order in the affected areas, and removing the counterrevolutionary kulak activists from the village the GPU organs of Ukraine ... HAVE ARRESTED, between 1 February and 15 March, 25,000 people, including the EXPOSURE AND LIQUIDATION of: a) counterrevolutionary organizations—36; b) counterrevolutionary kulak and terrorist groups—256. THE SUCCESS of the operative measures employed in connection with the operational blow against the counterrevolutionary activists is expressed in the following figures: SHOT—656 people, IMPRISONED IN CONCENTRATION CAMP—3,673, IN ADMINISTRATIVE EXILE-5,580 ..."[81]

In conclusion, the report discusses the exiling of kulaks that began in Ukraine on 20 February and should have been completed on 29 March. The figures as of 17 March were 17,602 families exiled, numbering 88,656 people. On 25 March a "decisive purge ... of bandit, counterrevolutionary, and kulak elements," expected to account for 15,000 people, was planned for regions along the Ukrainian borders.[82]

All this happened in the course of just a month and a half, and similar events were occurring in other regions of the country. A summary report on peasant disturbances in the Central Black-Earth region contained the following accounts.

· 142 ·

Excerpts from a summary report on peasant disturbances in the Central Black-Earth region, 1930. RGAE, f. 7486s, op. 37, d. 102, l. 198. Typewritten copy.

In the village of Pravo-Khava, on 31 March, a crowd of several hundred persons seized the seed fund and destroyed the kolkhoz administration. On

1 April, a forty-five-man strong GPU brigade arrived. Attempts to persuade [the crowd to disperse] failed. The crowd grew as people from neighboring villages kept arriving. After firing warning shots, a detached encircling force of ten Red Army soldiers let off a volley. Five people were killed and three wounded. After this the crowd dispersed. The bodies were transported to the hospital in the village of Rozhdestveno-Khava, and after a while they were buried in the cemetery.

[. . .] In the village of Krivopole, Ranenburg district, all the peasants went on strike and advanced the slogan "Down with the kolkhoz, and long live Soviet power."

In the village of Solntsevo, they closed the church and filled it with the communal oats. Under the leadership of the kulaks and semi-kulaks, all the peasants of the village went on strike, and it looked like a counterrevolution. Now in Ranenburg district it is forbidden to go from one village to another, and guards stand everywhere.

Trade has now stopped. In Ranenburg, kulaks and kulak-lackeys stand with sawed-off shotguns and are killing communal officials. The counter-revolutionary mood in these villages must be ended.

The following letter to the Central Committee, allegedly written by a Party member, discusses the situation in Siberia.

· 143 ·

An anonymous letter to the Central Committee of the VKP(b) from a self-described Party member describing anti-collectivization violence in Siberia, 1930. RGAE, f. 7486s, op. 37, d. 102, l. 215. Typewritten copy.

Siberia is in revolt. A situation has been created the likes of which has never been seen. There are uprisings in Shitka, Taishet,, and Taseevo districts. An uprising is expected at any moment in Taseevo district, where sixteen rebel organizers were sent. On either 30 June or 1, 2, 3, or 4 July, an uprising began one kilometer from the city of Kansk, in the village of Kansko-Perevoz. [Against] the uprising, all available communists were mobilized. The local authorities tried to maintain total secrecy, by the way, [but] the people know everything, and both the workers and the peasants are in a state of agitation. The only hope [rests] with Party and Komsomol members, but there have been cases where Party and Komsomol members have gone over to the side of the rebels. The workers and clerical staffs are also dissatisfied. Flour costs twenty-five rubles—there's never been such a price. Wages for a seasonal worker are lower than [they were] last year. And life has become much more expensive. They won't give food to seasonal workers. The workers are also talking about a general strike, except that they are waiting for an uprising or a war in order to get . . . and go drive out all the communists and then go to defend . . . power.

The pressure on the peasants now is the same as under the landlords—nothing is voluntary, everything is coerced. Is this really imaginable? From one village in Minusinsk region, they expelled about two hundred "kulak" families, who are not given any work and are not fed anything. Every day in every single district ten children and adults die. And of all those who have been exiled, a very, very small percentage are kulaks, since the real kulaks were already dekulakized in 1928–1929. Right now, they are contracting for livestock where they are taking the last of the peasants' cows. Representatives of different organizations are sent out just for show; they never read the newspapers, but they arrive somewhere and tell lie after lie that only antagonize the peasants. Right now, under the sovkhoz, they are dividing up the land, but the peasants are being sent off to unworkable tracts. Everyone says that Stalin is to blame for everything, that Stalin should be killed . . .

On top of that, the central authorities in Kansk are aware of the situation but aren't doing anything. A governmental commission and the old leaders of the partisan brigades who could explain [things] to the peasants must be sent out immediately. The Yakovenko brothers are leading the rebellious partisan brigades.

I beg you, take my letter very seriously. Otherwise, every day of delay will cost dearly.

Written by a Party member.

Accounts of violence and direct peasant resistance came in from across the country. A report on peasant uprisings in the North Caucasus stated: "In the Gostagaevka stanitsa, Anapa district, the peasants attacked the GPU and shouted, 'Take the rifles.' For ten days soldiers stood guard. Anastasievka stanitsa also rose up. The peasants slaughtered livestock. They went crazy." From Kherson region in Ukraine it was reported: "The militia is shooting peasants. The peasants are abandoning the fields, and an unprecedented harvest is being lost as a result of this." Rumors circulated among the peasants that "the Red Army is on the side of the discontented." For their part, those involved in suppressing the uprisings reported: "On the Chinese-Eastern Railroad, they greeted us like White bandits . . . the peasants are helping the rebels and not the communists . . . they haven't fed us, [so] we have to scavenge."[83]

Blame

On 2 March, *Pravda* published Stalin's article "Dizzy with Success," sig-
naling a halt to the first stage of collectivization and dekulakization. The
article was a direct response to the information flooding Moscow on the
scope and intensity of peasant resistance to collectivization and the violence and excesses by which collectivization was being enacted. In the article, while claiming great successes for collective farm construction, Stalin
attributed "distortions" in Party policy to overzealous elements within the
Party that had lost "all sense of proportion and the capacity to understand realities." Stalin specifically criticized forcible collectivization—
especially in grain-consuming and remote regions; the premature
organization of highly socialized kommunas in place of artels; and antireligious activity, which he ridiculed as pseudo-revolutionary.[84] The article's
purpose was plain enough: to place blame for the chaos unleashed by
collectivization squarely on the shoulders of lower-level officials responsible for policy implementation while shielding central policy makers
from any criticism whatsoever and thereby upholding the essential correctness of the "Party line" on agriculture.

There was a large grain of truth in Stalin's charges. Local officials and
their subordinates had indeed exceeded their formal instructions and
had employed measures not sanctioned or explicitly prohibited by central authorities. Nevertheless, in light of the previous two years' experience with forcible grain collections, the measures that local officials
employed could have surprised no one, especially given the time pressure under which they were forced to work.

For the implementers of collectivization, "Dizzy with Success" made
for demoralizing reading. Subsequent denunciations, investigations, and
punishments of those accused of having committed "excesses" kindled
the resentment of lower-level cadres still further. They plainly understood that all this amounted to scapegoating. "They always blame the
little guys," one of them wrote, "[but] they said [to] apply pressure
where you can." Other activists pointedly disagreed with the policy reversal. P. I. Chumachenko, from Berezovka, Odessa region, said, "The
devil take Comrade Stalin with his foolish views about how they took
all the productive livestock and draft power." Another local Party secretary expressed a similar sentiment, "Comrade Stalin is wrong. It's too
late to retreat."[85]

After the publication of Stalin's article, many activists and soviet officials complained that it had become impossible to work. One propagandist in a Mordovinian village complained that he could no longer
conduct antireligious propaganda because "the women have stopped

coming to the literacy school [*likpunkt*]."[86] Their authority undermined, collective farm activists were left helpless against those out to disband the kolkhozes. One activist-correspondent described the scene at one meeting of irate peasants.

· 144 ·

Letter to *Pravda* from I. Sakharov, a collective farm activist, describing a meeting of peasants seeking to disband a collective farm, 1930. RGAE, f. 7486s, op. 37, d. 102, ll. 197–98. Typewritten copy.

The former church, now a school, was packed. The women sat on the left—they were the obvious and overwhelming majority—and the men on the right.

"We are r-u-u-u-i-n-e-d! We are dying of hunger!" The face of the shouting woman was fat and flushed.

"They made off with my cart!"

"My reins are missing!"

"They busted the wheel of my wagon!"

"I'm cleaned out. Cleaned out!"

The [Party] activists remained solemnly quiet.

They shouted and cursed for a long time. The women let loose their feelings on us. The men stubbornly kept quiet, voting neither "for" nor "against" the kolkhoz. Mentally, the soured activists crawled under the table. The bookkeeper, Osipov, a bitter drunk, started a rumor about the end of the kolkhozes throughout the entire USSR.

By midnight, many had lost their voice and proposed that, instead of sowing, the land and the crops be divided by mouths to feed.

The honest labor of the workers had not been recorded or paid for. Their labor was gotten for a thank-you—and there wasn't even any thank-you. And the ones who did no work and didn't want to work got the crops. They only came [to the meeting] to systematically pull off this partition, and they tried to carry it off as best they could—quickly—so as not to damage the crops in the remaining field. That's how the kulaks operate.

IGOR SAKHAROV, postal district, Olga village, Vladivostok region.

Peasants who took "Dizzy with Success" and its emphasis on voluntary collectivization as a sanction to abandon the kolkhoz quickly encountered an array of legal and less formal obstacles and disincentives to withdrawal, however. The revised Model Statute on collective farms, published the same day as Stalin's article, stipulated that departing peasants forfeit a good deal of the livestock and property they had brought to the collective and that their new allotments be located off the kolkhoz, invariably on poorer land than they had previously worked. They were denied seed, hayfields, and pasture for livestock.[87]

If the law did not discourage peasants from leaving, officials reverted to other means. In a letter to *Pravda* under the title "The Left Opportunists in Klimovshchin, Mogilev region, Have Gone Completely Stupid," a group of peasants wrote: "Having read the new regulations on agricultural artels and Comrade Stalin's article, they [local officials] began to convene the village assemblies and to tell the peasants that Stalin didn't write this article, but some kulak, and even [if it was] Comrade Stalin, then he is not the leader [*vozhd*] of the proletariat but its destroyer [*podryvshchik*]. Whoever reads the newspaper and correctly explains it to the peasants, then these Sergeant Prishibeevs call them kulak-lackeys and threaten them with Solovki." In Kuban region, those leaving the kolkhoz were threatened with court action. Another letter described how a member of one kolkhoz administration threatened to "baptize" the women seeking to leave his farm with a log while he screamed at them, "Your mother, I'll give you the crops! Here's your allotment, mother of God! Clear out, [or] I'll shoot all of you!"[88]

The lengths to which kolkhoz administrations were prepared to go to prevent the withdrawal of farm inventory, regardless of "Dizzy with Success," is described in the following scene from the summer of 1930.

· 1 4 5 ·

An account sent to *Pravda* of one kolkhoz administration's efforts to recover the inventory of peasants who had left the collective farm, summer 1930. RGAE, f. 7486s, op. 37, d. 102, l. 219. Typewritten copy.

────────────

On 13 June of this year, a fourteen-man detachment from the kommuna of Neftianka stanitsa under the command of the head of transport for the kommuna, Comrade Pavlenkov, appeared at mechanical installation No. 2. They were all armed with rifles. Several comrades asked the detachment who they were and what they wanted, to which Comrade Pavlenkov answered that some members of the kommuna had fled, stealing livestock and equipment, and that they had headed for mechanical installations Nos. 2 and 3 to [work at] hauling materials. "We have come to search their carts." On 14 May, when the carts returned from [word missing], where they had hauled railroad ties, Comrade Pavlenkov demanded the return of the stolen livestock and equipment, to which the carters answered that they had left the kommuna and had taken the livestock and equipment that were their property. "If we've done something wrong, make us answerable in the people's court. Since they're our property, we won't give back the livestock and equipment."

Pavlenkov and his detachment seized a pair of horses and began to take a bridle from a boy, but the boy wouldn't give up the bridle. The assembled workers were outraged by the behavior of the detachment and especially at Comrade Pavlenkov. They observed that such actions on the part of the detachment were out of line. Then Pavlenkov shouted: "Disperse, I'll shoot . . ."

MACHUR

This sort of highhandedness discredited all authority in peasant eyes, as seen in the following excerpt from a letter to *Krestianskaia gazeta*.

· 146 ·

Excerpt from a letter to *Krestianskaia gazeta* written by the middle peasant Belotserkovsky on the obstacles encountered when attempting to leave a collective farm, 1930. RGAE, f. 7486s, op. 37, d. 102, l. 111. Typewritten copy.

Surrounded by lies.

[. . .] Everything that the Central Committee writes is unreliable. There was a Central Committee instruction to return horses and equipment to those not wishing to be in the collective. This didn't happen at all. On the contrary, Comrade Poliakov comes from the Kanevsky GPU toting a revolver and [says]: "*Now go and take back [your] appeal to leave the kolkhoz, and if you don't take [it back], we'll drive you in like cats.*"

We middle peasants all see that in soviet construction they steal and live off the labor of others. They took absolutely everything away from the peasant and say, "We've socialized."

I, losing the last drop of my blood, have gone about barefoot and naked, scrimped and saved for a horse—they took her.

They've turned everyone into a batrak. We walked four kilometers to the kitchen garden to water the cabbages—they don't give out horses. Let it be damned! And we left it to the whims [of fate]. [Stalin] wrote about dizziness. No doubt our leaders are dizzy. Now we see the lies, lies all around.

BELOTSERKOVSKY, Chelbasskaia stanitsa, Kanevsky district, Kuban region, North Caucasus territory.

From Ostrogozhsky region, Central Black-Earth region, M. Kvasov wrote of local officials' attempts to intimidate and discourage the peasants who took "Dizzy with Success" too literally as a policy reversal.

· 147 ·

Letter to *Krestianskaia gazeta* from M. Kvasov describing the violence suffered by peasants in the central Black-Earth region who attempted to act on "Dizzy with Success" and leave the kolkhoz, 1930. RGAE, f. 7486s, op. 37, d. 102, ll. 77–78. Typewritten copy.

. . . On 27 March there was a village assembly at which it was resolved to distribute seed. The next day the peasants went to get the seed, but drunken [Party] activists met them with bullets, a ramrod, and clubs.

. . . For two days it was impossible to walk or ride down the street. People with clubs rode up and down, seriously beating [people] and dragging [them] off to the district militia as well, where they were subjected to tortures. As a result, three were killed and many more were wounded . . .

. . . If they saw someone with a newspaper, they beat them harder and condemned [them]: "So, you're reading Stalin's article."

When the peasants showed the Party cell secretary, Petrov, Stalin's article, they declared, "You are concealing the Party line." But Petrov answered coldly: "You, comrades, are non-Party, and this does not concern you. Don't believe everything in the newspapers."

In a number of cases, when peasants who had been forced to join the kolkhozes justified their withdrawal by claiming that Stalin's article had "abolished force," they were informed "that Comrade Stalin himself is a right deviationist." One local leader maintained that "the article was written not by Stalin but by a kulak." According to another report, "Peasants were beaten for reading Stalin's article" and told "we don't acknowledge the central newspapers." One letter said that "on 4 April, in Starye Brody village, Chernushka district, Sarapul region, they arrested the peasant Kotov for reading *Krestianskaia gazeta* to his fellow villagers," and they brought him to trial. The court sentenced him to four months' forced labor. "The case consisted in the fact," the letter continued, "that Kotov on his own had led a discussion about collectivization with seven peasants. But Burmatov, the plenipotentiary, was very unhappy about this, especially since he had been told that he had prevented the gathering. In the opinion of the people's court, Kotov's talk brought about a withdrawal from the kolkhoz."[89]

Similarly, those who felt that the publication of "Dizzy with Success" would provide the opportunity to reverse their dekulakization found the task difficult. The following complaint to *Krestianskaia gazeta* arrived from Fedorovka village, Prigorodnovo district, Viazma region, Western region.

· 148 ·

Excerpt from a letter to *Krestianskaia gazeta* from F. F. Filatov on his inability to obtain his release from jail and to reestablish his right to vote, 1930. RGAE, f. 7486s, op. 37, d. 102, l. 222. Typewritten copy.

Where is the law, where is the truth?

[. . .] I, Filatov, Fedor Filatievich, had a landholding: one horse, one cow, one house, and three hectares of land. There are four people in my family: myself, sixty-two years old; my wife, an invalid; my son, underage; and a fourteen-year-old daughter, also an invalid. Because of her injury, my wife had a license to sell baskets. As a private trader, she was assessed an individual tax, and they took away our right to vote. After three days, they fined her fifty-four rubles for failing to pay the individual tax and took her to jail for one year and five months. Since I have a poor holding and cannot pay the aforementioned sum, I have been in detention since 13 October. After Stalin's article "Dizzy with Success," my holding was exempted from the individual tax, and my voting rights were restored. They told me personally that I retained my voting rights and was not deprived [of them]. Yet I find myself in confinement since 13 October 1929. Where is the law?

Even after the appearance of "Dizzy with Success," middle peasants continued to suffer dekulakization and arrest. For their part, local officials complained that kulaks and speculators were having their rights restored on appeal, and called on their superiors not to make this "second blunder."[90]

This following letter illustrates the bureaucratic process and the conflicting interests involved in getting one's civil rights and property restored. It also shows that minor officials could still wield a great deal of authority despite Stalin's article.

· 149 ·

Account of a local official's efforts to prevent the rescinding of a peasant family's dekulakization in Mordovia, 1930. RGAE, f. 7486s, op. 37, d. 102, ll. 44–45. Typewritten copy.

In the village of Tazneev, Kozlovka district, [there is an] eighty-year-old peasant, Onufrii Glukhov. His insured property is valued at 942 rubles. He was never assessed the individual tax. He never engaged in trade. His sons, Andrei and Elizar, were shepherds, and both served in the Red Army. In his youth, the old man was a batrak. Harmonious family labor helped them advance to the middle-peasant level. As a result, [they were] fully dekulakized and reduced to poverty. The chief perpetrator of the dekulakization was the village soviet chairman, Iliushkin.

The family has submitted requests to the village soviet on rescinding their dekulakization from [the following]: 1) the Kozlovka district executive committee; 2) the senior inspector of Mordova province; 3) the prosecutor of Mordova province; and 4) the prosecutor of the republic, Andreev. For ignoring the requests to rescind the dekulakization, the prosecutor of Mordova province and the republican prosecutor are threatening to hold the [soviet] chairman responsible. Despite all this, the chairman corrected "the exaggeration" reluctantly. He returned the property and let them back into the house, but said: "It doesn't matter. You won't hold on to it. You'll just live [here] for now."

After returning the property, Iliushkin came and declared: "The garden and the farmstead are not yours." He took the horse. He presented no papers and left no inventory list. When we lodged a protest and pointed to the papers from the center, he said: "I don't recognize anyone and I'll do what I want." We said that our dekulakization was rescinded, but he said: "I'm starting all over again." When will our tortures end, and who will stop Iliushkin? All the same, we still believe in justice.

Despite such cases, after "Dizzy with Success" kolkhoz enthusiasts found themselves on the defensive. During the campaign, two communists, Abol and Trambovetsky, who worked in the Kamenets-Podolsky region of Ukraine, had threatened peasants with exile. Their actions provoked the slaughter and sale of livestock and a mass uprising of the villagers against the kolkhoz. In the subsequent investigation, they referred to their instructions from the district executive committee: "Either 100 [percent] collectivization or [your] Party card on the table."[91]

The following is a vivid and even poignant appeal by another kolkhoz organizer, entitled "You can't measure everyone by the same yardstick."

· 150 ·

Letter from Anton Pavlovich Dervoedov to his brother explaining the
reasons for the excesses he committed in the course of collectivization. His
brother forwarded the letter to *Pravda*, 1930. RGAE, f. 7486s, op. 37, d. 102, ll.
204–203. Typewritten copy.

From Dervoedov, a VKP(b) member since 1927, [Party card] no.
0635590. I am now on active military duty in the GPU forces. I request
[that you] consider my brother's letter. They sentenced him to a year of
forced labor [. . .]

"Dear brother! You ask, How could I have done this excess? It's not out
of any mercenary motives. I carried out the directive of the Party. The
VKP(b) regional committee was given instructions to socialize the livestock
in the kolkhoz, down to the last chicken and kitchen garden seeds in forty-
eight hours. In February of this year, I cursed a certain woman in the
Maslianka village soviet of Tavda district [Ural province]—she refused to
turn over [her] seed. Her husband got ten years in prison as a bandit, and
her two sons [were sentenced] for raiding and murder. The members of the
grain procurement commission also got ten years. I compromised someone,
but not my poor-peasant brother or middle-peasant ally. In the Nosorevka
village soviet, the chairman of the district executive committee himself,
Comrade Martynov, showed me the ropes—how one has to go to the granaries and confine the peasants who refuse to turn over seed in the sauna. The
entire village of Tandashkovo, in the Nosorevka village soviet of Tavda district, can confirm this. In this village soviet I had already put the peasants
[in the sauna], but in a different village, Tyomnaia Rechka, [I put them] in a
hut. At first it was cold [inside]—[no one] had lived in it for two weeks—
but there was a small stove. I divided up the hut, since there were ten to
twelve people who stubbornly refused to turn over the seed, and I sent for
them all the time. Whoever refused [to hand over seed] would be sent back
in there, and so I held six to eight men like this for about a whole day. Of
course, I treated them crudely, cursed at them, but there was no hitting or
anything else. On 24 March, they removed me from the ranks of the
VKP(b). At the regional Party conference, the regional committee realized
its mistakes in the kolkhoz movement. The mistakes I permitted were due to
confusion, even though I worked as the secretary of the Komsomol district
committee, but you see I was a promotee, [I used to be] a worker-batrak.
My natural family, the VKP(b), destroyed me, and for what, for my first
mistake, [even though] I have been a Komsomol member since 1925 and in
the VKP(b) from 1928. I'm twenty-one years old, and [now] I work as a
stevedore for the state steamship line on the Tura River in Tiumen. Everyone is a cheat, all are convicts. You wouldn't want to and couldn't do what
they do, but you'd do it if fate had forced you to come here in such circumstances. In this situation, brother, I also will become a conman and criminal.

I need to put a bullet in my chest. You know that I'm tough, and it's not out of faintheartedness, but it's not possible to go on living. I was a batrak so probably I have to labor. But now there aren't any kulaks, and there is nowhere to labor; now you need a scam. When I went to Comrade Markus, the secretary of the VKP(b) provincial committee, I wanted to talk to him, and he told me, "You wanted to provoke the peasants into rebellion with your methods. Now you are an enemy, and you have no place with us. Get lost and don't talk to me anymore." I went to the secretary of the Komsomol provincial committee, Comrade Pekarevich, and started to say that I had committed a crime, that [I] am admitting it myself, but I should remain in the Komsomol. Then Comrade Pekarevich told me, "You are a Trotskyist. You need to follow Trotsky [i.e., leave the country] and not admit your mistakes." Pekarevich is the son of a former gendarme or sergeant-major and now he already works as the secretary of the provincial committee of the Komsomol. This came out recently. Look who leads us and look who they drive to ruin. If you speak to anyone, they write "right deviation." Let them investigate my case, and then they will see if I am guilty or not. I am not the guilty party but the victim. [...]"

DERVOEDOV, ANTON PAVLOVICH, city of Tiumen.

Caught in the switches of a dramatic policy reversal, those who ended up in prison or on trial for their actions expressed bewilderment and often attributed their plight to kulak intriguers. Convinced that they had acted correctly, many refused to show even a hint of repentance. They boldly pointed accusing fingers at the highest Party leadership. In a letter sent to *Pravda* and intended for Stalin, six workers—members and candidates of the Communist Party—from the agricultural machine factory Selmashstroi, wrote, "You accused many prominent Party members of right deviation—Comrade Bukharin and others. At the present time you yourself have condescended to right deviation." A. P. Podarsky, a member of the "Bolshevik" kommuna in Chashi district, Kurgan region, wrote: "No one foresaw the disorganization. Everyone was sure that there would be difficulties the first year but we'd get through them. Now, all the servants [*cheliad*] of counterrevolution have come back to life. They put on trial those who created the kommunas. It is painful to watch. [It is] spring, the sun is shining, but my spirit is dark."[92]

The disruption of agricultural production in the first half of 1930 threatened many regions of the country with famine. Numerous reports spoke of crops standing unharvested in the fields. A group of youths from the Lower Volga Territory in the summer of 1930 wrote that both the kolkhozniks and individual farmers were eating gophers and that the children got up early and ran to the swamp to pull up grass. Famine was reported in Kazakhstan, in Leningrad, and in the Ivanovo regions. In Kalmyk villages there was no produce, and a difficult situation was shaping up in the famous North Caucasian sovkhoz "Gigant." According to other reports, hungry peasants were fleeing to the cities to find work in production.[93]

That all this was occurring at a time when newspapers and the leadership of the country were trumpeting the successes of socialism and the crisis of the capitalist economic system did not escape notice. The following letter to *Pravda* is a sharp commentary on the official version of the international economic situation.

· 151 ·

Excerpt from a letter to *Pravda* from the peasant Stepanishchev comparing the crisis of world capitalism and the situation in the Soviet Union, after July 1930. RGAE, f. 7486s, op. 37, d. 102, ll. 172–73. Typewritten copy.

On behalf of the peasants, but not in the peasant way.

[. . .] Having read Comrade Stalin's report to the Sixteenth Party Congress, I feel I must say a few words about this report. Comrade Stalin says that the current economic crisis is a crisis of overproduction in all the capitalist countries. That is, the crisis is characterized not, as with us, by a shortage of a great number of different commodities and grain but by the fact that the capitalists sink their ships with [loads of] corn, feed rye to their pigs, throw coffee into the sea, and have everything in abundance, and of course [everything] is very cheap. That's what I call a crisis! The peasants say that if we should have such a crisis, then all the people would lie down and eat bread. They would spit at the ceiling while lying on their backsides. But instead you work day and night and die of hunger. You walk around in rags like a dog. I wonder why our government doesn't buy surplus coffee from the capitalists, because it would be a lot more profitable [for them] to sell it, even at a very low price, than to toss it as an offering to the sea. Russia certainly needs real coffee. If we have reached and surpassed all the prewar [levels], then why is there a shortage of all sorts of goods for 150 million people when up to the war in 1914 there was enough for 165 million people?

If we are smarter than the capitalist economists, then why are they invited to Russia to our construction site[s] and [why does] their work bring about improvements? It turns out that the clever communists are learning from the imbecile capitalists, and the imbeciles do things better then the clever ones. Now, it's clear that the stupid one always learns from the intelligent one—what's going on here?

It's a shame that I am not at the congress. I could give a three- to four-hour speech if the congress was open and non-Party people had the right to speak. I know that no one from the Party will say what I would. Some don't say it because they're blind and others because they are looking out for themselves, looking for warm places.

What do you call it when the communists come to us and, instead of taking from the kulaks, pull the shoes right off the legs of the laboring middle peasant? What is it called when they took dresses, blankets, and pillow cases from peasant women, even the poor ones? In response, you will no doubt say that these are simply mistakes, but no one here is saying that.

With a comradely greeting, STEPANISHCHEV. Bezvodny settlement.

The following evaluation of the economic situation entitled "You are all lying to the peasants" is startling in its grasp of the general direction that the still-unfolding situation was taking. It quite accurately describes the prevailing chaos in the industrial and agricultural sectors that resulted from the employment of feverish production methods. In traditional fashion, the letter was submitted as a resolution of the entire village.

· 152 ·

Letter to *Pravda* from the peasants of Guliaevo, Western region, on the effects of the policies of crash industrialization and wholesale collectivization, 1930. RGAE, f. 7486s, op. 37, d. 102, ll. 230–231. Typewritten copy.

We, the peasants of the Western region, Gzhatsk district, Guliaevo village soviet, decided to convene an assembly. There were 150 individual citizens and fifteen kolkhozniks (in six villages). They elected a presidium made up of Nikolai Stepanov, the kolkhoz chairman; Mikhail Rysenkov, a member; [and] Ivan Nikolaev as secretary. It was decided to discuss the current situation of the Soviet republic. After a long argument and discussion, we came to be convinced that our country is heading to ruin and poverty. Editors of *Pravda,* don't get angry that we give this answer. We will prove to you that our [answer] is the truth.[94] Here it is, the second year of the five-year plan, and there is nothing good to be seen. You all proclaim that this has improved, and that the other has increased, that everything is going neatly and smoothly, that the factories and plants are working at a shock-work [pace], that the kolkhozes and sovkhozes have expanded the area sown, that all that remains is to organize wholesale collectivization and destroy the independent middle peasants, who, in your opinion, [practice] a primitive [form] of agriculture. And then there will be heaven in the land of Soviets. No, you are greatly mistaken. We peasants see that you all lie, especially about agriculture. The first question is about industry and construction. You write that in some city they are putting up a factory. That by next year this factory will turn out 100,000 tractors. In another city a giant factory will turn out all the machinery required for agriculture—that's an awful lot, an infinite amount, etc. But if one looks now at those factories and plants that you are predicting, what will you see? In one, the factory foundation is being laid, and in another, they are putting up the walls, and as you will see, the factory is half built, and there aren't enough materials to complete it in a short time. And you are already promising that the factory will turn out 200,000–300,000 machines. True, there are a few genuine achievements, but only a few. Some of the achievements are quite unnecessary at the present time. We peasants see the whole picture well. Industry has advanced, but no one will buy such a machine because of its poor quality. Compared to the foreign machines, all our machines are good for nothing, and other products are of lesser quality next to the foreign ones. Now let's take the second question— farming (agriculture) and what we peasants see and hear from our government. It writes: "The 'Gigant' kolkhoz in the Caucasus exceeded the sowing plan, and in the Crimea, where the sowing area [increased] by two thousand hectares, [there was] the same expansion, etc., etc. In all branches of agriculture the harvest has attained the proper heights, machines have lightened the workers' labor, and the seredniak peasant saw the great advantage the kolkhozniks have in individual cultivation and is entering the kolkhoz willingly." This is also false. You are quite mistaken about this, as well, and at times lie. We peasants see how sovkhoz after sovkhoz is not expanding the sown area, and they did not raise the harvest; instead we see that the immense crisis of agriculture production is growing not by the day but by the hour. This has caused the ruin of the large peasant holding, but the poor, small-scale, middle-level peasants can't even feed themselves, never mind hand over produce to the state. The kolkhozes and sovkhozes produce a lot, but only on paper; in reality, in fact, there is nothing there, and the crisis grows and will grow; even though you have forced all the independent middle-peasant holdings, using all possible pressures, to enter the kolkhozes, the crisis will come anyway. Until the rural population goes voluntarily, willingly, hopefully into the kolkhoz, the authorities won't be able to straighten things out. You can't escape the crisis of rural production by breaking up large peasant holdings before the kolkhozes and sovkhozes can actually replace the production of these holdings; only then can they be liquidated. Until then, they should be left intact and not partitioned with their property; they should remain intact. At the present time, the seredniaks [with] holdings that survived dekulakization are not improving their holding and are not developing them. No amount of agitation can force [them] to expand their holdings, only to reduce [them], and they are reducing [them]. They are selling cows and slaughtering [them], the same with horses, and they are not breeding them. It's the same with the expansion of sown area; the local government is expanding [it] more on paper than in fact. We repeat again that the rural population does not trust any decree of yours that says the seredniak will not be dekulakized. We say this in the west, but the rural population is of the same mind in the east, the south, and the north since the country is dying. Now it only remains for the authorities immediately to issue some decrees to the effect that the seredniak will not be dekulakized according to the limits—that for two family members should be allowed one cow, one horse, one piglet, and two sheep, four colonies of bees, four hectares of land, or else drive everyone into the kolkhoz immediately and see what happens there. But you can shut down some capital construction. We'll manage without it. And then, so long!

> Signatures of the presidium: NIKOLAI STEPANOV, chairman;
> MIKHAIL RYSENKOV, member; I. NIKOLAEV, secretary.

For all the practical wisdom and common sense expressed in this letter, the measured approach it advocates for industrial and agricultural development already belonged to another era. Caution, once thrown to the winds, proved irretrievable. During the next decade, collectivization would proceed until it encompassed almost the entire peasantry, and industrialization plans, though toned down, remained highly ambitious, wringing ever greater sacrifices from the common people, workers and peasants alike. The resulting social turmoil, not to mention inevitable mistakes, false starts, and failures arising from the furious transformations, helped create an atmosphere of recrimination and dread that soon engulfed the entire country and added new categories of victims to those already classified as "kulaks."

Like the peasants from Guliaevo in the Smolensk countryside, who had gathered in time-honored fashion to discuss pressing issues, Nikolai Bukharin, the foremost advocate for economic equilibrium and balanced development, had also sensibly warned against trying to build today's factories with tomorrow's bricks. This position, rooted as it was in Soviet reality, proved out of step with the times and could not withstand the building pressure for an immediate breakthrough to the socialist future. Reality meant limits, but forcible collectivization and crash industrialization were revolutionary assaults on reality in an effort to obliterate limits. Of all the limitations that reality imposed on Soviet development, peasant Russia, with its muzhiks and babas, its strip farming and dirty huts, its folk traditions and superstitions, its village priests and healers, was the most burdensome. Now, in the new reality of the kolkhoz and the five-year plan, the village would be firmly bound to the city, and socialist construction could begin in earnest. This truly was a profound revolution, a leap from an old to a new qualitative state, as proclaimed in the official Stalin-era history of the Party. Tragically, for many who placed their hopes in the new configuration, the promised joyous life failed to materialize.

Conclusion

Collectivization and dekulakization destroyed the Russian village and, with it, what remained of the peasants' traditional life. Together with rapid industrialization and the implementation of economic planning, the two campaigns helped bring about the reconstitution of Soviet society along highly centralized and authoritarian lines. The ruling elite, with Stalin and his supporters firmly at the helm, now employed every means at their disposal to mobilize all necessary resources—human and otherwise—in their mission to build socialism. Proceeding at breakneck speed, this cruel and ruthless process reinforced the hardness, if not callousness, of officials at all levels of the state, Party, and economic hierarchies. Officials placed never-ending demands on their subordinates and on the Soviet people to produce more, faster, while making do with less in the way of compensation and material comforts. Large numbers of ordinary citizens responded to these challenges resolutely, even enthusiastically; they identified their personal interests and the national well-being with the Soviet Union's economic advance and were willing to sacrifice for it. But for many millions the forced march from the realm of necessity to that of freedom proved to be oppressive and deadly. Individuals now faced increased social pressures, together with harsh forms of discipline and punishment, to meet ever-higher quotas at the workplace. In the rush to expand the industrial base, accidents (often fatal), bottlenecks, and production errors commonly occurred and were criminalized, exposing more people to arrest and to terms in the camps.[1]

Apart from the direct violence perpetrated against the rural population in the course of collectivization and dekulakization, the most tragic result was the famine of 1932–1933, which engulfed large portions of Ukraine, the North Caucasus, and the Lower and Middle Volga, and Central Black-Earth Region, all of them traditionally rich grain-producing areas. The cause of the catastrophe rests in state officials' aggressive procurement of grain despite poor harvests in 1931 and 1932, leaving peasants with little or no reserves and no markets to turn to. Collection officials themselves were responding to the center's unrelenting demand for grain arising from industrialization (which required food for an expanding workforce and grain exports to obtain hard currency) and the needs of collective farms and the military. The growing labor-camp population also needed to be fed, if only minimally. By mid-1932, state grain reserves, which had never been sufficient, dropped to dangerous levels in Ukraine, the Lower Volga, and the North Caucasus. Reductions (though not cessation) of grain exports and the reopening of private markets could not avert the catastrophe. Grain reserves from other areas of the country were not released.[2]

Dekulakization also occurred in all the Central Asian republics and, as elsewhere, led to the mass slaughter of livestock and the disruption of agricultural production, which contributed to famine in 1933. In Kazakhstan, however, the forcible transformation of millions of nomadic and semi-nomadic herders into collective grain farmers produced a demographic and agrarian catastrophe that, in terms of percentage of population affected, surpassed any of the other calamities of the time. Nearly six million head of cattle were lost and twenty million of sheep. Famine and flight reduced the Kazakh population by approximately one and a half million.[3]

Precisely determining famine-related deaths is a notoriously difficult task. Quantitative information is often incomplete and open to interpretation. The malignant effects of hunger are not always immediate and typically leave victims susceptible to diseases incurred later—perhaps years or decades later—that may count as the actual cause of death. Earlier deprivations may make it impossible to withstand later rigors of exile or life in the camps. Before the opening of Soviet archives, reliable data were hard to come by, and scholars employed different methods of analysis and arrived at estimates of losses that diverged by many millions. Access to Soviet statistical information since the early 1990s, particularly the previously suppressed 1937 census, has been a boon to demographers, who, as a result, have been able to provide a statistically based range of losses. Still, a universally accepted figure expressing famine-related deaths has remained elusive. Low-end estimates place famine-related deaths in the affected Russian and Ukrainian regions at four million to five million, to which must be added the numbers from Kazakhstan. Higher-end estimates advance a range of just over seven million to just over eight million in the USSR as a whole for 1933 alone. All researchers agree that Ukraine suffered the greatest losses by far. Losses in the Kazakh republic may have amounted to more than 40 percent of its total population.[4]

Such cataclysmic events in the countryside could not but have a negative impact on urban and industrial areas. Food shortages arising from the 1928 grain crisis had necessitated the introduction of a ration-card system for bread in Soviet cities in late 1928 and early 1929. Other food items, like meat, sugar, tea, and eggs, were soon included. The ration system, which remained in place until 1935, was based on social class and favored factory workers in important industrial centers (the "special list"). The remainder of the workforce received partial rations depending on which of three other categories they belonged to. Individuals who had been deprived of civil rights received no rations. By 1931, a special "closed distribution" system for high government and Party officials had also come into being, entitling their recipients not only to higher rations than special-list workers received but access to a greater assortment of foodstuffs. Thus, the campaign to socialize food production created greater inequalities in food consumption and provided the rationale for a new system of social stratification.[5]

Documents

For many people, the deteriorating urban situation called to mind the hardships endured during the revolutionary and war-communism periods. One person who remembered earlier periods of hunger wrote: "Famine in Saratov. It's like 1921–1922 when they piled up the corpses of typhus and famine victims in barracks by the wagonload." Another wrote, "The year [1930] resembles the end of 1916 and the beginning of 1917," and a third declared, "1920 is coming back." Reports of food shortages came in from cities across the country.[6] Workers wrote of speculation and high prices at the markets. They also complained about the lack of truthful information in the newspapers, and warned of the consequences should the situation continue to deteriorate as in the following letter.

· 153 ·

Letter to *Pravda* from Ya. Buslaev, a worker in Mordova province, on the hardships imposed by shortages and rising costs, 1930. RGAE, f. 7486s, op. 37, d. 102, l. 239. Typewritten copy.

The department of trade and the consumer union have forgotten the Vindrei worker settlement, Zubovo-Poliana district, Mordova province. They've forgotten the Vindrei sawing mill in particular.

For their part, the workers have expended every effort to fulfill the industrial-financial plan, and they have fulfilled it by more than 100 percent. But how are they being supplied? Only the worker in production gets rations, while his wife and small children get nothing, apart from rye flour. The workers and their families are in worn-out clothes, the kids are raggedy, and their [the kids'] bellies have started to bloat up. [They] need to buy [clothes], but where? In the workers' cooperative! [They] don't want to think about that. There are only empty shelves and bottles of perfume. The market? The speculators skin the worker alive. A meter of calico print used to cost thirty-five kopeks. Now the speculators in the market sell the same print for one ruble, thirty-five kopeks a meter. The workers are infuriated that the speculators have calico and other [things], but there is nothing in the workers' cooperative. I think, naturally, that such a [negative] atmosphere may take hold at the mill, but this must not be permitted. All that's needed is to push the right person in the right place.

A large number of previously secret letters date from 1930 and 1931. These read as an unending and painful lament. The writers describe deprivations and shortages, rotten food in workers' canteens, disorders and misconduct in lines for food and other goods, the hoarding of provisions and the failure to receive rations, the depreciation of paper money and the disappearance of small coins from circulation. Some writers sought scapegoats and blamed the desperate situation on the machinations of Jews and other groups. One letter presented "facts" to prove that at the Moscow railroad stations and in the nearby city of Noginsk, "Yids" were arrested from whom large batches of silver and copper had been seized: "The Noginsk militia seized six puds and then released the 'Yids' to go off to the four winds." A worker from the Moscow factory "Manometr" wrote of the catastrophic fall in real wages. He blamed this on an increase in the salaries of office workers.[7]

Accounts of shortages came in from the far-flung corners of the country. In Melenki, Vladimir province, lines for "inedible bread" were reportedly three hundred to four hundred strong. Stores in the Dagestanian capital, Makhachkala, on the Caspian Sea, carried nothing but rotten fish. In Tarashch, Ukraine, the lines for meat began forming at five o'clock in the morning. Railroad workers in Kiev warned that clothes and footwear were lacking, and claimed to be walking about without boots. From the Motovilikha plant in Perm came reports that workers relieved "those who weaken" in the lines where it was necessary to stand for fifteen hours at a clip. When two or three pair of shoes became available, twenty people lined up. A letter from Golutvin, in Moscow province, claimed that people stood in line all night for shoes "and in the morning were told that there wouldn't be any."[8]

Through all this, the financial situation continued to place burdens on the people. A group of workers living near Irkutsk complained: "They smother you with different collections and taxes. If you earn twenty rubles a month, then from this they give three rubles to the cooperative, to the trade union—one ruble, to the insurance office—fifty kopeks, to MOPR [International Organization Aiding Revolutionary Fighters] — twenty kopeks, for the orphans—ten kopeks, [and] one *grivennik* [ten kopeks] for every ruble to the insurance fund, [plus] house taxes for those who have their own hut, [then] to the Society "Down with Illiteracy!"—twenty kopeks, [and to] Osoaviakhim [Society Assisting Defense, Aviation and Chemical Production]—twenty kopeks.[9] This amounts to ten rubles. In addition, there are the various voluntary collections [amounting to] five rubles a month. Only this is not voluntary, but obligatory."[10]

A large number of letters discussing material difficulties came from the industrial Donbass. This excerpt is taken from a worker's letter entitled "The Donbass Needs Emergency Help."

· 154 ·

Excerpt from a letter to *Pravda* from the Donbass worker Bosenko on food shortages and their effects on miners, 1930. RGAE, f. 7486s, op. 37, d. 102, l. 100. Typewritten copy.

————————

[. . .] In the Donbass there are no matches: a box costs twenty kopeks. There is no soap, and the miners have nothing to wash off the coal dust, nothing to take to the bath. There's no fish, nothing edible. There has never been such a situation in the Donbass as there is now.

The Donbass needs emergency help as fast as possible. The miners go into the surrounding villages in search of grub. It is disgraceful and criminal how the mine cafeterias are dealing with this especially important matter.

Up to now they have been making borscht with very sour pickled cabbage that brings on stomachaches, diarrhea, and vomiting.

How much coal can one hew on such grub when, after this belly wash, you have no strength left? If you spend money on a smoke, you and your family go hungry. Someone is creating a provocation to upset the workers. A delegation will be dispatched to the center. Our work is being disrupted, and then they will blame us workers.

BOSENKO, Lugansk region.

The socialist offensive was all-encompassing; the country's spiritual life came under assault during these years together with its material foundations. The antireligious campaign struck a chord among urban youth, and young people's conduct in the cities may have been even more vehement than in the village. In Samara, for example, according to one letter, only seven of twenty-five churches remained. Churches were closed down to make clubs, according to another letter, because schoolchildren demanded this, "even those who are practically infants of seven to ten years." The authorities in Dedovsk, Moscow province, decided to turn a church into an electrical power station. A workers' rally was organized in support of the decision. Allegedly, thirty-five hundred people attended in order to block the attempt by a group of "kulaks and the disenfranchised" to reopen the church.

The antireligious movement was part of a larger "cultural revolution" that accompanied the socialist offensive in the years of the First Five-Year Plan. Many professions, especially those in the arts, education, and the sciences, were subjected to an ideological assault that aimed to purge them of moderation and political neutrality. The movement was sanctioned from above and drew energy from the young and other groups impatient with the slow advance to socialism and the indulgences granted noncommunists during the NEP years. Ideas and individuals identified as "bourgeois" or "nonproletarian" were at particular risk from Party hardliners, who laid siege to what in their view were intellectual and cultural Winter Palaces. Wrath was also rained down on Party members labeled "rightists," like Bukharin and Rykov, who questioned the wisdom of such fortress storming. The cultural revolution helped reenergize the utopianism and revolutionary optimism of the early Soviet era, but many academic, scientific, and artistic disciplines never recovered from the havoc and the political controls forced on them.[11]

The condition of city life further deteriorated thanks to an influx of peasants seeking to escape the turmoil created by collectivization and dekulakization. Their sudden arrival, en masse, in the country's large cities and towns exacerbated the already-acute urban problems of housing and transport. Housing construction and the creation of model "socialist cities" lagged far behind the urban population explosion, and finding adequate living space became desperately difficult. Rumors about the solutions being considered by the authorities to solve the overcrowding problem caused understandable anxiety among some urban dwellers, as this letter describes.

· 155 ·

Letter to *Pravda* expressing fear that the government was preparing to seize workers' houses, 1930. RGAE, f. 7486s, op. 37, d. 102, l. 97. Typewritten copy.

———————

A rumor is circulating among the workers and employees in the Urals that the Soviet government is going to seize their tiny houses even though they have only one to two rooms. This has created a depressed mood, since, after years of stubborn labor, many have built themselves a corner so that when they can no longer work, they can rest in their corner.

[. . .] Putting up cooperative housing satisfies no one since [they expect nothing from it] but squabbles and unpleasantness from the common communal kitchens and the like.

A worker in a communal apartment does not feel it's "his" corner and does not take to it. It needs to be explained to the worker that Soviet power does not intend to rob the toilers and that there is no basis for them to abandon their quarters and allow them to decay before their time.

———————

Crash industrialization and collectivization combined to create a highly mobile population. Torn from a familiar environment, this declassed, disoriented mass wandered the country seeking some foothold in what Moshe Lewin termed the "quicksand society." Factory shops and other work sites filled with migrants from the villages who were new to production and lacked the habits and skills of experienced workers. The two capitals, Moscow and Leningrad, were the most attractive destinations, and it is not surprising that they were the first cities to set limits on the issuance of resident permits. Despite the obstacles, however, people found creative ways to circumvent the restrictions in order to remain in these centers. Thus, in an unanticipated way, the final victory of the city over the village actually led to the "ruralization" of the city.[12]

By the same token, these migrants also provided the recruits for the new working class that moved in to fill the vacancies created by Stalin's industrial revolution. Rapid industrialization solved the problem of NEP unemployment and led to a large expansion in the ranks of both white- and blue-collar workers. During the First Five-Year Plan (1928–1932), the industrial workforce nearly doubled to six million. By its end, workers who had entered industry during the plan years outnumbered their pre-plan comrades by three to one. Of these new workers, nearly 60 percent claimed peasant origins. Although the simultaneous forced expansion of the industrial base and creation of a large, inexperienced working class from recent peasants eliminated the scourge of unemployment and provided countless opportunities for social advancement, the process was not smooth or pretty. A leading historian of Stalinist industrialization has likened its brutalities and exploitation of labor to that of early capitalism. Thanks to the decline in agricultural production and the capital it generated, workers—new and old—found themselves paying for industrialization through reduced living standards.[13]

Naturally, workers reacted to this situation with resentment and even outrage. The lack of basic necessities could dampen the spirits of the most ardent and enthusiastic supporters of socialist construction. One worker, sympathetic to the transformations under way, expressed the feelings of many when he wrote, "We understand everything, but the stomach wants something to eat." Sharply worded messages also made their way to the leadership. The worker Vladimirov from the Zubovsk factory in Briansk region, wrote: "The Central Committee has brought us to ruin. For what did we battle in the swamps of Karelia and the mud of Perekop?"[14] The following letter was sent to *Pravda* in the name of the workers from the Ukrainian mining and metallurgical center Krivoi rog.

· 156 ·

Letter to *Pravda* in the name of the miners at Krivoi rog on the catastrophic situation there due to collectivization, 1930. RGAE, f. 7486s, op. 37, d. 102, l. 131. Typewritten copy.

"HASTE MAKES WASTE"[15]

You cannot rule the state, because you've allowed a famine. For a rotten salt herring you have to stand in line for twelve hours. [. . .] [Before,] the priests deceived us, and now who? We ourselves.

This year the land is barren, uncultivated; the pigs and cattle were all killed, destroyed when they started to put the peasants in the kolkhozes. The kulak must be replaced by the collective, the private trader by the cooperative, but don't forget the proverb "Haste makes waste."

During the civil war, they told us that we would work eight hours, sleep eight hours, play on the harmonica, attend clubs, theaters, but that's not how it is now. We don't work eight hours but seven. As a result, the intestines play a twelve-hour march in your gut.

Let the Sixteenth Party Congress in Moscow attest to my letter. It is a letter from a group of miners at the Krivoi rog Soviet mine. The author of the letter is Tolmachev, barracks No. 5.

As far as the collective goes, we must create more of them, only on voluntary principles, and don't take in idlers. [. . .]

We ask that an explanation appear in the press to the workers of Krivoi rog as to why criticism is suppressed among us.

In a more hostile vein, one Moscow worker denounced the increased piecework norms and tempos imposed on factory workers.

· 157 ·

Letter to *Pravda* from the Moscow worker P. Skatov denouncing piecework and increased tempos imposed on factory workers, 1930. RGAE, f. 7486s, op. 37, d. 102, l. 131. Typewritten copy.

[. . .] Both the industrial managers and others are still evidently "dizzy" from the successful exploitation and robbing of the workers. I, a worker, have now ceased to believe our leaders [when they say] that someday the life of the worker will get better.

The worker is not greedy, not self-seeking, and despite what some bastards say about the worker, he is not lazy. Just give him what the bourgeois gave him. I call labor penal servitude because it has become harder than before. I have worked for sixteen hours and longer but was never as worn out as now, [when I am working] eight hours.

In no fascist country do they cut wages like this, and if they did, then he [the worker] has the right to fight. I don't remember that the [old] bosses treated us like this, causing us to go elsewhere.

I am a highly skilled shoe worker of the eighth rank. Thanks to the bunglers, the bureaucrats, I am paid between one ruble, twenty kopeks, and two rubles, fifty kopeks a day.

This is what is forcing [workers] to leave production—unpaid labor, the stupid approach to piecework. The more you work, the more they lower the wage, and they write a higher norm.

Now I have to overpay for everything. I bought five eggs for eighty ko-peks, and there's practically nothing to eat. Our factory buffet costs more than the cooperative. Here or there, there is practically nothing.

I fought the Whites. Twice we kicked Shkuro's Whites out of Kislo-vodsk.[16] We captured thirty million rubles for Soviet power in Kislovodsk; five hundred head of cattle were taken from the Whites near Devichii mon-astery in 1918. We fought for the Bolsheviks, but what are you doing for us, the workers?

In some cities workers, having reached the breaking point, took direct action against the political authorities. A report from Berdiansk, in Ukraine, detailed an assault on the city soviet instigated by the complete breakdown of supplies and the rise in prices on the black market. Win-dows were broken and office workers were beaten because the chairman was nowhere to be found. The peasants brought rural forms of protest to the urban areas, and a number of cities experienced their own ver-sions of babi bunty—violent resistance by women. As a result of all the time that workers needed to spend in shopping lines, discipline and good factory organization began collapsing. According to one report, workers stopped coming to factory meetings, complaining that time wasted in the meetings was better spent seeking provisions. The food shortages also encouraged workers to call for an end to grain sales abroad. In response to the food crisis, the Politburo reluctantly approved reductions in collec-tion quotas for grain and meats in May 1932 and granted collective-farm and other peasants freedom to trade their produce at "market prices." As R. W. Davies notes, these concessions resembled the early measures of the NEP.[17]

As the situation in the country worsened, many mourned the passing of the NEP period, which now seemed to have been a time of relative abundance and good order. The Donbass worker F. Sobol, for example, wrote, "In the past, '25, '26, and '27, both the workers and the peasants felt like citizens in their own right. Since '28 we have been stumbling in the wrong direction and by the end of the five-year plan will have tum-bled so far that we will be panting like dogs."[18] Similar sentiments are expressed in this collective letter to A. I. Rykov from the workers of one of the largest armament plants in the country, located in the city of Izhevsk. The letter was received on 17 August 1930.

· 158 ·

Letter to the chairman of the Council of People's Commissars, A. I. Rykov, in the name of the workers of the Izhevsk factory, 17 August 1930. RGAE, f. 8043, op. 11, d. 26, l. 13. Typewritten copy.

Comrade Rykov.

In the name of the 50,000 workers of the Izhevsk factory, we beg you to save us from starvation. The cafeterias are closing They serve us a watery oatmeal and a little bit of bread. In the stores they give a half-funt of black bread or flour per person; they already haven't given [us] any more for a month. We are swelling up from hunger and have no strength to work. Before, they gave the soldier-parasites three funts of bread, a quarter-funt of meat, and a half-funt of porridge for supper—and they didn't do anything, while we perform heavy labor, especially in the foundry. The workers are abandoning production [work] and are selling everything they have just to feed their children. The children are condemned because you cannot provide anything. What good is your heavy industry to us when there is no small [industry]. Heavy [industry] won't fill [our bellies]. First, you should provide the basic necessities [and only] then think about heavy industry or else hunger will turn everything to dust.

Comrade Rykov, heed what we are suggesting to you. Don't be stubborn any longer. You have enjoyed yourself, to be sure, but, you see, it's come to nothing. You must turn down another road. First, you must open the private factories and plants and [allow] private trade. If not loaded down with taxes, the private trader will find everything. Then, create model landholdings; that is, allow [. . .] the middle peasants to have three to four horses, ten cows, fifty sheep and chickens and geese [. . .]. When there is private trade, the peasant will try to get more output from his labor, and then he will cheaply sell bread, and wool, and flax, and meat. But when the peasant is not interested in buying, he will sell nothing—then there will be less agricultural produce, and it will be expensive, and if agricultural produce is expensive, then [goods from] the city will also be expensive. [. . .] [The supply of] grain and fruit depends on this.

We beg you, don't compel us to [take] drastic measures. We are turning to you as comrades. There are many communists among us, and we have all come to the same decision. The situation is catastrophic and can no longer be tolerated.

Of course, there is still a way to save the country from destitution. That is to get everything from abroad. Don't accept this support. Save us from hunger. [. . .]

The signatures of the representatives of the workshops follow.

Signed: SKVORTSOV and TAMARNIKOV.

That Rykov had not supported the extreme policies of Stalin's revolution from above is either lost on or of little moment to these factory workers, who only want an end to hardships. Other citizens—including rank-and-file communists—now openly expressed a preference for Stalin's political opponents, whether of the right or the left. An anonymous letter writer from Orenburg declared, "Stalin is exchanging raw materials, leather, and horse tails overseas for a lot of tractors. They are humming along this bumpy path straight to the abyss, but with Rykov, Bukharin, Tomsky, and the others we on our horses are getting to the bright, Leninist path. With Rykov and Bukharin our paradise maybe won't be so great, but at least we'll be fed and clothed." A similar letter writer maintained, "When Rykov, Tomsky, and Bukharin were in charge, everything was in abundance, but they got rid of them, and there is nothing left. That's the real reason they're purging them at the congress."[19] A certain Borovikov from Semipalatinsk wrote, "The press considers any realistic correspondence the scribbling of a kulak or NEPman. Trotsky and the Trotskyists sided with Marxism—that in one country it is impossible to build socialism—and they were gagged and exiled like the Mensheviks. Bukharin, Rykov, and the others suffered the anathema of the Stalinist papal throne; their mouths were gagged, and the press no longer carries their articles. If you really want to build socialism, then you must revive the newspapers and purge them of communist arrogance." Others felt that the great leap forward was premature. "They started socialism and communism too early. When we build automobiles, then maybe it will go faster . . . they've set too high a tempo . . . a horse can't carry the excess." Another wrote, "The leaps are too huge. Lenin would not have allowed such things . . . the right [oppositionists] are right and not the Party. The two years of the five-year plan [mark] total deterioration . . . There are no goods for provisions."[20]

Whatever nostalgia the people now expressed for the difficult but comparatively calm NEP years, there could be no turning back. With collectivization, dekulakization, and rapid industrialization, Stalin and the Communist Party he headed turned socialist transformation into a forced march and a race against time. Only the future mattered. The present would have to be sacrificed to that future, and in the present little comfort would be found by lamenting the past.

Notes

INTRODUCTION

1. N. N. Sukhanov, *Zapiski o revoliutsii*, 3 vols. (Moscow, 1992), vol. 3, pp. 356–357. John Reed does not refer to this episode in his account of the congress; see Reed, *Ten Days That Shook the World* (New York, 1989), pp. 133–138.

2. These letters have previously appeared in a Russian-language edition; see Andrei K. Sokolov, ed., *Golos Naroda: Pis'ma i otkliki riadovykh sovetskikh grazhdan o sobytiiakh 1918–1932* (Moscow, 1997).

3. On the role of the press in the 1920s, see Matthew Lenoe, *Closer to the Masses: Stalinist Culture, Social Revolution, and Soviet Newspapers* (Cambridge, Mass., 2004), pp. 1–45; Jeffrey Brooks, *Thank You, Comrade Stalin! Soviet Public Culture from Revolution to Cold War* (Princeton, N.J., 2000), pp. 3–18. For the congress resolution "On the Press," see *Kommunisticheskaia Partiia Sovetskogo Soiuza v rezoliutsiiakh i resheniiakh s"ezdov, konferentsii i plenumov TsK* (Moscow, 1984), vol. 3, p. 254 (hereafter *KPSS v rezoliutsiiakh*).

4. *KPSS v rezoliutsiiakh*, vol. 3, p. 257.

5. See Lenoe, *Closer to the Masses*, pp. 70–100. Jeffrey Brooks traces the emphasis the Bolsheviks placed on readers' letters to the revolutionary underground era when, as he writes, "Workers' letters to *Pravda* were highly prized"; see Brooks, *Thank You, Comrade Stalin!* p. 8.

6. On the workers' and peasants' correspondent movement, see Peter Kenez, *The Birth of the Propaganda State: Soviet Methods of Mass Mobilization, 1917–1929* (Cambridge, England, 1985) pp. 233–237; Brooks, *Thank You, Comrade Stalin!* pp. 8–9.

7. Brooks, *Thank You, Comrade Stalin!* pp. 52–53.

8. See, for example, Sarah Davies, *Popular Opinion in Stalin's Russia: Terror, Propaganda and Dissent, 1934–1941* (Cambridge, England, 1997), pp. 49–53.

Lynne Viola has written most extensively and in detail on the subject of Soviet peasant resistance; see Viola, *Peasant Rebels under Stalin: Collectivization and the Culture of Resistance* (New York, 1996); see also the essays in Lynne Viola, ed., *Contending with Stalinism: Soviet Power and Popular Resistance in the 1930s* (Ithaca, N.Y., 2002). Viola's introduction and her essay entitled, "Popular Resistance in the Stalinist 1930s: Soliloquy of a Devil's Advocate" are especially thoughtful and instructive contributions to this debate.

9. William Henry Chamberlin, *Soviet Russia: A Living Record and a History* (London, 1930), p. 189. The Constitution of the Russian Soviet Federated Socialist Republics (RSFSR) adopted in 1918 denied the franchise to anyone who profited through the use of hired labor, including peasants, and weighted a worker's vote five times greater than that of a peasant: in elections to the All-Russian Congress of Soviets representatives from city soviets were elected at a ratio of one to every 25,000 electors, and representatives from provincial soviets, at one to every 125,000.

10. On the RSFSR agricultural apparatus, see James W. Heinzen, *Inventing a Soviet Countryside: State Power and the Transformation of Rural Russia, 1917–1929* (Pittsburgh, Pa., 2004).

11. See Lynne Viola, "The Peasants' Kulak: Social Identities and Moral Economy in the Soviet Countryside in the 1920s," *Canadian Slavonic Papers*, 42, no. 4 (2000), pp. 431–460.

12. This overt conflict around the issue of social identity is one notable instance of peasant-state/Party engagement that may be considered in light of Michel Foucault's concept of "subjectivity," according to which modern individuals are produced, and produce themselves, through a dynamic and continuous interaction with institutions of power. Examples of the application of Foucault's ideas to Soviet history include Stephen Kotkin's paradigmatic study of workers in the 1930s, *Magnetic Mountain: Stalinism as Civilization* (Berkeley, Calif., 1995); Igal Halfin, *From Darkness to Light: Class, Consciousness, and Salvation in Revolutionary Russia* (Pittsburgh, Pa., 2000).

13. Jeffrey Brooks, *Thank You, Comrade Stalin!* pp. 5–11.

14. See, for example, George Rudé, *Revolutionary Europe, 1783–1815* (New York, 1964), p. 89.

15. See James C. Scott, *Domination and the Arts of Resistance: Hidden Transcripts* (New Haven, Conn., 1990).

16. Hugh D. Hudson Jr. has examined one such attempt on the part of officialdom to set the terms of acceptable discourse in peasant letter-writing in 1926; see Hudson, "Shaping Peasant Political Discourse during the New Economic Policy: The Newspaper *Krestianskaia gazeta* and the Case of 'Vladimir Ia.,' " *Journal of Social History*, 36, no. 2 (2002), pp. 303–317.

17. In a study of workers in the 1930s, Stephen Kotkin captures the essence of this form of discourse with the phrase "speaking Bolshevik." In Kotkin's words, "It was not necessary to believe. It was necessary, however, to participate as if one believed." See Kotkin, *Magnetic Mountain*, pp. 198–237. The emulation of official and journalistic styles in letter writing became more widespread in the 1930s; see, for example, Sheila Fitzpatrick, "Suppliants and Citizens: Public Letter-Writing in Soviet Russia in the 1930s," *Slavic Review*, 55, no. 1 (1996), pp. 78–105. Examples of letters written during the 1930s can be found in Lewis Seigelbaum and Andrei Sokolov, *Stalinism as a Way of Life: A Narrative in Documents* (New Haven, Conn., 2000).

18. This reading of the letters is in line with James C. Scott's critique of the Italian communist Antonio Gramsci's concept of "hegemony." According to Scott, Gramsci erred when he attributed successful class rule to the "false-consciousness" instilled in subordinate classes via elite control and manipulation of the cultural-ideological superstructure of a given society. On the contrary, Scott argues, it is precisely in the realm of consciousness that subordinate classes are most free, as evidenced by their ability to see through, or "penetrate," to the reality of domination that elite ideology seeks to conceal. What appears to be acquiescence in a given power structure, on the other hand, is a rational evaluation by members of the subordinate class that the chance of overturning that structure is unlikely, and a very reasonable desire to avoid the retribution that would follow a failed revolt. Subordinate class resistance, therefore, is more likely to be located, not in direct struggle with power holders, but in more indirect or "passive" acts and in thought. See James C. Scott, *Weapons of the Weak: Everyday Forms of Peasant Resistance* (New Haven, Conn., 1985), pp. 314–322.

19. An extremely helpful analysis of lower-class writing styles is provided by Ekatarina Betekhtina in Mark D. Steinberg, *Voices of Revolution, 1917* (New Haven, Conn., 2001), pp. 309–338.

20. Orlando Figes, *Peasant Russia, Civil War: The Volga Countryside in Revolution, 1917–1921* (Oxford, 1991), pp. 30–46.

21. See Andrea Graziosi, "The Great Soviet Peasant War: Bolsheviks and Peasants, 1917–1933," *Harvard Papers in Ukrainian Studies* (Cambridge, Mass., 1996). As commentators on peasant behavior have noted, eruptions of violence among peasants is frequently an indication that other, passive forms of struggle are failing to achieve their purposes; see, for example, James C. Scott, *Weapons of the Weak*, p. 37.

22. See Rosa Luxemburg, "The Russian Revolution," in *Rosa Luxemburg Speaks* (New York, 1970), p. 378.

23. For Lenin's ideas on this subject, see "On Cooperation," written in early January 1923, in V. I. Lenin, *Polnoe sobranie sochinenii,* 5th ed. (Moscow, 1964), vol. 45, pp. 369–377 (hereafter *PSS*).

24. On socialist propagandizing among the peasants, see Orlando Figes and Boris Kolonitskii, *Interpreting the Russian Revolution: The Language and Symbols of 1917* (New Haven, Conn., 1999), pp. 148–152.

25. Some notion of an obligatory paternalism held a constant place in Russian peasant conceptualizations of the just society. Sheila Fitzpatrick traces it to an idealized image of the "good master" that existed before the revolution and finds it continuing after collectivization in the kolkhozniks' attempts to negotiate welfare-state-like conditions on collective farms. See Sheila Fitzpatrick, *Stalin's Peasants: Resistance and Survival in the Russian Village after Collectivization* (New York, 1994), pp. 9–10.

26. Among the most thorough and valuable treatments of the subject are Moshe Lewin, *Russian Peasants and Soviet Power: A Study of Collectivization* (New York, 1975); E. H Carr and R. W. Davies, *Foundations of a Planned Economy, 1926–1929* (New York, 1969), vol. 1; Merle Fainsod, *Smolensk under Soviet Rule* (New York, 1963); R. W. Davies, *The Socialist Offensive: The Collectivisation of Soviet Agriculture, 1929–1930* (Cambridge, Mass., 1980); V. P. Danilov, *Rural Russia under the New Regime* (Bloomington, Ind., 1988); Fitzpatrick, *Stalin's Peasants;* V. Danilov, R. Manning, and L. Viola, eds., *Tragediia sovetskoi derevni. Kollektivizatsiia i*

raskulachivanie. 1927–1939. Dokumenty i materially, 5 vols. (Moscow, 1999 and 2000), vols. 1–2 (hereafter *TSD*); Lynne Viola et al., eds., *The War against the Peasantry, 1927–1930: The Tragedy of the Soviet Countryside* (New Haven, Conn., 2005), vol. 1; N. A. Ivnitskii, *Kollektivizatsiia i raskulachivanie (nachalo 30-kh godov)* (Moscow, 1996). Among the firsthand accounts of the transformation of the Soviet countryside, the one by Maurice Hindus, *Red Bread* (New York, 1931), remains highly engaging and rewarding. The present volume draws heavily on these and other scholarly works.

27. In addition to works already cited, letters, documents, and other materials relating to the pre- and post-revolutionary countryside may be found in *Krest'ianskie istorii: Rossiiskaia derevnia 20-kh godov v pis'makh i dokumanentakh* (Moscow, 2001). The oral history *Golosa krest'ian: Sel'skaia Rossiia XX veka v krest'ianskikh memuarakh* (Moscow, 1996) contains valuable anecdotal evidence on 1920s village life, the collectivization process, and rural conditions after collectivization.

28. See Jeffrey Brooks, "Public and Private Values in the Soviet Press, 1921–1928," *Slavic Review,* 48, no. 1 (1989), appendix B, pp. 30–31.

29. Ibid., appendix C, p. 32.

30. *KPSS v rezoliutsiiakh,* vol. 3, pp. 109–110. In fact, during the civil war, anywhere from 50 to 75 percent of the issues of *Bednota,* which had a print run of 750,000 copies in 1919, were distributed among the Red Army; see Brooks, "Public and Private Values"; and the *Bednota* entry in *Bol'shaia Sovetskaia Entsiklipediia* (Moscow, 1970), vol. 3; *The Smolensk Archive,* WKP 209, l. 57. Circulation figures are from Brooks, "Public and Private Values," appendix A, p. 29.

31. The Russian designations are, respectively, *sekretno, sovershenno sekretno, ne podlezhit oglasheniiu,* and *dlia sluzhebnogo pol'zovaniia.*

32. For a detailed discussion of *svodki,* see Sarah Davies, *Popular Opinion,* pp. 9–14.

CHAPTER 1. REVOLUTION AND WAR COMMUNISM

1. This figure is for the territory falling within the USSR's 1926 borders. See *Naselenie Rossii v XX veke: Istoricheskie ocherki,* 3 vols.(Moscow, 2000), vol. 1, p. 95. If extended to include 1922, when famine along the Volga River was still severe, then the figure is closer to thirteen million, but, as the authors of *Naselenie Rossii,* note, other credible estimates place the figure as high as fifteen million. If deficit births are included in the calculations, the total demographic catastrophe in these years (1917–1922) may be as high as twenty-five million.

2. General histories of the 1917 revolution and its aftermath are numerous and varied in their approaches and sympathies. Leon Trotsky, a prominent participant in the events of that year, set a high literary and analytical standard for revolutionary studies with his *History of the Russian Revolution* (New York, 2003). Other notable works include E. H. Carr, *The Bolshevik Revolution, 1917–1923,* 3 vols. (Middlesex, England, 1975); Robert V. Daniels, *Red October: The Bolshevik Revolution of 1917* (New York, 1967); Richard Pipes, *The Russian Revolution* (New York, 1990); Sheila Fitzpatrick, *The Russian Revolution* (Oxford, England, 1994); James D. White, *The Russian Revolution, 1917–1921: A Short History* (London, 1994); Orlando Figes, *A People's Tragedy: The Russian Revolution, 1891–1924* (New York, 1996).

3. On the political and social evolution of the Bolshevik Party over the course of 1917, see Alexander Rabinowitch, *Prelude to Revolution: The Petrograd Bolsheviks*

and the July 1917 Uprising (Bloomington, Ind., 1968); Rabinowitch, *The Bolsheviks Come to Power: The Revolution of 1917 in Petrograd* (New York, 1976).

4. For an eyewitness account of Lenin addressing the congress, and texts of the proclamation and decree in English, see Reed, *Ten Days That Shook the World*, pp. 123–136.

5. During the last year of the old regime, the amount of currency in circulation increased by nearly two-thirds. Under the Provisional Government, it increased a further 70 percent. In the first five months of Bolshevik rule, the money supply grew yet again, by almost three-fourths. In 1918, retail prices doubled between January and March, then doubled again over the next three months. See R. W. Davies' essay, "Economic and Social Policy in the USSR, 1917–41," in Peter Mathias and Sidney Pollard, eds., *The Cambridge Economic History of Europe: The Industrial Economies* (Cambridge, England, 1989), vol. 8, p. 993.

6. See William G. Rosenberg, "Russian Labor and Bolshevik Power after October," *Slavic Review*, 44, no. 2 (1985), pp. 213–238.

7. On the factory committee movement see, for example, S. A. Smith, *Red Petrograd: Revolution in the Factories, 1917–18* (Cambridge, England, 1983); Vladimir Brovkin, "The Mensheviks' Political Comeback: The Elections to the Provincial City Soviets in Spring 1918," *Russian Review*, 42 (1983), pp. 1–50.

8. RGASPI, f. 5, op. 1, d. 1500, l. 18, and d. 1501, l. 39. For a list of archives and explanations of archive abbreviations, see the Note on the Documents. In the second letter, the writer addresses Lenin with the familiar *ty* (you), not the more formal *Vy* (You). In other documents writers use one or the other, as we occasionally note, or, as in Document 2, shift between the two. Using *Vy* is a standard Russian practice when writing a formal letter or addressing someone who is not a close friend.

9. Probably a reference to a site along the Toropa River in neighboring Pskov province.

10. "Northern Army" probably refers to the Northern Corps created by the German military command in the Pskov and Dvinsk regions from the remnants of Russian units in October 1918.

11. S. M. Zav'ialov, ed., *Neizvestnaia Rossiia, Sbornik, Dokumentov* (Moscow, 1993), vol. 2, p. 216.

12. Maxim Gorky, *Untimely Thoughts: Essays on Revolution, Culture and the Bolsheviks, 1917–1918* (New Haven, Conn., 1995), pp. 139–140.

13. RGASPI, f. 5, op. 1, d. 1500, l. 47.

14. Richard Stites, *Revolutionary Dreams: Utopian Vision and Experimental Life in the Russian Revolution* (New York, 1989), p. 61.

15. Count Wilhelm Mirbach served as German ambassador to the Soviet republic at this time. Mirbach was an object of vituperation in the heightened anti-German atmosphere following the signing of the Brest-Litovsk Treaty in March 1918. On 6 July, one week after this letter was written, Mirbach was, in fact, assassinated in Moscow by the Left Socialist-Revolutionary and Chekist Yakov Bliumkin in the course of the Left SR uprising against the Bolsheviks.

16. A reference to state grain requisitioning then under way.

17. "Little Russians" (*Malorossy*) refers to Ukrainians. Orel province borders Ukraine, and that may be the source of the reference.

18. The presentation of bread and salt was a traditional peasant display of welcome and hospitality.

19. Created in 1866, the Okhrana was the security police force of the tsars.

20. A reference to the fourth century B.C. Greek philosopher Diogenes of Sinope, whose renunciation of worldly possessions and extreme asceticism led him to live in a wooden barrel or tub.

21. RGASPI, f. 5, op. 1, d. 1500, l. 44.

22. On the creation of the Red Army, see Mark Von Hagen, *Soldiers in the Proletarian Dictatorship: The Red Army and the Soviet Socialist State, 1917–1930* (Ithaca, N.Y., 1990), pp. 13–66.

23. For the Decree on Land, see *Resheniia partii i pravitel'stva po khoziaistvennym voprosam, 1917–1967,* 5 vols. (Moscow, 1967), vol. 1, pp. 15–17. The Decree on Land's antecedents may be traced to the agrarian resolution of the first All-Russian Congress of Soviets held in June and dominated by the SRs, and the "model decree" published in August by the All-Russian Peasants' Congress, where the SRs also enjoyed a majority. For symbolic reasons, the Central Executive Committee passed "On the Socialization of Land" on 19 February 1918, the anniversary of serf emancipation. On the evolution of Lenin and the Bolsheviks' actions in relation to these and other agrarian measures, see Carr, *The Bolshevik Revolution,* vol. 2, pp. 35–53.

24. See, for example, Figes, *Peasant Russia, Civil War,* pp. 30–69.

25. The lower number is given in Oliver H. Radkey, *Russia Goes to the Polls: The Election to the All-Russian Constituent Assembly, 1917* (Ithaca, N.Y., 1989), pp. 148–151. A more recent Russian publication includes 4.2 million more total votes than Radkey does and a much higher figure of over seven million votes for socialists other than the SRs, Bolsheviks, or Mensheviks, see L. G. Protasov, *Vserossiiskoe Uchreditel'noe sobranie: Istoriia rozhdeniia i gibeli* (Moscow, 1997), especially the table following p. 362. Deputies were elected not only from the main socialist parties (SRs, Bolsheviks, and Mensheviks) but also from the lists of the Popular Socialists, the Constitutional Democrats, and various national parties (of Muslims, Bashkirs, Kirghizians, Armenians, Jews, Poles, Letts, Estonians, and Cossacks). The revolution had clearly opened the possibilities for a vibrant political life throughout the former empire that expressed the varied interests of a multinational population; see Radkey, *Russia Goes to the Polls,* p. 23.

26. Cheka is a Russian acronym formed from the first two words of the body's official title, Extraordinary Commission (*Chrezvychainaia komissia,* or *ChK*). As the Cheka's reach expanded, so did its official title, which ultimately read, "The All-Russian Extraordinary Commission for Combating Counterrevolution, Speculation, Sabotage, and Misconduct in Office"; see George Leggett, *The Cheka: Lenin's Political Police* (Oxford, 1981), pp. 30–31. On the conflict between popular revolutionary expectations and Bolshevik policies, see Christopher Read, *From Tsar to Soviets: The Russian People and Their Revolution, 1917–21* (New York, 1996).

27. On the diverse intellectual sources of and conflicts within early Bolshevism, see Robert C. Williams, *The Other Bolsheviks: Lenin and His Critics, 1904–1914* (Bloomington, Ind., 1986). On Vikzhel and the Bolshevik leadership's position on a coalition government, see Rabinowitch, *The Bolsheviks Come to Power,* pp. 308–310.

28. See Roger A. Clarke and Dubravko J. I. Matko, *Soviet Economic Facts, 1917–81* (London, 1983), pp. 83–89, 113; Alec Nove, *An Economic History of the USSR* (Middlesex, England, 1969), pp. 63–68.

29. See Daniel R. Brower, "'The City in Danger': The Civil War and the Russian Urban Population," in Diane Koenker et al., eds., *Party, State and Society in the Rus-*

sian Civil War: Explorations in Social History (Bloomington, Ind., 1989), pp. 58–65; William J. Chase, *Workers, Society, and the Soviet State: Labor and Life in Moscow, 1918–1929* (Urbana, Ill., 1987), pp. 31–33.

30. On the influence of ideology on Bolshevik rural policy at this time, see Bertrand M. Patenaude, "Peasants into Russians: The Utopian Essence of War Communism," *Russian Review*, 54, no. 4 (1995), pp. 552–570; see also Lars T. Lih's response to Patenaude's article in *Russian Review*, 55, no. 3 (1996), pp. 494–496.

31. For background on war communism, see Nove, *An Economic History of the USSR*, pp. 46–82; Carr, *The Bolshevik Revolution*, vol. 2, pp. 151–268. For the non-military impulses behind war communism, see, for example, Figes, *A People's Tragedy*, pp. 612–615.

32. Lenin, *PSS*, vol. 36, p. 360.

33. Figes, Peasant Russia, Civil War, pp. 188–199.

34. Lars T. Lih, "Bolshevik *Razverstvka* and War Communism," *Slavic Review*, 45 (1986), pp. 673–688.

35. Chase, *Workers, Society, and the Soviet State*, pp. 26–27. On bagging, see Figes, *A People's Tragedy*, pp. 611–612; Carr, *The Bolshevik Revolution*, vol. 2, pp. 241–245.

36. On Narkomprod's relationship with other food-supply organs, see Mauricio Borrero, *Hungry Moscow: Scarcity and Urban Society in the Russian Civil War, 1917–1921* (Peter Lang, 2003). For the relevant decrees, see *Dekrety Sovetskoi Vlasti* (Moscow, 1959), vol. 2, pp. 261–267, 307–312.

37. *Dekrety Sovetskoi Vlasti*, vol. 2, pp. 264–266; Leggett, *The Cheka*, p. 211; Zav'ialov, *Neizvestnaia Rossiia*, vol. 2, p. 212. The Russian phrase used here is "Pridetsia deistvovat' zatvorom."

38. In March 1919, the Eighth Russian Communist Party Congress instructed local Communist Party organizations to cease using coercive methods against middle peasants.

39. This refers to provincial-level congresses of soviets.

40. "Rural bourgeoisie" was the Party's label for rich peasants, or kulaks.

41. The Extraordinary Committee for the Collection of Hay (Chrezkomzagsen) was an organ of Narkomprod.

42. The labor-conscription departments and district labor committees were local organs of the Main Administration of Labor (Glavkomtrud).

43. Literally, "play Lazarus," which means "complain."

CHAPTER 2. THE OLD VILLAGE AND THE NEW ECONOMIC POLICY

1. On the peasant wars, see Robert Conquest, *The Harvest of Sorrow: Soviet Collectivization and the Terror-Famine* (New York, 1986), pp. 49–53; Figes, *Peasant Russia, Civil War*, pp. 321–353; Figes, *A People's Tragedy*, pp. 753–758; Read, *From Tsar to Soviets*, 258–272. On the rebellion led by Makhnov, see Peter Arshinov, *History of the Makhnovist Movement, 1918–1921* (Detroit, Mich., 1974); Nestor Makhno, *Vospominaniia* (Moscow, 1992). Estimates of Antonov's forces vary from twenty thousand to fifty thousand men. On the Antonov movement, see Erik C. Landis, *Bandits and Partisans: The Antonov Movement in the Russian Civil War* (Pittsburgh, Pa., 2008); Oliver H. Radkey, *The Unknown Civil War in Soviet Russia: A Study of the Green Movement in the Tambov Region, 1920–1921* (Stanford, Calif., 1976).

2. On the Kronstadt rebellion, see Paul Avrich, *Kronstadt 1921* (New York, 1974). For documents relating to the mutiny, see V. P. Naumov and A. A. Kosakovskii, eds., *Kronshtadt 1921* (Moscow, 1997).

3. See Lenin's report on the tax-in-kind to the Tenth Congress, in *Desiatkii s"-ezd RKP(b), Mart 1921 goda, Stenograficheskii otchet* (Moscow, 1963), pp. 403–415.

4. V. I. Lenin, "The Tax in Kind," in *Collected Works* (Moscow, 1973), vol. 32, p. 343. In this pamphlet, Lenin owned up to the fact that peasant "necessities" had been taken along with "surpluses."

5. Subsequent legislation in 1923 and 1925 lifted restrictions on land leasing even further. On the 1922 Land Code, see Dorothy Atkinson, *The End of the Russian Land Commune, 1905–1930* (Stanford, Calif., 1983), pp. 234–246; Teodor Shanin, *The Awkward Class: Political Sociology of Peasantry in a Developing Society: Russia, 1910–1925* (Oxford, 1972), pp. 225–226; *Istoriia sovetskogo krest'ianstva*, 5 vols. (Moscow, 1986), vol. 1, pp. 250–256.

6. Nove, *An Economic History of the USSR*, p. 84. By March 1922, 233 million puds had actually been collected; see *Istoriia sovetskogo krest'ianstva*, vol. 1, p. 229. In Moscow province, for example, before the change, individual taxes were assessed on rye, oats, potatoes, oil seeds, hay, dairy products, eggs, natural fibers, wool, vegetables, honey, domestic fowl, and fur pelts; see, D. V. Kovalev, *Podmoskovnoe krest'ianstvo v perelome desiatiletie, 1917–1927* (Moscow, 2000), pp. 70–71. Active Red Army soldiers, commanders, and students in military schools (*kursanty*) were included in each household's head count. Soldiers demobilized after April 1922 and working an allotment no bigger than two and a half desiatinas were exempted from the tax altogether; see *Istoriia sovetskogo krest'ianstva*, vol. 1, p. 232; *KPSS v rezoliutsiiakh*, vol. 3, pp. 75–79.

7. Pyotr Stolypin, Russian prime minister from 1906 to his assassination in 1911, was the major political proponent, if not the actual author, of these reforms. For a thorough treatment of the agrarian reforms that followed the 1905 Revolution, see Geroid Tanquary Robinson, *Rural Russia under the Old Regime* (Berkeley, Calif., 1972), pp. 208–242. On the complexities of Soviet agriculture and Party policy in 1924 and 1925, see Edward Hallett Carr, *Socialism in One Country, 1924–1926*, 3 vols. (New York, 1958), vol. 1, pp. 189–329.

8. Carr, *The Bolshevik Revolution*, vol. 2, pp. 275–279. On the dichotomy between "large" Bolshevik theories and "small" peasant realities, see Roger Pethybridge, *The Social Prelude to Stalinism* (New York, 1974), pp. 196–242.

9. T. V. Osipova, *Rossiiskoe krest'ianstvo v revoliutsii i grazhdanskoi voine* (Moscow, 2001), p. 237; D. V. Kovalev, *Agrarnye preobrazovaniia i krest'ianstvo stolichnogo regiona v pervoi chetverti XX veka (po materialam Moskovskoi gubernii)* (Moscow, 2004), p. 161; V. V. Kabanov, *Krest'ianskoe khoziaistvo v usloviiakh "voennogo kommunizma"* (Moscow, 1988), pp. 83–84. For Lenin's comments, see *Desiatkii s"ezd RKP(b)*, p. 404; N. Bukharin and E. Preobrazhensky, *The ABC of Communism* (Ann Arbor, Mich., 1983), p. 301.

10. Moshe Lewin, "Rural Society in Twentieth Century Russia," in *The Making of the Soviet System: Essays in the Social History of Interwar Russia* (New York, 1985), p. 51; Danilov, *Rural Russia*, pp. 109–110; Shanin, *The Awkward Class*, pp. 164–169.

11. On denunciation letters, see Sheila Fitzpatrick, "Signals from Below: Soviet Letters of Denunciation of the 1930s," in Sheila Fitzpatrick and Robert Gellately, eds., *Accusatory Practices* (Chicago, Ill., 1997), pp. 85–120.

12. On the central agricultural establishment's reception of the national plan for electrification (GOELRO), see Jonathan Coopersmith, *The Electrification of Russia, 1880–1926* (Ithaca, N.Y., 1992), pp. 163–167.

13. For reports on the initial peasant reactions to the announcement of the NEP, see *Sovetskaia derevnia glazami VChK-OGPU-NKVD: Dokumenty i materialy,* 4 vols. (Moscow, 1998), vol. 1, pp. 400 ff. The report from Kaluga province is on p. 403.

14. Ibid., pp. 401, 434, 514, 528, 530, 535.

15. The population of the areas experiencing crop failure at this time was approximately thirty-five million; see *Naselenie Rossii XX veka,* vol. 1, pp. 129–131; Figes, *A People's Tragedy,* pp. 776–780. Figes analyzes the relationship between grain requisitioning and the magnitude of the famine along the Volga in *Peasant Russia, Civil War,* pp. 267–273.

16. See A. A. Kurenyshev, *Krest'ianstvo i ego organizatsii v pervoi treti XX veka* (Moscow, 2000), pp. 174–178.

17. On the role of Narkomzem officials in drafting the Land Code, see Heinzen, *Inventing a Soviet Countryside,* pp. 70–72.

18. *Istoriia sovetskogo krest'ianstva,* p. 251.

19. A Russian proverb meaning "Back where we started," popularized by Alexander Pushkin in his "Tale of the Fisherman and the Golden Fish."

20. P. A. Mesiatsev (1889–1938), an economist and professor at the Petrovsk Aricultural Acadamy and Land-Survey Institute. At the time of this letter, he was a collegium member of Narkomzem RSFSR.

21. In fact, the congress did not consider recognizing private land ownership. It did propose to give over to individuals all the land parcels they were currently working rather than undertake further redistributions. This flew in the face of most peasants' support for equalization of holdings and was taken by them as a rejection of land reorganization and a step in the direction of privatization.

22. In the original the author writes "prisech' sebe kak sviatye kliki" but probably intended "prichislit' sebe kak sviatye liki."

23. Vasily Grigorievich Yakovenko (1889–1938). A former peasant, Yakovenko had been named people's commissar of agriculture of the RSFSR in early 1922.

24. Lev Semenovich Sosnovsky (1886–1937), a political ally of Trotsky's in the 1920s. At the time of this letter he served as editor of *Bednota.*

25. For generations the peasants referred to the longed-for redistribution of land from the gentry to the peasantry as the "black repartition."

26. N. P. Rudin, a land reform specialist and former member of the Socialist-Revolutionary Party. In the early 1920s, he taught courses on land reform for Narkomzem in Ekaterinoslav.

27. In the original, "*liudi-spetsy*"—that is, prerevolutionary technical and other experts.

28. RGAE, f. 396, op. 2, d. 16, ll. 49–50.

29. Before the revolution, landowners counted how many "souls" they owned to work their land.

30. RGAE, f. 396, op. 2, d. 18, l. 326.

31. Atkinson, *The End of the Russian Land Commune,* pp. 246, 249.

32. The date refers to the proclamation of serf emancipation during the reign of Tsar Alexander II.

33. The term "revision soul" dates to prerevolutionary censuses that counted adult male serfs as a measure of a landowner's wealth. After serf emancipation in 1861, it was used to determine communal tax obligations.

34. This is one example of the astronomical inflation rate. Because of the unrestricted printing of money, prices in 1922 were 200,000 times higher than in 1913.

35. The report is contained in *Sovetskaia derevnia glazami,* vol. 2, pp. 198–199.

36. Here the writer uses the word *muzhichek,* a diminutive form of *muzhik.*

37. RGAE, f. 396, op. 4, d. 21, l. 239. These figures are cited in Danilov, *Rural Russia,* p. 160 (the acres—400,000 to 1,900,000—are here converted to hectares). On p. 165, Danilov also reports that some peasants believed grain would not grow on reorganized land. In Moscow province, land reorganization cost one ruble, ten kopeks per desiatina (one desiatina equals 2.7 acres or 1.09 hectares); see Kovalev, *Podmoskovnoe krest'ianstvoe,* p. 79. Beginning in 1925, the state provided more funds to poor peasants for land reorganization; see Danilov, *Rural Russia,* pp. 170–172; *RKI v sovetskom stroitel'stve* (Moscow, 1926), pp. 36–37; Atkinson, *The End of the Russian Land Commune,* p. 246.

38. *RKI v sovetskom stroitel'stve,* p. 37. Strictly speaking, *barshchina* refers to the labor obligations serfs owed their gentry landlords before emancipation, the Russian equivalent of corvée.

39. RGAE, f. 396, op. 2, d. 16, l. 390. The exact number of former gentry that remained in the countryside in the 1920s remains unknown. For the best account in English of this and related questions, see John Channon, "Tsarist Landowners after the Revolution: Former Pomeshchiki in Rural Russia during NEP," *Soviet Studies,* 39, no. 4 (October 1987), pp. 575–598.

40. Parentheses in original.

41. A tarantas is a type of low-slung four-wheeled horse-drawn carriage named for its resemblance to a tarantula. Note the letter writer's use of the spider analogy.

42. Danilov, *Rural Russia,* p. 98. The figure of 108,000 acres is here given in hectares.

43. Committees of Indigent Peasants (*Komitety nezamozhnikh selian* or *Komnezamy*), also abbreviated KNS, were the successors to the *kombeds.* The committees were created exclusively in Ukraine in May 1920 as an extension of Communist Party authority. Initially they carried out food and grain requisitioning, and then they regulated the allotment of land and inventory among poor peasants. Unlike kombeds, committees included middle peasants among their members. A republic-level Central Commission of Indigent Peasants (TsKNS) oversaw the committees' activities and was headed by G. I. Petrovsky, chairman of the Ukrainian central executive committee. The committees were disbanded in 1933. On the political significance of the committees, see Robert Conquest, *The Harvest of Sorrow,* esp. p. 40.

44. Artemis (Diana), the Greek goddess of hunting and defender of women, was at times identified as the goddess of nature.

45. Lazar Volin, *A Century of Russian Agriculture: From Alexander II to Khrushchev* (Cambridge, Mass, 1970), pp. 112–113; Lenin, *Collected Works,* vol. 32, p. 349. Lenin began expressing his interest in cooperatives soon after the October Revolution; see V. P. Danilov, ed., *Kooperativno-kolkhoznoe stroitel'stvo v SSSR, 1917–1922: Dokumenty i materialy* (Moscow, 1990), pp. 9, 13–14. On Soviet cooperatives, see Lewin, *Russian Peasants,* pp. 93–102; Carr, *Socialism in One Country,* vol. 1, pp. 275–282; Carr and Davies, *Foundations,* vol. 1, pp. 144–157.

46. V. V. Kabanov, *Kooperatsiia, revoliutsiia, sotsializm* (Moscow, 1996), pp. 158–162; Carr and Davies, *Foundations,* vol. 1, pp. 154–156.

47. G. Ye. Zinoviev coined the "face to the countryside" slogan in July 1924. On the 1924 harvest, see Carr, *Socialism in One Country,* vol. 1, pp. 189–209.

48. *Lineiki* and *drozhki* are types of wagons.

49. The remainder of the letter provides details on the way the local authorities prevent the peasants from obtaining free firewood during the ten-day forest-clearing periods, and the failures of the local literacy campaign.

50. RGAE, f. 478, op. 7, d. 564, l. 241.

51. Evidently a reference to a provincial (*guberniia*) conference of soviets that deeply impressed the author.

52. A reference to the last tsar, Nicholas II, who granted a limited constitution in October 1905, then gradually restricted its provisions.

53. See Danilov, *Rural Russia,* pp. 260–269.

54. In the original, *kulaki-miroedy. Miroedy* is a prerevolutionary term of abuse for kulaks that literally means "those who devour the commune."

55. On the question of defining the kulak, see Viola, "The Peasants' Kulak"; Moshe Lewin, "Who Was the Soviet Kulak?" in *The Making of the Soviet System,* pp. 121–141.

56. In the original: "black," translated here as "degrading."

57. The writer uses *rossiiskie,* a word signifying all residents of the Russian republic, rather than *russkie,* the term for ethnic Russians.

58. Here, the author of the letter wrote *kazhdyi* (each) but probably intended *vazhnyi* (important).

59. Elise Kimerling, "Civil Rights and Social Policy in Soviet Russia, 1918–1936," *Russian Review,* 41, no. 1 (January 1982), pp. 30–35.

60. A reference to the sections of the 1918 RSFSR Constitution that denied the franchise to anyone employing the labor of another for profit or engaging in commercial activities.

61. *Krestianskii komitet obshchestvennoi vzaimopomoshchi* (KKOV), or Peasant Committee of Social Mutual Aid. Created by a Council of People's Commissars decree in May 1921, the committees were attached to district (*volost*) and village-level soviets. They provided various forms of social and material aid to families of Red Army soldiers, famine victims, poor peasants, and others in dire straits.

62. In the original, "za dlinnymi chervontsami."

63. The word *farsivyi,* used here in the phrase "s farsivyim spaseniem svoego zdorov'ia," is not entirely clear.

64. A *baranka* is a ring-shaped loaf of white bread.

65. In the original: *band net.* This is probably a reference to armed anti-Soviet groups.

66. Danilov, *Rural Russia,* pp. 219, 214–216, 211–212. The total of twenty-five million households is based on information gathered by the Central Statistical Administration (TsSU). For reasons that he explains, Danilov believes this figure may overestimate the total by as much as one million, but the lesser figure would still represent a substantial increase.

CHAPTER 3. *SMYCHKA:* THE BOND BETWEEN CITY AND VILLAGE

1. *PSS,* vol. 45, pp. 76–77.

2. Ibid., pp. 369–377; N. I. Bukharin, "Put' k sotsializmu i raboche-krest'ianskii soiuz," in *Put' k sotsializmu. Izbrannye proizvedeniia* (Novosibirsk, 1990), pp. 35–36.

3. When plotted on a single graph, rising prices for industrial goods and falling prices for agricultural products resembled the open blades of a scissors, hence the name.

4. For a valuable discussion of the competing developmental theories of the left and right wings of the Party, see Lewin, *Russian Peasants*, pp. 132–171.

5. The Moscow figures are from Kovalev, *Podmoskovnoe krest'iantsvo*, table 3, p. 129. The national figures are from L. Kritsman, *Geroicheskii Period Velikoi Russkoi Revoliutsii*, as quoted in E. H. Carr, *The Bolshevik Revolution*, vol. 2, p. 171. The percentages have been rounded.

6. On the difficulties of peasant classification, see Lewin, *Russian Peasants*, pp. 41–80.

7. Published in English as F. Panferov, *Brusski: A Story of Peasant Life in Soviet Russia* (Westport, Conn., 1977), pp. 104–105.

8. For the debates over the concessions, see Carr, *Socialism in One Country*, vol. 1, pp. 240–282; Charles Bettelheim, *Class Struggles in the USSR: Second Period, 1923–1930* (Sussex, England, 1978), pp. 89–91.

9. See *Vos'moi s"ezd RKP(b), mart 1919g.: Protokoly* (Moscow, 1953), pp. 406.

10. Figures on the rural Communist Party are from Daniel Thorniley, *The Rise and Fall of the Soviet Rural Communist Party, 1927–39* (New York, 1988), pp. 11–17 and from the tables on pp. 200–204. Fainsod, *Smolensk*, p. 44.

11. Lynne Viola, *The Best Sons of the Fatherland: Workers in the Vanguard of Soviet Collectivization* (New York, 1987), p. 23. On shefstvo, see Chase, *Workers, Society and the Soviet State: Labor and Life in Moscow*, pp. 301–302; Carr, *Socialism in One Country*, vol. 2, pp. 342–344.

12. For an overview of Soviet efforts to eliminate illiteracy in the 1920s and 1930s, see Pethybridge, *The Social Prelude*, pp. 132–186. On the civil war campaigns, see Peter Kenez, *The Birth of the Propaganda State: Soviet Methods of Mass Mobilization, 1917–1929* (Cambridge, England, 1985), pp. 70–83; V. A. Kozlov, *Kul'turnaia revoliutsiia i krest'ianstvo, 1921–1927* (Moscow, 1983), p. 45.

13. The Society "Down with Illiteracy!" (ODN) operated between 1923 and 1936 to eliminate adult illiteracy and was especially active in rural areas; see Kenez, *The Birth of the Propaganda State*, pp. 153–156.

14. Kozlov, *Kul'turnaia revoliutsiia*, pp. 42–45. On the methods and results of literacy work in the 1920s, see Kenez, *Birth of the Propaganda State*, pp. 145–166. On the findings of the 1926 census, see Pethybridge, *The Social Interlude*, pp. 155–156; Danilov, *Rural Russia*, pp, 43–44. Pethybridge and Danilov provide slightly different figures.

15. See Jeffrey Brooks, "The Breakdown in Production and Distribution of Printed Material, 1917–1927," in Abbott Gleason et al., eds., *Bolshevik Culture: Experiment and Order in the Russian Revolution* (Bloomington, Ind., 1985), pp. 151–171.

16. In February 1924, the gold-backed *chervonets*, which had been in circulation since 1922, completely replaced the existing, highly inflated paper ruble known as the *sovznak*.

17. Though not a Communist Party member, Ignatiev is addressed here as "comrade."

18. On communist ceremonies, see Christopher A. P. Binns, "The Changing Face of Power: Revolution and Accommodation in the Development of the Soviet Ceremonial System, Part 1," *Man*, 4 (1979), pp. 585–606.

19. Zakotnov is referring to organized labor activity in Moscow-area textile factories. On the extent of strike activity in Moscow at this time, see Diane Koenker, *Moscow Workers and the 1917 Revolution* (Princeton, N.J., 1981), pp. 90–91.

20. In the original, "svobodnyi put' sol'et mashiny bol'she, chem liudei."

21. Frances L. Bernstein examines one manifestation of this problem and the medical profession's response to it, in "Panic, Potency, and the Crisis of Nervousness in the 1920s," in Christina Kiaer and Eric Naiman, eds., *Everyday Life in Early Soviet Russia: Taking the Revolution Inside* (Bloomington, Ind., 2006), pp. 156–182.

22. *Gudok* (The Whistle) was the daily newspaper of the railroad workers' union.

23. A reference to Kalinin's speech to the Twelfth Communist Party Congress, April 1923.

24. On the All-Russian Peasant Union during the 1905 Revolution, see Teodor Shanin, *Russia 1905–07: Revolution as a Moment of Truth* (New Haven, Conn., 1986), pp. 114–124. For a history of the union movement, see A. A. Kurenyshev, *Vserossiiskii krest'ianskii soiuz, 1905–1930 gg. Mify i real'nost'* (Moscow and St. Petersburg, 2004), pp. 264–274 (for the unions' role in the Tambov rebellion) and 284–287 (for information on a "Soviet" peasant union); *Sovetskaia derevnia glazami,* vol. 2, pp. 192–193; *TSD,* vol. 1, p. 575.

25. As the author correctly notes later, this famous manifesto was actually published in October, not December, 1905.

26. In fact, the composition of both the first (April–July 1906) and second (February–June 1907) Dumas (parliaments) was quite radical, and each was dissolved by Nicholas II. Only after the government, led by Pyotr Stolypin, drastically (and unconstitutionally) changed the requirements for voter eligibility in June 1907 was a conservative, third Duma (1907–1912) elected and allowed to sit for its entire term.

27. Probably a local cooperative organization.

28. Yakov Arkadeevich Yakovlev (1896–1939) was editor of *Krestianskaia gazeta* and a Party spokesman on rural affairs.

29. The reference is to Tsar Ivan IV's paramilitary force—the *oprichnina*—that terrorized and displaced the sixteenth-century Muscovite ruling class of boyars. The Russian root *oprich* means "apart," and the author is making an obvious play on the meaning.

30. See Carr, *The Bolshevik Revolution,* vol. 1, pp. 150–154. On the actions of peasant assemblies in the revolution, see Figes, *Peasant Russia, Civil War,* esp. 70–153.

31. Viola, *The Best Sons,* pp. 65–66.

32. On changes in the tax regulations, see Carr and Davies, *Foundations,* vol. 1, pp. 746–755. The total agricultural tax collected in 1925/1926 amounted to 252 million rubles; the 1926/1927 figure was 358 million.

33. On the elections, see ibid., vol. 2, pp. 274–281.

34. In the original: *splaatatory* for *ekspluatatory,* and *izplaatirovat'* for *ekspluatatirovat'.*

35. A reference to article 107 of the RSFSR Criminal Code punishing speculation. Beginning in 1927, this article was used to prosecute "kulaks" alleged to be hoarding grain.

36. "State fund" (*gosudarstvennyi fond*) refers to the state land reserve from which state farms (*sovetskie khoziaistva* or *sovkhozy*) were created. The fund also

consisted of meadows, forests, and orchards that generated money for the state. The author evidently wants to see the land from this reserve made available to the peasants at affordable prices.

37. For some reason this letter ended up in the *Krestianskaia gazeta* files.

38. *Kommunkhozy* is the author's own neologism combining the Russian words *kommuna* (commune) and *khoziaistvo* (farm).

39. *Krasnyi voin* (The Red Warrior) was the chief publication of the political department (*politotdel*) of the Revolutionary-Military Council (*Revvoensovet*) of the Second Army fighting on the eastern during the civil war.

40. There may be some confusion in Kuznetsov's timeline. Minsk, which had been initially taken by the Red Army in December 1918, fell to the Polish army in August 1919. It was then recaptured by Soviet forces under General Tukhachevsky's command in July 1920. See Evan Mawdsley, *The Russian Civil War* (Boston, 1987), pp. 118–119, 253.

41. I.e., born lucky.

CHAPTER 4. WAS SOCIETY TRANSFORMED?

1. On this subject, see Pethybridge, *The Social Prelude.*

2. Leon Trotsky, "Chtobyi perestroit' byt, nado poznat' ego," in *Sochineniia* (Moscow-Leningrad, 1927), vol. 21, p. 21.

3. The acronym GOELRO is derived from *Gosudarstvennaia Komissiia po Elektrifikatsii Rossii* (The State Commission for the Electrification of Russia). See Eugène Zaleski, *Planning for Economic Growth in the Soviet Union, 1918–1932* (Chapel Hill, N.C., 1971), pp. 35–40.

4. *PSS*, vol. 42, pp. 160–161.

5. Only 84,000 out of 18,000,000 Russian peasant households had electricity by 1926. See Coopersmith, *The Electrification of Russia*, pp. 243.

6. See Larry E. Holmes, *The Kremlin and the Schoolhouse: Reforming Education in Soviet Russia, 1917–1931* (Bloomington, Ind., 1991), pp. 7–9.

7. Ibid., pp. 91–96.

8. Ibid., p. 97. On the Party's anti-illiteracy efforts, see Kenez, *The Birth of the Propaganda State*, pp. 70–83.

9. For Grachev's account, see *Golosa krest'ian*, pp. 51–52. The figures are taken from Kenez, *The Birth of the Propaganda State*, pp. 146–147.

10. Neil B. Weissman, "Origins of Soviet Health Administration: Narkomzdrav, 1918–1928," in Susan Gross Solomon and John F. Hutchinson, eds., *Health and Society in Revolutionary Russia* (Bloomington, Ind., 1990), pp. 97–107; Christopher Davis, "Economic Problems of the Soviet Health Service, 1917–1930," *Soviet Studies*, 35, no. 3 (July 1983), pp. 343–361.

11. On the Komsomol's early years, see Kenez, *The Birth of the Propaganda State*, pp. 84–94. On the relations between the peasants and the Komsomol during the NEP years, see Isabel A. Tirado, "The Komsomol and Young Peasants: The Dilemma of Rural Expansion, 1921–1925," *Slavic Review*, 52, no. 3 (1993), pp. 460–476; Kenez, *The Birth of the Propaganda State*, pp. 177–185.

12. See, for example, Alexandra Kollantai, "Sexual Relations and the Class Struggle" and "Communism and the Family," in *Selected Writings* (London, 1977); Barbara Evans Clements, "The Utopianism of the Zhenotdel," *Slavic Review*, 51, no. 3 (1992), pp. 486–487.

13. Clements, "The Utopianism of the Zhenotdel," pp. 490–496.

14. In the original Russian: *Kirianikha-znakharka.*

15. In the 1920s and 1930s, the Communist Party used meetings of "women delegates" (*delegatki*) as vehicles for organizing women at workplaces and through local soviets.

16. RGASPI, f. 89, op. 3, d. 152, ll. 6, 8. The word *duma* shares the root *dum* with the Russian verb "to think" (*dumat'*) and usually refers to elected legislative assemblies or city councils. On reading huts and their effectiveness as agents of indoctrination, see Kenez, *The Birth of the Propaganda State,* pp. 137–143.

17. For a history of Russian taverns and drinking practices, see I. V. Kurukin and E. A. Nikulina, *Povsednevnaia zhizn' russkogo kabaka ot Ivana Groznogo do Borisa El'tsina* (Moscow, 2007).

18. On state efforts to regulate alcoholic beverages, see Neil Weissman's highly informative article "Prohibition and Alcohol Control in the USSR: The 1920s Campaign against Illegal Spirits," *Soviet Studies,* 38, no. 3 (July 1986), pp. 349–368.

19. S. E. Panin, "Krest'ianstvo i samogon v Srednem Povolzh'e v 1920-e gg.," in V. A. Yurchenkov, ed., *Krest'ianstvo i vlast' Srednego Povolzh'ia: Materialy VII mezhregion. nauch.-prakt. konf. istorikov-agrarnikov Srednego Povolzh'ia* (Saransk, 2003), pp. 322–323.

20. Susan Gross Solomon, "David and Goliath in Soviet Public Health: The Rivalry of Social Hygienists and Psychiatrists for Authority over the *Bytovoi* Alcoholic," *Soviet Studies,* 41, no. 2 (April 1989), pp. 259–260; Weissman, "Prohibition," pp. 355–356, 359, 364; Hindus, *Red Bread,* p. 196.

21. Local troikas usually consisted of the Party chief, the soviet chairman, and the head of the militia.

22. RGAE, f. 396, op. 3, d. 391, l. 3.

23. Weissman, "Prohibition," p. 355; Leon Trotsky, "Vodka, tserkov i kinomatograf," in *Sochineniia* (Moscow-Leningrad, 1927), vol. 21, pp. 22–25; TsGAOD g. Moskvy, f. 3, op. 11, d. 310, ll. 30–31.

24. See Ben Eklof, *Russian Peasant Schools: Officialdom, Village Culture, and Popular Pedagogy, 1861–1914* (Berkeley, Calif., 1986), pp. 292–293; Holmes, *The Kremlin and the Schoolhouse,* p. 96.

25. On Soviet education policy and its effects during the NEP and the First Five-Year Plan, see Sheila Fitzpatrick, *Education and Social Mobility in the Soviet Union, 1921–1934* (Cambridge, England 1979). These figures are taken from the table on p. 62.

26. Holmes, *The Kremlin and the Schoolhouse,* pp. 28–29.

27. Ibid., pp. 97–98 and 113–114.

28. The overwhelming majority of demobilized soldiers returned to their home villages; see Von Hagen, *Soldiers in the Proletarian Dictatorship,* p. 299.

29. The "Higher Party School" (*Vysshaia partinaia shkola*), or Sverdlov Communist University, had been created in Moscow in June 1918 to train Party propagandists and instructors.

30. On worker faculties, see Fitzpatrick, *Education and Social Mobility,* pp. 49–51 ff.

31. See Tirado, "The Komsomol and Young Peasants."

32. On the complexities of Soviet social identities, see Sheila Fitzpatrick, "Ascribing Class: The Construction of Social Identity in Soviet Russia," *Journal of Modern History,* 65, no. 4 (December 1993), pp. 745–770; Fitzpatrick, "The Two Faces of

Anastasia: Narratives and Counter-Narratives of Identity in Stalinist Everyday Life," in Kiaer and Naiman eds., *Everyday Life,* pp. 23–34.

33. Pavel Karlovich Renenkampf (1854–1918), a Russian imperial cavalry commander. Following the 1905 Revolution, he led punitive expeditions to suppress the revolutionary movement in eastern Siberia.

34. The SRs are the Socialist-Revolutionaries, and the SDs are the Social Democrats. In the Russian text, the author uses the acronyms phonetically: *esery* and *esdeki.*

35. "Always prepared!" was a Komsomol slogan.

36. See Peter Gooderham, "The Komsomol and Worker Youth: The Inculcation of 'Communist Values' in Leningrad during NEP," *Soviet Studies,* 34, no. 4 (October 1982), pp. 506–528. The report is quoted in Fainsod, *Smolensk,* p. 412.

37. Gooderham, "The Komsomol and Worker Youth," p. 513.

38. Olga Semyonova Tian-Shanskaia describes such gatherings in the late nineteenth century in *Village Life in Late Tsarist Russia* (Bloomington, Ind., 1993), pp. 51–57.

39. RGAE, f. 396, op. 3, d. 391, l. 93.

40. RGAE, f. 396, op. 7, d. 14, ll. 99–100.

41. See Wendy Z. Goldman, *Women, the State and Revolution: Soviet Family Policy and Social Life, 1917–1936* (Cambridge, England, 1993), pp. 1–58, 144–183.

42. The Communist Party sponsored conferences of peasants as a means of organizing cohorts of noncommunist rural activists.

43. In the 1920s, Narkompros organized children's labor colonies to provide care and vocational training to orphans. These colonies should not be confused with penal institutions.

44. Goldman, *Women, the State, and Revolution,* pp. 103–109. The original quatrain is "Sovetskaia vlast': muzha ne boiusia, esli plokho budem zhit', voz'mu—razvedusia."

45. The original: "my rebiata slavnye . . . a potom v kusty." Here the writer is implying that the man may be inclined to forget his responsibilities to wife and child, abandon them, and seek out another woman.

46. The Kirghiz Autonomous Soviet Socialist Republic created in 1920, was renamed the Kazakh ASSR in 1925.

47. Usually "old style" is used to distinguish the Julian calendar in use before 1918 from the Gregorian "new style" calendar. In the twentieth century the Julian calendar was thirteen days behind the Gregorian calendar.

48. Anna Louise Strong, *I Change Worlds: The Remaking of an American* (New York, 1937), pp. 191–205.

49. To lower the costs of the state apparatus, the Soviet government initiated the "regime of economy" in June 1926. During the campaign, most state institutions were forced to cut budgets and reduce the size of their staffs.

50. According to a 1925 study of local budgets cited by Christopher Davis, urban per capita health expenditures averaged 3.53 rubles; rural, only 29 kopeks. See Davis, "Economic Problems," p. 350; Chamberlin, *Soviet Russia,* pp. 188–189.

51. On the Bolshevik response to religious belief, see Stites, *Revolutionary Dreams,* esp. pp. 101–123.

52. See ibid., pp. 109–111; Leon Trotsky, "Sem'ia i obriadnost'," in *Sochineniia* (Moscow-Leningrad, 1927), vol. 21, pp. 39–42.

53. The Young Sparticists (*Iunye spartakovtsy*) was a children's organization of the 1920s.

54. Hindus, *Red Bread,* pp. 186–192.

55. Stites, *Revolutionary Dreams,* p. 122. The peasants' "dual faith" (*dvoeverie*), a combination of pagan and Christian beliefs, is the best example of this.

56. Like the Russian word for christening (*krestiny*), *oktiabriny* is a plural-only noun. Following the usual custom and for clarity's sake, it has been rendered here as a singular.

57. For a more extensive list and discussion, see Stites, *Revolutionary Dreams,* pp. 111–112.

58. This is a derogatory reference to artistically inferior icons attributed to painters from the cities of Vladimir and Suzdal.

59. On official perceptions of peasant women, see Lynne Viola, "Bab'i Bunty and Peasant Women's Protest during Collectivization," in Beatrice Farnsworth and Lynne Viola, eds., *Russian Peasant Women* (New York, 1992).

60. RGAE, f. 396, op. 3, d. 391, l. 94.

61. The father evidently slaughtered an animal in preparation for the wedding feast.

62. This is probably a reference to changes in admissions policy in higher education that went into effect in 1926. The new policy emphasized academic performance over class and political criteria; see Fitzpatrick, *Education and Social Mobility,* pp. 105–107.

63. *Sovetskaia derevnia glazami,* vol. 2, p. 414.

64. Ibid.

65. Three-quarters of all prison sentences imposed in 1926 were for hooliganism. See Peter H. Solomon Jr., *Soviet Criminal Justice under Stalin* (Cambridge, England, 1996), p. 58.

66. On this and on the antihooligan campaign, see Neil Weissman, *Russia in the Era of NEP: Explorations in Soviet Society and Culture* (Bloomington, Ind., 1991), pp. 186–187; *Sovetskaia derevnia glazami,* vol. 2, p. 415.

67. The people's court (*narodnyi sud* or *narsud*), the lowest-level Soviet court, was presided over by a judge and two lay assessors.

68. Moshe Lewin, "Society, State and Ideology during the First Five-Year Plan," esp. pp. 216–218, and "The Social Background to Stalinism," in *The Making of the Soviet System.*

Chapter 5. People and Power

1. Yury Libedinsky, *A Week* (New York, 1923), p. 153.

2. On the size of the rural Communist Party in the 1920s, see Daniel Thorneily, *The Rise and Fall of the Soviet Rural Communist Party, 1927–39* (New York, 1988), pp. 11–23; Fainsod, *Smolensk,* pp. 44–45, 141.

3. On the difficulties in determining the exact number of soviets, see D. J. Male, *Russian Peasant Organisation before Collectivisation: A Study of Commune and Gathering, 1925–1930* (Cambridge, England, 1971), pp. 87–97; Carr, *Socialism in One Country,* vol. 2, pp. 321–323. On the relations between rural soviets and peasant communes, see Atkinson, *The End of the Russian Land Commune,* pp. 295–312.

4. See V. V. Kuibyshev's report in *RKI v sovetskom stroitel'stve* (Moscow, 1926), p. 34.

5. *Sovetskaia derevnia glazami,* vol. 2, pp. 263–265.

6. On the various stages of soviet revitalization, see John Slatter, "Communes with Communists: The *Sel'sovety* in the 1920s," in Roger Bartlett, ed., *Land Commune and Peasant Community in Russia: Communal Forms in Imperial and Early Soviet Society* (New York, 1990), pp. 272–282.

7. On the soviet revitalization campaign, see Carr, *Socialism in One Country*, vol. 2, pp. 304–340.

8. Atkinson, *The End of the Russian Land Commune*, p. 298; Male, *Russian Peasant Organization*, p. 89 and the table on pp. 92–93; *RKI v sovetskom stroitel'stve*, p. 35. There were 13,897 rural Party cells in 1924 and 20,719 in 1928. Between 1925 and 1930, however, average cell membership increased by only one—from eleven to twelve; see Thorneily, *The Rise and Fall*, p. 17.

9. *Sovetskaia derevnia glazami*, vol. 2, pp. 266, 638–639.

10. *Istoriia sovetskogo krest'ianstva*, vol. 1, p. 302; S. N. Ikonnikov, *Sozdanie i deiatel'nost' ob"edinennykh organov TsKK-RKI v 1923–1934gg.* (Moscow, 1971), p. 151. On the Dymovka Affair, see Carr, *Socialism in One Country*, vol. 1, pp. 196–198.

11. See Steven Coe, "Struggles for Authority in the NEP Village: The Early Rural Correspondents Movement, 1923–1927," *Europe-Asia Studies*, 48, no. 7 (1996), pp. 1151–1171. In July 1925, the Political Administration of the Red Army established courses designed to train soldiers for various positions in the countryside, including as selkors; see Von Hagen, *Soldiers in the Proletarian Dictatorship*, p. 303.

12. *KPSS v rezoliutsiakh*, vol. 3 pp. 256–257.

13. Menshevism had historically been stronger in Georgia than Bolshevism. On the rebellion, see Ronald Grigor Suny, *The Making of the Georgian Nation* (Bloomington, Ind., 1994), pp. 223–225. A telegram sent by Stalin on behalf of the Central Committee to the Transcaucasian Party committee, dated 2 September, demanded a stop to the shootings of arrestees that had evidently been undertaken by the Transcaucasian Cheka; see *Sovetskaia derevnia glazami*, vol. 2, p. 239; Stalin, *Works* (Moscow, 1954), vol. 7, pp. 20–21.

14. Stalin, *Works*, vol. 7, pp. 22–23.

15. Ibid. vol. 6, pp. 320–323.

16. Nina Tumarkin, *Lenin Lives! The Lenin Cult in Soviet Russia* (Cambridge, Mass., 1997), pp. 134–143. On the antecedents and early expressions of the Lenin cult, see Figes and Kolonitskii, *Interpreting the Russian Revolution*, pp. 71–103. In Mikhail Chiaureli's 1946 film *The Vow* (*Kliatva*), peasants trek to Gorki to seek redress for their treatment at the hands of malicious kulaks but on arrival find that Lenin has already expired. On actual peasant pilgrimages to Gorki, see Tumarkin, *Lenin Lives!* pp. 136–137. For letters see, for example, *Serdtsem i imenem* (Moscow, 1967).

17. The writer may be referring to Vyshne-Volotsky county in Tver province.

18. Probably a reference to panic in connection with savings banks.

19. The state inspectorate, the People's Commissariat of Worker-Peasant Inspection, known by its Russian acronym, Rabkrin, was responsible for investigating and eliminating mismanagement and bureaucratization in governmental agencies and economic enterprises. Just what role the author assumes it is playing in this context is not entirely clear.

20. RGAE, f. 396, op. 3, d. 100, ll. 49, 43.

21. S. M. Zav'ialov, ed., *Neizvestnaia Rossiia, Sbornik Dokumentov* (Moscow, 1993), vol. 4, pp. 12–20.

22. In the early years, non-Muslim central appointees dominated Party and government institutions in the Kazakh republic; see Richard Pipes, *The Formation of the Soviet Union: Communism and Nationalism, 1917–1923* (Cambridge, Mass., 1997), pp. 172–174.

23. Stalin, *Sochineniia* (Moscow, 1953), vol. 6, pp. 302–312.

24. *The ABC of Communism,* the 1919 primer cowritten by N. I. Bukharin, was aimed at these new recruits. See, for example, Isaac Deutscher, *The Prophet Unarmed: Trotsky, 1921–1929* (New York, 1965), pp. 135–136. E. H. Carr emphasizes the bureaucratizing effects of Party growth and purging; see Carr, *The Interregnum, 1923–1924* (Baltimore, Md., 1969), pp. 358–364. According to Catherine Merridale, the stultifying effects of membership expansion were not immediately evident in the Moscow Party organization; see Merridale, *Moscow Politics and the Rise of Stalin: The Communist Party in the Capital, 1925–32* (New York, 1990), pp. 25–26.

25. *Gaidamaki* or *haidamaki* originally referred to bands of eighteenth-century right-bank Ukrainian peasants and Cossacks who rebelled against Polish rule. Here, the author is evidently referring to Ukrainian nationalists who fought the Red forces during the civil war. The terms are anglicized in the text.

26. Anton Ivanovich Denikin (1872–1947), imperial Russian army general during World War I and, following the death of General L. G. Kornilov in April 1918, commander of the anti-Bolshevik "Volunteer Army" during the civil war.

27. The Devil's Hundreds was a special detachment of the First Officers' Regiment of the Volunteer Army. Denikin renamed the regiment in honor of its former commander, General S. L. Markov, after his death in June 1918.

28. Baron Petr Nikolaevich Wrangel' (1878–1928), imperial Russian army general during World War I, who replaced General Denikin as commander of the White forces in southern Russia in April 1920.

29. The nature of these jobs—fighting "counterrevolution and speculation"—indicates that Burtsev may have worked with or for the Cheka.

30. The people's court (*narodnyi sud,* or *narsud*), the lowest-level Soviet court, was presided over by a judge and two lay assessors.

31. See Fitzpatrick, "Supplicants and Citizens," pp. 78–105.

32. Viola, *The Best Sons,* p. 21. Georgy Gapon, an Orthodox priest who ministered to St. Petersburg's working class, became influential in the turn-of-the-century police-union movement and led the 9 January demonstration. Sergei Zubatov, a tsarist police official, is credited with conceiving the police unions. For these examples, see *Itogi proverki chlenov i kandidatov RKP(b) neproizvodstvennykh iacheek* (Moscow, 1925), p. 70.

33. Fainsod, *Smolensk,* p. 48.

34. At the time of this letter, Trotsky was already a politically isolated and disgraced figure within the Party leadership. As this honorific salutation attests, however, for many outside leading Party circles, he remained the renowned revolutionary leader and civil war hero of old.

35. The appearance of the "Miting" column in the *Krestianskaia gazeta* initiated a serious discussion about rural problems and elicited a flood of letters from readers.

36. The horn of plenty was an often-encountered Soviet symbol of agricultural abundance.

37. A reference to rumors of imminent war with Poland and England.

38. From the Russian proverb "From righteous labors, you will not get palaces of stone" (*Ot trudov pravednykh ne nazhevesh' palat kamennykh*), expressing the belief that hard work is not rewarded with wealth.

39. RGAE, f. 396, op. 6, d. 114, l. 745.

40. On peasant manipulation of the "myth of the good tsar" in the nineteenth century, see Daniel Field, *Rebels in the Name of the Tsar* (Boston, 1976).

41. A reference to state bonds (*obligatsii krest'ianskogo zaima*), for which peasants were often pressured to subscribe.

42. Volkhovstroi, near Leningrad, and Dneprostroi, in Ukraine, were large-scale projects to construct electrical-generating stations.

43. TsGAOD g. Moskvy, f. 3, op. 11, d. 310, l. 74.

44. *Itogi proverki*, pp. 85, 91; TsGAOD g. Moskvy, f. 3, op. 11, d. 310, l. 46.

45. The precise reference is unclear. In January 1925, Stalin delivered two speeches in which he criticized local Party officials, but he did not explicitly call for a purge, Stalin, *Works*, vol. 7, pp. 19–33.

46. It would appear that the writer was renting land in the sovkhoz mentioned in the letter.

47. In peasant dwellings the large stoves, which often took up to a third of the living space, also served as desirable sleeping spots.

48. *Itogi proverki*, p. 85; RGAE, f. 396, op. 6, d. 28, l. 35.

49. RGASPI, f. 89, op. 3, d. 152, ll. 5–8.

50. The author uses the attractively alliterative phrase *partiets-p'ianitsa*.

51. An ironic reference to a well-known verse from the communist anthem, *The International*.

52. On the Shakhty Affair, see Kendall E. Bailes, *Technology and Society under Lenin and Stalin: Origins of the Soviet Technical Intelligentsia, 1917–1941* (Princeton, N.J., 1978), pp. 69–94; Hiroaki Kuromiya, *Stalin's Industrial Revolution: Politics and Workers, 1928–1932* (Cambridge, England, 1988), pp. 12–17.

53. Probably a reference to Rykov's speech on the Shakhty Affair that appeared in *Pravda* on 11 March 1928.

54. In early 1928, Ibraimov (Ibragimov) was expelled from the Party and arrested for collusion with kulaks. In April, the RSFSR Supreme Court sentenced him to death in connection with the murder of a witness who threatened to expose his relationship with former counterrevolutionaries. On the Ibraimov affair, see Carr and Davies, *Foundations*, vol. 2, p. 136; *Kak lomali NEP: Stenogrammy plenumov TsK VKP(b), 1928–1929 gg.*, 5 vols. (Moscow, 2000), vol. 3, pp. 458, 630.

55. Article 76 of the RSFSR Criminal Code details the punishments for "publicly insulting" governmental representatives in the course of their duties.

CHAPTER 6. WHITHER SOCIALISM?

1. *KPSS v rezoliutsiiakh*, vol. 3 p. 427.

2. On Lenin's political activity during his illness, see Moshe Lewin, *Lenin's Last Struggle* (New York, 1970). A number of classic accounts detail the factional conflicts of the 1920s. Among these are Deutscher, *The Prophet Unarmed: Trotsky;* Robert V. Daniels, *The Conscience of the Revolution: Communist Opposition in Soviet Russia* (New York, 1969); Roy Medvedev, *Let History Judge: The Origins and Consequences of Stalinism* (New York, 1989).

3. For details, see Carr and Davies, *Foundations*, vol. 1, pp. 271–312.

4. Chase, *Workers, Society, and the Soviet State*, pp. 136, 173–188, 200–204.

5. See *Sovetskaia derevnia glazami*, vol. 2, pp. 416–418, 427–432.

6. On Soviet tax policy at this time, see Carr and Davies, *Foundations*, vol. 1, pp. 752–763. On the Fifteenth Party Congress, see Lewin, *Russian Peasants*, pp. 198–213.

7. Carr and Davies, *Foundations*, pp. 752–763. On the self-assessment, see Atkinson, *The End of the Russian Land Commune*, pp. 308–311.

8. For these and other accounts, see *Sovetskaia derevnia glazami*, vol. 2, pp. 473–474, 496–497.

9. Alexander Herzen, *Childhood, Youth and Exile* (Oxford, 1980), p. 217.

10. On the evolution of the repartitional commune, see Jerome Blum, *Lord and Peasant in Russia: From the Ninth to the Nineteenth Century* (Princeton, N.J., 1971), pp. 508–514. For the 1905 Socialist-Revolutionary Party program, see Basil Dmytryshyn, *Imperial Russia: A Source Book, 1700–1917*, 2nd ed. (Hinsdale, Ill., 1974), pp. 399–405.

11. Figes and Kolonitskii, *Interpreting the Russian Revolution*, esp. pp. 127–152.

12. See Eric R. Wolf, *Peasants* (Englewood Cliffs, N.J., 1966), pp. 107–109.

13. The writer uses the phrase *polirovannye zveri*, literally "polished beasts."

14. As noted in the text at the Fourteenth Party Congress (December 1925), Stalin, Bukharin, and their supporters routed the opposition headed by Zinoviev and Kamenev. Trotsky, however, did not speak at the congress.

15. On the Bolshevik conceptualization of suicide and "eseninism," see Kenneth M. Pinnow, "Violence against the Collective and the Problem of Social Integration in Early Bolshevik Russia," *Kritika: Explorations in Russian and Eurasian History*, 4, no. 3 (2003), pp. 653–677, esp. p. 669 n. 48.

16. TsGAOD g. Moskvy, f. 3, op. 11, d. 440, ll. 88, 111, 93, 87. A. V. Lunacharsky (1875–1933) served as people's commissar of enlightenment from 1917 to 1929.

17. Ibid., ll. 86 and 87.

18. TsGAOD g. Moskvy, f. 3, op. 11, d. 440, d. 310, l. 55 and d. 440, l. 18.

19. In addition to working as head of the Central Institute of Labor from 1920 to 1938, Gastev (1882–1941) penned collections of verse, including *The Poetry of the Worker's Stroke* (1918), and wrote the study *How One Must Work* (1921).

20. Burkma is the acronym for the Administration of Boring Works of the Kursk Magnetic Anomaly.

21. *Na ispol*, from *ispol'shchina*, is a form of sharecropping in which one-half (*polovina*) of the harvest is paid out as rent.

22. See Scott, Domination and the Arts of Resistance, pp. 87–103.

23. By "associations" the author has in mind the TOZ.

24. TsGAOD g. Moskvy, f. 3 op. 11, d. 440, l. 88.

25. "Pruzhinka" comes from the Russian *pruzhina* (spring); that is, the teeth of the harrow are on springs.

26. The author spells out the word *tochka* (period) in the parentheses for emphasis.

27. TsGAOD g. Moskvy, f. 3, op. 11, d. 440, l. 87.

CHAPTER 7. THE GREAT BREAK

1. The stenographic reports of the critical Central Committee plenary sessions of this period are published in *Kak lomali NEP*.

2. On the Chinese events, see Helmut Gruber, *Soviet Russia Masters the Comintern: International Communism in the Era of Stalin's Ascendacy* (Garden City, N.Y., 1974), pp. 324–500; Michael Reiman, *The Birth of Stalinism: The USSR on the Eve of the Second Revolution* (Bloomington, Ind., 1987), pp. 14–15. In the immediate aftermath of the assassination on 7 June, Stalin, from his vacation home in Sochi, telegraphed Molotov with instructions to declare all "monarchists" in Soviet custody "hostages" and "to shoot five or ten" of them. He further ordered the OGPU to seek out and arrest all "monarchists" and "White Guardists" throughout the entire country. The actual Politburo decision, taken the next day, exceeded Stalin's instructions. Twenty "monarchists" were shot, and the OGPU was granted the right to carry out "extrajudicial sentences up to shooting." Based on reports in the German press, Reiman concludes that at least fifteen thousand were arrested and estimates that hundreds were ultimately shot. For a detailed account of the telegram exchanges between Stalin and Molotov, and the Politburo's decisions, see V. P. Danilov's introduction in *TSD*, vol. 1, pp. 22–23.

3. *TSD*, vol. 1, pp. 23–27.

4. On the decision to lower prices for industrial goods, see Carr and Davies, *Foundations*, vol. 1, pp. 684–696. State prices for these items had been raised in the summer; see p. 46.

5. For grain collection figures, see ibid., table 7, p. 943. For the relevant documents, see Lynne Viola et al., *The War against the Peasantry*, vol. 1, pp. 32–34, 45–47, 55–56.

6. Viola et al., *The War against the Peasantry*, vol. 1, pp. 46, 55–56. The 5 January directive was signed by Stalin. For excerpts of Stalin's Siberian speeches, see Stalin, *Sochineniia*, vol. 11, pp. 1–9. Stalin claimed that procuracy and judiciary officials were rooming in kulak households and, therefore, were reticent to prosecute kulaks. Stalin's Siberian speeches were not published at the time.

7. For Zagumennyi's objections, see Viola et al., *The War against the Peasantry*, vol. 1, pp. 71–74.

8. Ibid., pp. 87, 91. On the conduct of the procurement campaign, see Lewin, *Russian Peasants*, pp. 214–249.

9. See *Piatnadtsatyi s"ezd VKP(b), Stenograficheskii otchet*, 2 vols. (Moscow, 1962), vol. 2, pp. 1462–1468; Stalin, *Sochineniia*, vol. 11, pp. 2–5. Officials determined to be abetting speculation were liable to prosecution under article 105 of the RSFSR Criminal Code. Lower-level Soviet officials, because of their proximity to the peasantry, were especially vulnerable. After Molotov's arrival in the Urals in January, more than eleven hundred local officials were removed from their posts; see *TSD*, vol. 1, p. 39; Lewin, *Russian Peasants*, p. 229.

10. These and other incidents of excesses (*peregiby*) are contained in an OGPU report compiled on 3 June 1928; see *TSD*, vol. 1, pp. 278–280.

11. Ibid., pp. 283–284.

12. See, for example, A. I. Mikoian's speech in *Kak lomali Nep*, vol. 1, pp. 37–58; *TSD*, vol. 1, pp. 261–262.

13. These figures are from an OGPU report compiled in March 1931; see *TSD*, vol. 1, p. 63.

14. Stalin, *Sochineniia*, vol. 11, pp. 159–160. On the right deviation, see Medvedev, *Let History Judge*, pp. 192–210; Stephen F. Cohen, *Bukharin and the Bolshevik Revolution: A Political Biography, 1888–1938* (New York, 1975), pp. 270–336.

15. R. W. Davies emphasizes the role played by adverse weather conditions, as does Moshe Lewin, but Lewin also stresses the effects of governmental policy: Davies, *The Socialist Offensive,* pp. 41–47; Lewin, *Russian Peasants,* pp. 385–386. On rationing, see E. A. Osokina, *Ierarkhiia potrebleniia: O zhizni liudei v usloviiakh stalinskogo snabzheniia, 1928–1935 gg.* (Moscow, 1993), p. 15.

16. Davies, *The Socialist Offensive,* pp. 47–48, 56; Lewin, *Russian Peasants,* pp. 384–385.

17. For valuable explications of the Ural-Siberian method, see James Hughes, "Capturing the Russian Peasantry: Stalinist Grain Procurement and the 'Ural-Siberian Method,'" *Slavic Review,* 3, no. 1 (1994), pp. 76–103; Y. Taniuchii, "A Note on the Ural-Siberian Method," *Soviet Studies,* 33, no. 4 (1981), pp. 518–547; Atkinson, *The End of the Russian Land Commune,* pp. 318–326.

18. Davies, *The Socialist Offensive,* p. 57 and table 8, p. 428; Lewin, *Russian Peasants,* pp. 389–391; Viola et al., *The War against the Peasantry,* vol. 1, pp. 119–120, 126–133.

19. See Davies, *The Socialist Offensive,* pp. 63–71.

20. For a detailed account of the 1929 grain collection campaign, see ibid., pp. 56–108. The figures are taken from *TSD,* vol. 1, p. 63.

21. See Davies, *The Socialist Offensive,* pp. 104–108.

22. *Piatnadtsatyi s"ezd VKP(b),* vol. 1, pp. 43–55.

23. Probably a reference to her monitoring and denouncing the activities of individual peasants.

24. The uprising in Serezh village, in late 1920, was part of the general peasant rebellion against Bolshevik rule and grain requisitioning in Siberia at the end of the civil war.

25. This letter was placed in the files of the People's Commissariat of Trade (Narkomtorg).

26. "Soiuzkhleb" (The All-Union State Joint-Share Association of Grain) was the grain-collecting agency of the People's Commissariat of Trade (Narkomtorg).

27. Probably E. I. Yurevich (1878–1958) who became a Central Committee member in 1929.

28. In the original, the author uses the phonetic transcription of the abbreviation SR (Socialist-Revolutionaries), *esery.* In this particular case, the writer may be applying the term to anyone who speaks on the peasants' behalf or against Party policy.

29. On the measures taken in these years and on the difficulty in determining aggregate numbers, see Carr and Davies, *Foundations,* vol. 1, pp. 158–166 and, for figures, table 9a, p. 944; Danilov, *Kooperativno-kolkhoznoe stroitel'stvo v SSSR,* p. 327; R. W. Davies, *The Soviet Collective Farm, 1929–1930* (Cambridge, Mass., 1980), table 3, p. 184.

30. The references here to bast sandals are probably an effort to expose superficial attempts by their wearers to show solidarity with the peasants. "Land and liberty!" was a longstanding revolutionary slogan of the peasants, which probably explains the phrase's somewhat awkward use in this sentence.

31. A discussion of workers' wages has been omitted by the editors.

32. For a discussion of this question, see Lewin, *Russian Peasants,* pp. 433–435; Davies, *The Socialist Offensive,* p. 111.

33. Shanin, The Awkward Class, pp. 190–192; Von Hagen, *Soldiers in the Proletarian Dictatorship,* pp. 300–304.

34. Davies, *The Socialist Offensive*, pp. 109–110 and table 17, pp. 442–443; Lewin, *Russian Peasants*, p. 435.

35. Viola, *Peasant Rebels*, pp. 46–47.

36. RGAE, f. 396, op. 7, d. 14, ll. 241–242.

37. See Carr and Davies, *Foundations*, vol. 1, pp. 158–181 and table 9, pp. 944–945.

38. In this period the number of households in collective farms in North Caucacus increased from 7.3 to 19.1 percent of households; the number in the Middle Volga, from 3.9 to 8.9; and the number in the Lower Volga, from 5.9 to 18.1. See Davies, *The Socialist Offensive*, p. 133 and tables 16–17, pp. 441–444; Stalin, *Sochineniia*, vol. 12, pp. 118–135. The phrase "the great break," taken from the title of Stalin's article, has become a shorthand reference for the turn away from the NEP to the Stalinist policies of rapid industrialization and collectivization.

39. See excerpts of Molotov's address in Viola et al., *The War against the Peasantry*, vol. 1, pp. 158–159. For the plenum stenographic report and other materials, see *Kak lomali NEP*, vol. 5.

40. R. W. Davies provides a valuable and detailed account of the December Commission's deliberations based on published information available to him at the time, in *The Socialist Offensive*, pp. 185–194. Archival materials from the commission can be found in Viola et al., *The War against the Peasantry*, vol. 1, pp. 171–197.

41. For examples, see Davies, *The Socialist Offensive*, pp. 177–180.

42. For the history of the 25,000ers, see Viola, *The Best Sons*,

43. For figures, see Davies, *The Socialist Offensive*, table 17, pp. 442–444, and pp. 238–243; Ivnitskii, *Kollektivizatsiia i raskulachivanie*, pp. 86–92.

44. For example, see the report dated 15 January 1930, in *TSD*, vol. 2, pp. 131–134; RGAE, f. 7486s, op. 37. d. 102. ll. 264, 265. According to figures published in 1931, by 1 June 1930 more than a third of the value of the property comprising the indivisible capital fund of collective farms in the USSR came from dekulakized households; see N. A. Ivnitskii, *Sud'ba raskulachennykh v SSSR* (Moscow, 2004), p. 23.

45. Evidently a reference to the tensions in USSR-China relations at this time.

46. RGAE, f. 396, op. 5, d. 14, l. 216.

47. "Red rooster" is a colloquialism for arson.

48. Viola et al., *The War against the Peasantry*, vol. 1, pp. 279–288.

49. Ibid., pp. 283, 286, 288.

50. Both OGPU reports are dated 24 January 1930 and are labeled "top secret." See *TSD*, vol. 2, pp. 138–141.

51. RGAE, f. 7486s, op. 37. d. 102. l. 260–261.

52. In January 1930, the military high command committed the Red Army to train 100,000 men for service in the countryside, many as tractor drivers. The practice continued in 1931, though at reduced numbers, for military leaders worried about the effect that collectivization was having on soldiers' morale; see Von Hagen, *Soldiers in the Proletarian Dictatorship*, pp. 317–324.

53. RGAE, f. 7486s, op. 37. d. 102. l. 221. On the significance of women's hair, see Fitzpatrick, *Stalin's Peasants*, pp. 52–53.

54. Article 28 of the RSFSR Criminal Code actually established terms of incarceration (*lishenie svobody*).

55. For the OGPU report, see *TSD*, vol. 2, pp. 87–88; RGAE, f. 8043, op. 11, d. 26, l. 109.

56. Stalin, *Sochineniia,* vol. 12, pp. 166–172.

57. *TSD,* vol. 2, pp. 195–196.

58. Viola et al., *The War against the Peasantry,* vol. 1, pp. 205–216, 218–220, 208. See also Lynne Viola, "The Role of the OGPU in Dekulakization, Mass Deportations, and Special Resettlement in 1930," in *The Carl Beck Papers in Russian and East European Studies* (Pittsburgh, Pa., 2000).

59. *TSD,* vol. 2, pp. 127, 129–130. The Party members were to come from regional (*okrug*) or higher-level organizations.

60. Ibid., p. 128.

61. Ibid., pp. 128–129.

62. Ibid., pp. 131, 155, 169–170.

63. The precise meaning of *zimogor* (a compound of *zima,* "winter," and *gora,* "mountain") in this context is not entirely clear. In some areas the name is used for people who spend the winter in the mountains. Here, it possibly refers to Stalin's Georgian origins and the "wildness" attributed to people from the Caucasus Mountains.

64. *TSD,* vol. 2, pp. 128–129, 200, 205–206, 231, 276; RGAE, f. 7486s, op. 1. d. 100. l. 127.

65. All four Central Committee members openly opposed forced collectivization and were thus branded "rightists" by the majority. Deputy People's Commissar of Finance M. I. Frumkin (1878–1938) had written a letter to the Politburo on 15 June 1928 in which he criticized the Party's rural policy. Stalin and Molotov both responded to Frumkin with their own letters to the Politburo. For Frumkin's and Molotov's letters, see *TSD,* vol. 1, pp. 290–294 and 297–301, respectively. For Stalin's letter, see Stalin, *Sochineniia,* vol. 11, pp. 116–126.

66. On this and other forms of peasant resistance, see Viola, *Peasant Rebels,* esp. pp. 67–99; *TSD,* vol. 2, pp. 157, 161–162.

67. A hero of the civil war, Kliment Yefremovich Voroshilov (1881–1969) was serving as people's commissar of military and naval affairs at this time and was a Stalin ally on the Politburo.

68. For examples, see *TSD,* vol. 2, pp. 209–210, 200; RGAE, f. 7486s, op. 1. d. 100. l. 47.

69. Ibid., ll. 48, 46.

70. RGAE, f. 7486s, op. 37. d. 102. ll. 219, 218.

71. Mosselprom is the acronym for the Moscow Amalgamated Enterprise for the Processing of Agricultural Products.

72. Probably Riabovsk district, Leningrad province.

73. *TSD,* vol. 2, pp. 593, 745–746; Ivnitskii, *Sud'ba raskulachennykh,* pp. 40–46. Although twice as many families were dekulakized in 1930 as in 1931, Ivinitsky attributes the higher deportation rate in the latter year to the fact that in 1930 more than 200,000 families remained in their home regions after initial dekulakization.

74. *TSD,* vol. 2, pp. 27, 809, 594–595.

75. Ibid., pp. 168–169, 166.

76. RGAE, f. 7486s, op. 37. d. 102. l. 209.

77. On this point and for available documentation, see Oleg V. Khlevniuk, *The History of the Gulag: From Collectivization to the Great Terror* (New Haven, Conn., 2004), pp. 287–327; RGAE, f. 7486s, op. 1. d. 102. l. 208, and d. 100, l. 47.

78. RGAE, f. 7486s, op. 1. d. 100. l. 49. Tracy McDonald has analyzed one such mass rebellion in which peasants, significantly, exhibited class solidarity

across economic lines and had the support of the village soviet; see McDonald, "A Peasant Rebellion in Stalin's Russia: The Pitelinskii Uprising, Riazan, 1930," *Journal of Social History,* 35, no. 1 (2001), pp. 125–146.

79. See Viola, *Peasant Rebels,* pp. 181–204; RGAE, f. 7486s, op. 1. d. 100. l. 48.

80. RGASPI, f. 85, op. 1s, d. 118, l. 48.

81. Ibid., l. 49.

82. Ibid. For additional material on peasant resistance during collectivization, see Valerii Vasil'ev and Lynne Viola, eds., *Kollektivizatsiia i krest'ianskoe soprotivlenie na Ukraine, noiabr' 1929–mart 1930 gg.* (Vinnitsa, Ukraine, 1997); Lynne Viola, Sergei Zhuravlev et al., eds, *Riazanskaia derevnia v 1929–1930 gg.: Khronika golovokruzheniia. Dokumenty i materialy* (Moscow, 1998).

83. RGAE, f. 7486s, op. 37, d. 102, l. 216

84. See *TSD,* vol. 2 n. 101; Stalin, *Sochineniia,* vol. 12, pp. 191–199.

85. For figures on the investigations and sentencing, see Danilov, Manning, and Viola, *TSD,* vol. 2, pp. 479–481; RGAE, f. 7486s, op. 37, d. 102, ll. 64, 62.

86. RGAE, f. 7486s, op. 37, d. 102, l. 65.

87. On the Model Statute and its various applications, see Davies, *The Socialist Offensive,* pp. 269–278.

88. RGAE, f. 7486s, op. 37, d. 102, ll. 204, 258–263. The title character of an Anton Chekhov short story, Sergeant Prishibeev, a former noncommissioned officer and obsequious martinet, acted servilely before his social superiors and tormented the peasants around him. Despite lacking any formal authority, the rigid Prishibeev took it upon himself to stamp out activity not explicitly permitted by law—like singing—in the name of social order. His name, derived from the verb meaning "to dispirit or depress" (also "to kill"), is a byword for narrowness and authoritarianism. In 1923, the Bolsheviks turned the monastery located on the remote Solovetsky Islands in the White Sea into a labor camp, colloquially known as Solovki.

89. RGAE, f. 7486s, op. 37, d. 102, ll. 65, 78, 266.

90. Ibid., l. 45

91. Ibid., l. 225.

92. Ibid., ll. 65, 64.

93. Ibid., ll. 166, 165.

94. The writers here are making a play on the word *pravda,* "truth."

CONCLUSION

1. On the development of the Stalinist state apparatus in the course of rapid industrialization, see David R. Shearer, *Industry, State, and Society in Stalin's Russia, 1926–1934* (Ithaca, N.Y., 1996). On popular reaction to Stalinist policies, see Davies, *Popular Opinion;* Kotkin, *Magnetic Mountain.* John Scott, an American who worked at the Magnitogorsk construction site and steel mill in the 1930s, provides a firsthand account of working conditions there in *Behind the Urals: An American Worker in Russia's City of Steel* (Bloomington, Ind., 1989).

2. See Conquest, *The Harvest of Sorrow;* Fitzpatrick, *Stalin's Peasants,* pp. 69–76; N. A. Ivnitskii, *Repressivnaia politika sovetskoi vlasti v derevne, 1928–1933 gg.* (Moscow, 2000), pp. 288–323; R. W. Davies, M. B. Tauger, and S. G. Wheatcroft, "Stalin, Grain Stocks and the Famine of 1932–1933," *Slavic Review,* 54, no. 3 (1995), pp. 642–657. Conquest's thesis that Stalin used the famine to terrorize and decimate the Ukrainian and other national-ethnic populations remains highly controversial.

For the state of the debate on the famine as genocide, see R. W. Davies and Stephen G. Wheatcroft, *The Years of Hunger: Soviet Agriculture, 1931–1933* (London, 2009), pp. xiii–xx.

3. Conquest, The Harvest of Sorrow, pp. 189–198; Ivnitskii, *Repressivnaia politika sovetskoi vlasti v derevne*, pp. 297–299.

4. For a clear and helpful discussion of these and other problems, see Massimo Livi-Bacci, "On the Human Costs of Collectivization in the Soviet Union," *Population and Development Review*, 19, no. 4 (December 1993), pp. 746–750. Davies and Wheatcroft provide an invaluable analysis of the archival and demographic evidence in *The Years of Hunger*. The authors settle on an estimate of 5.7 million excess deaths from famine during the period 1930–1933; see pp. 409–431. See also Nove, "Victims of Stalinism: How Many?" in J. Arch Getty and Roberta T. Manning, eds., *Stalinist Terror: New Perspectives* (New York, 1993); Stephen G. Wheatcroft, "More Light on the Scale of Repression and Excess Mortality in the Soviet Union in the 1930s," in ibid.; Michael Ellman, "A Note on the Number of 1933 Famine Victims," *Soviet Studies*, 43, no. 2 (1991), pp. 375–379.

5. On the food shortages and food rationing of these years, see R. W. Davies, *Crisis and Progress in the Soviet Economy, 1931–1933* (London, 1996), pp. 176–192 and pp. 530–533 (table); Osokina, *Ierarkhiia potrebleniia*, pp. 15–21, 63–64.

6. See the letter *svodki* in RGAE, f. 7486s, op. 37, d. 102, ll. 119–127.

7. Ibid., ll. 212, 97.

8. Ibid., ll. 92–94.

9. Osoaviakhim was a union of several organizations whose primary purpose was giving military training to civilians in preparation for their service as reserve armed forces.

10. Ibid., l. 115.

11. Ibid., l. 223. See Sheila Fitzpatrick, ed., *Cultural Revolution in Russia, 1928–1931* (Bloomington, Ind., 1978).

12. See Moshe Lewin, "Society, State and Ideology," in *The Making of the Soviet System*, pp. 218–221.

13. Vladimir Andrle, *Workers in Stalin's Russia: Industrialization and Social Change in a Planned Economy* (New York, 1988), pp. 34–35; Donald Filtzer, *Soviet Workers and Stalinist Industrialization: The Formation of Modern Soviet Production Relations, 1928–1941* (Oxford, 1986).

14. RGAE, f. 7486s, op. 37, d. 102, ll. 99, 98. The references are to civil war battles.

15. In the original: "Tishe edesh', dal'she budesh'."

16. A reference to Lieutenant General Andrei Grigorievich Shkuro (1886–1947), a leader of the White armies in the North Caucasus during the civil war.

17. RGAE, f. 7486s, op. 37, d. 102, ll. 216, 127, 97; Davies, *Crisis and Progress*, pp. 209–215.

18. Ibid., l. 98.

19. The Sixteenth Communist Party Congress, June–July 1930.

20. RGAE, f. 7486s, op. 37, d. 102, ll. 67–68.

Documents

CHAPTER 3. *SMYCHKA*

37. Letter to *Krestianskaia gazeta* from the peasant D. T. Yesipov describing the shefstvo relationship between the People's Commissariat of Foreign Trade and his village, 27 December 1924.

38. Letter to *Krestianskaia gazeta* from the peasant Fedor Morozov demanding the formation of a peasant union, June 1924.

39. Letter to *Krestianskaia gazeta* from the peasant P. I. Bazhin on the economic burdens placed on the peasant by the Soviet state, 13 June 1925.

40. Letter to *Krestianskaia gazeta* from the peasant Sergei Gogoi requesting an explanation of the peasants' role in the proletarian dictatorship, 13 June 1926.

41. Letter to *Krestianskaia gazeta* from the peasant G. Masiura on the Communist Party's exploitation of the peasantry, 1925.

42. Letter to *Krestianskaia gazeta* from the peasant I. L. Chibutkin on the Soviet state's "sons" and "stepsons," 12 March 1927.

43. Letter to *Krestianskaia gazeta* from the village correspondent Leonid Yarovoi describing the condescending behavior of Komsomolists toward the peasants, 27 June 1925.

44. Letter to the VTsIK chairman, M. I. Kalinin, care of *Krestianskaia gazeta*, from the peasant T. V. Shevchenko requesting a definition of a kulak, 21 April 1926.

45. Letter to *Krestianskaia gazeta* from the peasant S.M. on the detrimental effects of the state's lack of faith in the middle peasant, 11 August 1928.

46. Letter to *Krestianskaia gazeta* from the peasant A. F. Shklinov on the relative economic condition of the worker and the peasant, 4 January 1928.

47. Letter to *Krestianskaia gazeta* from an unknown poor peasant on the favoritism the state shows toward the working class, 27 January 1928.

48. Letter to *Krestianskaia gazeta* from the peasant A. A. Shchipakin calling for equality between workers and peasants, 14 October 1927.

49. Letter to the VTsIK chairman, M. I. Kalinin, care of *Krestianskaia gazeta*, from the peasant I. A. Rusov on the conflict between bureaucrats and those who labor, 30 March 1927.

50. Letter to *Gudok* from the peasant A. I. Sechko on the proper way to organize a collective farm, 14 July 1928.

51. Letter to *Krestianskaia gazeta* from the worker A. N. Kuznetsov detailing his civil war experience and his later efforts to organize collective farms, 15 March 1928.

Chapter 4. Was Society Transformed?

67. Letter to *Krestianskaia gazeta* from V. S. Goncharenko, a member of the "Red Rose" kommuna, describing Communist Party members' abuse of divorce laws, 25 March 1925.
68. Letter to *Krestianskaia gazeta* from the peasant A. I. Pukhov on the need for children's nurseries in the countryside, no date.
69. Letter to *Krestianskaia gazeta* from the peasant Klavdia Pliusnina describing the workings and benefits of a children's nursery, fall 1926.
70. Letter to *Krestianskaia gazeta* from the peasant and Privodino village soviet commission member V. A. Khokhlov describing peasant opposition to nurseries and fire brigades, 18 October 1926.
71. Letter to *Krestianskaia gazeta* from the peasant Kolosovsky describing the mismanagement of the Lezno village children's nursery, Novgorod province, 30 July 1926.
72. Letter to *Krestianskaia gazeta* from the peasant V. K. Kulikov describing his stay at the Usole sanatorium, 6 December 1926.
73. Excerpt from an account of an oktiabrina, *Moskovskaia pravda*, 28 June 1924.
74. Letter to *Krestianskaia gazeta* from the peasant S. A. Ganin describing an oktiabrina in his village, 25 November 1924.
75. Letter to *Krestianskaia gazeta* from the Komsomol cell secretary I. M. Gutsev on one family's adoption of the new rituals, 15 April 1925.
76. Excerpt from a letter to *Krestianskaia gazeta* from N. A. Bobkov on Komsomolists who choose religious marriage ceremonies, 18 August 1925.
77. Excerpt from the peasant I. I. Melnikov's letter to *Krestianskaia gazeta* describing youth-gang violence in his village, 27 November 1927.
78. Letter to *Krestianskaia gazeta* from the peasant and Committee of Indigent Peasants (KNS) member I. S. Chernoivanov describing the conflict between a kulak and the chairman of the KNS on the occasion of the May Day holiday, May 1924.
79. Letter to *Krestianskaia gazeta* from the peasant V. M. Turovtsev on criminality and how to deal with it, 24 September 1927.
80. Letter to *Krestianskaia gazeta* from the village correspondent S. Kugorev describing instances of hooliganism, 22 February 1927.

CHAPTER 5. PEOPLE AND POWER

81. Letter to *Krestianskaia gazeta* from the peasant and antireligious activist A. P. Poliakov explaining the harmful effects of the Lenin cult, 10 May 1927.

82. Letter to *Krestianskaia gazeta* from the village correspondent "A. Diletant" on rumors of a tsarist restoration following Lenin's death, early 1924.

83. An anonymous letter to *Krestianskaia gazeta* describing how local communists dominate village institutions, 19 May 1925.

84. Letter to Stalin care of *Krestianskaia gazeta* from T. G. Burtsev asking for reinstatement to the Communist Party, 4 July 1926.

85. Letter to Trotsky, care of *Krestianskaia gazeta,* requesting an explanation of his disagreements with the Central Committee from the peasant I. P. Vostryshov, 14 December 1927.

86. Letter to *Krestianskaia gazeta* from the peasant A. F. Sdobniak explaining the peasants' view of communist opposition groups, 5 January 1928.

87. Letter to *Krestianskaia gazeta* from the peasant N. F. Yelichev, who is responding to M. I. Kalinin and comparing Soviet agricultural policy unfavorably to that of the tsars, 10 October 1927.

88. Letter to *Krestianskaia gazeta* from the peasant M. V. Kiselkina in response to N. F. Yelichev's letter (no. 8), 11 October 1927.

89. A hostile anonymous letter to *Krestianskaia gazeta* condemning the Bolsheviks and their rural policies, 13 August 1928.

90. Excerpts from a letter to *Krestianskaia gazeta* from the peasant I. G. Shokin criticizing local officials and the unfair application of the "kulak" designation to productive peasants, 10 October 1927.

91. Letter to G. I. Petrovsky, chairman of the central executive committee of the Ukrainian republic, from the peasant Ya. Yu. Stepanov, 19 June 1928.

92. Letter to *Krestianskaia gazeta* from the peasant A. Grigoriev on the exploitation of the peasantry for the benefit of government bureaucrats, 4 July 1928.

93. Letter to *Krestianskaia gazeta* from the peasant I. F. Goloborodko detailing the corrupt and dissolute behavior of local communists, 4 May 1925.

94. Letter to *Krestianskaia gazeta* from the village correspondent Grigoriev on the opportunism and careerism of local communists, 7 March 1926.

95. An anonymous letter to *Krestianskaia gazeta* on the bureaucratic degeneration of the Communist Party, 15 August 1928.

CHAPTER 6. WHITHER SOCIALISM?

CHAPTER 7. THE GREAT BREAK

CONCLUSION

Index

BOOKS IN THE ANNALS OF COMMUNISM SERIES